D0151341

THE MEASURE OF MERIT

THE MEASURE OF MERIT

TALENTS, INTELLIGENCE, AND INEQUALITY IN THE FRENCH AND AMERICAN REPUBLICS, 1750–1940

John Carson

PRINCETON UNIVERSITY PRESS PRINCETON AND OXFORD

Published by Princeton University Press, 41 William Street, Princeton, New Jersey 08540
In the United Kingdom: Princeton University Press, 3 Market Place, Woodstock, Oxfordshire OX20 1SY

Library of Congress Cataloging-in-Publication Data

Carson, John, 1954–
The measure of merit : talents, intelligence, and inequality in the French and American republics, 1750–1940 / John Carson.
p. cm.
Includes bibliographical references and index.
ISBN-13: 978-0-691-01715-0 (hardcover : alk paper)
ISBN-10: 0-691-01715-8 (hardcover : alk paper)
1. Intelligence levels—Social aspects—France. 2. Intelligence levels—Social aspects—United States. 3. Intellect—Social aspects—France. 4. Intellect—Social aspects—United States. 5. Ability—France. 6. Ability—United States. 7. Equality—France. 8. Equality—France. 9. Affirmative action programs in school—France. 10. Affirmative action programs in schools—France. 11. Social stratification—France. 12. Social stratification—United States. I. Title.
BF431.5.F8C37 2006
305.0944—dc22 2006006604

British Library Cataloging-in-Publication Data is available

This book has been composed in Sabon

Printed on acid-free paper.∞

pup.princeton.edu

Printed in the United States of America

1 3 5 7 9 10 8 6 4 2

10 9 8 7 6 5 4 3 2 1

For Mom, and in memory of Dad and Betsy

Contents

Tables

Preface and Acknowledgments

ON JUNE 23, 2003, the U.S. Supreme Court handed down two landmark decisions about the right of American colleges and universities to consider race in their admissions procedures. Ruling on suits filed against affirmative action programs adopted by the University of Michigan for admissions to its undergraduate College of Literature, Science, and the Arts (LSA) and its Law School, the Court upheld the principle that race could be considered for admissions, at the same time as it struck down mechanical methods of applying specific race-weighting factors to admissions decisions.[1] The University and its many supporters in academia and industry justified their positions largely in the name of the value of diversity to institutions of learning and the existence of multiple ways of assessing a candidate's merit. Arrayed against them were a range of individuals and groups, most of whom also adopted the language of merit, though in their case to express their objections to the University's affirmative action policies. School grades, the plaintiffs contended, and especially the results of impartial instruments such as the SAT or LSAT, were the only valid criteria on which to make admissions decisions. Merit, they argued, was a matter of past performance and native ability, nothing more.[2]

Five months later, a French high court also considered the question of affirmative action. Confronted with persistent complaints that the French higher-education system of elite *grandes écoles* routinely discriminated against individuals from lower socioeconomic strata, in 2001 the Institut d'Etudes Politiques de Paris (Sciences Po)—among the most prestigious *écoles*—had inaugurated a small program of affirmative action for graduates of selected *lycées* (high schools) in economically depressed neighborhoods. A national student organization immediately challenged this alternative admissions method for violating the French Constitution's guarantee of nondiscrimination and for subverting the traditional system of merit-based selection via *concours* (examinations). Like the U.S. Supreme Court, the French high court ruled on November 6, 2003, in Sciences Po's favor, upholding the constitutionality of their affirmative action policy. To the policy's supporters, such as Sciences Po's director of communications Xavier Brunschvicg, the decision was a victory not only for diversity and equality of opportunity, but also for intellectual merit and intelligence. The traditional selection system, he argued, privileged those coming from good cultural backgrounds who were exposed throughout their lives to books and learning; the supplemental affirmative action procedure sought to find intelligence where learning had been more scarce.[3]

That two nations with such different educational and political histories should have arrived at approximately the same place—the need to embrace and vigorously defend the value of diversity within a language of merit—speaks volumes. Three features of this conjuncture stand out. First, it is striking that the parties on both sides of each dispute sought to anchor their claims at least in part by invoking some concept of merit. While there was little agreement about how merit should be determined and whether there was one standard or many, no one dismissed it as irrelevant to the issue at hand by arguing, for example, that the need for social justice might temper the demands of merit. Discussions of which individuals or groups deserved what social goods inevitably revolved around questions of merit, as indeed had been the case in both republics since their creation. Even when claims about social justice were pressed, they were most typically framed within the language of merit, with individual achievements and group histories of oppression or success mobilized in almost dizzyingly complex ways to sustain defenses of or attacks upon particular merit-based systems of selection.

Second, the parties supporting older admissions methods and hostile to affirmative action policies had much less work to do to represent their positions as embodying a commitment to merit. These "traditional" procedures—whether school grades and SAT scores or performance on comprehensive examinations—reflected what merit had come to mean; they had been put into place and sustained, in the French case for almost two centuries, precisely because they had been successfully tied to the ideology of merit. The struggle was on the other side, to expand and transform notions of merit sufficiently so that the claims of affirmative action could be made consonant with them. And third, there is the curious role of the concept of intelligence. In the American case, intelligence—particularly in the form of standardized test results—was a weapon of the plaintiffs, one of the means of trying to demonstrate that Michigan's admissions procedures were choosing less able candidates over more deserving ones. In the French case, however, it was the reverse. There, *supporters* of affirmative action appropriated the language of intelligence, using it to suggest that there was a criterion of merit not completely captured by the grueling admissions examinations that formed the usual route into the grandes écoles. What accounts for this divergence? And, for that matter, why was intelligence being discussed at all? What roles did it fulfill?

The Measure of Merit is my attempt to explore these questions and to understand how merit and intelligence came to be linked within French and American conversations about democracy and equality. Although these affirmative action cases were decided near the end of this project, they capture nicely the similarities and differences in the ways that two republics, born at approximately the same moment and with similar commitments to equality and opportunity, have at the same time confronted the issue of difference, especially "natural" inequality. Both in one way or another eventually turned to testing as a way of establishing merit, and in each the response was as much one of anxiety as approbation. Some worried that the tests might be wrong: inaccurate, ill-conceived, and doomed to choose the wrong people based on the wrong criteria. Others, however, were unsettled by the opposite: that the tests might conceivably be right, and thus that some people really were naturally better than others. What if inequality was the product not of poor environment or personal choices but the luck of the genetic draw, and what if scientists could "see" the difference?

The Measure of Merit explores what might be termed this shadow language of inequality paralleling the much more visible rhetoric of republican equality in France and America, and traces its changing history up to the moment where it assumed the form invoked within the affirmative action debates of the early twenty-first century. Both nations turned to seemingly impersonal and objective methods to make, or at least justify, decisions about merit, about who gets access to what opportunities. And in both nations these methods developed as thoroughly modern solutions to a similar problem raised by their shared revolutionary heritages: having toppled aristocratically organized societies in the name of natural rights and the people's sovereignty, what would be put in their place? How could a new elite be selected and justified within a political ideology also celebrating equality and universal rights? How, in other words, could inequality be rendered legitimate?

To answer these questions, I have looked at the work of mental philosophers and anthropologists, political theorists and popular pundits, psychologists and educators. My goal has been to provide a kind of genealogy of the politics of natural difference, capturing illuminating moments and broad trends in the two cultures I examine and compare. In large measure, it is a study of discourse, of languages of human nature and how they intersected with political and social languages. As in all such studies, causality remains murky.[4] Clearly, at moments certain actors, such as Samuel Morton or Alfred Binet or Lewis Terman, were critical to the trajectory of developments. Just as clearly, however, much broader cultural, social, scientific, and political ecologies were also crucial to determining when, how, and with what meanings a key concept such as intelligence might become something that mattered. What I assume throughout is that ideas do have consequences, whether when instantiated in concrete practices such as cranium measuring or intelligence testing, or as ontological elements that individuals rely on to make sense of the world. Examining and understanding the public discourses that have framed concepts such as talents, merit, and intelligence, I would argue, helps to unearth a culture's presuppositions, the boundaries within which individuals operate while persuading one another, or themselves, to act on or think about the world in particular ways. It is that story which *The Measure of Merit* seeks to tell.

I could not possibly construct such a story by working in isolation, and so one of my great pleasures in completing this book is the chance to thank publicly the many institutions and individuals that have helped make it possible or sustained me during the long period of its gestation. First, my scholarship has been generously supported by Princeton University, an NSF-NATO postdoctoral fellowship, the Wellcome Trust, the Cornell University Department of Science and Technology Studies, the Mellon Foundation, the University of Michigan, and the National Humanities Center. I also received small research grants from the Rockefeller Foundation, the Hagley Museum and Library, and the University of Michigan Department of History, and a generous publication subvention from the University of Michigan. I deeply appreciate their willingness to sustain my and others' scholarly research.

Second, the project has depended heavily on library resources and archival collections spread across three countries. While I cannot name all the individual librarians who helped me over the years, I would at least like to acknowledge some of the institutions that allowed me to use their collections. In the United States, these include the American Archives for the History of Psychology, Columbia University archives, Cornell University library, Educational Testing Service archives, National Archives, National Humanities Center librarians, National Research Council archives, Princeton University library, Rockefeller Foundation archives, Stanford University archives, University of Michigan library, and Yale University archives. In France, I made particular use of the Archives Nationale, Bibliothèque Nationale, Bibliothèque Nationale de Recherche Pédagogique, and Sorbonne library. Finally, in Britain I am grateful to the University College London archives, University College London library, University of London (Senate House) library, and Wellcome Institute for the History of Medicine library.

Third, among the most patient organizations in the world must be Princeton University Press, which has stuck by me for the ridiculously long time it has taken to get this book to them and has, at every juncture, done all that it could to support me and the manuscript on its road to production. I want to particularly thank my editors, Sam Elworthy and Emily Wilkinson, and their assistants, Hanne Winarsky and Kevin Downing. I have also benefited from the editorial assistance of Kate Washington, Marsha Kunin, Mary O'Reilly, and Karen Carroll, and from the research assistance of a number of undergraduates associated with the University of Michigan Undergraduate Research Opportunities Program (UROP): Sabrina Hiedemann, Megan Lambert, Erin Mays, and Joshua Wickerham. Parts of chapters 1, 6, and 7 have been published previously, and I thank the editors of and anonymous reviewers for those publications for their comments on my work.

Fourth, I have been lucky enough to receive friendship, encouragement, guidance, and the occasional whack upside the head from an extraordinary group of friends, teachers, and institutions. I want to begin by singling out a number of people who deserve special mention. Ken Arnold, Marybeth Hamilton, Susie Blumenthal, Tom Broman, Lynn Nyhart, Elspeth Carruthers, Betsy Clark, Jay Cook, Rita Chin, Matthew Countryman, Rosie Ceballo, Peter Dear, Pauline Dear, Greg Dowd, Ada Verloren, Kevin Downing, Paul Edwards, David Hancock, Gabrielle Hecht, Tom Green, Ellen Herman, Sarah Igo, Sheila Jasanoff, Martha Jones, Carol Karlsen, Liz Lunbeck, Jonathan Metzl, Ted Porter, Dan Rodgers, Hannah Rosen, Richard Turits, Andrea Rusnock, Paul Lucier, Charis Thompson, Emily Thompson, and Skip Weidner—each of you has been stalwart in your support of me and a constant reminder of the many values of friendship. My academic career began in earnest at Princeton, where a philosophy major got retrained as a historian through the patience and support of the faculty and graduate students in the Program in History of Science and the Department of History. My teachers there—particularly Dan Rodgers, Gerry Geison, Liz Lunbeck, John Servos, and Raine Daston—opened my intellectual horizons in any number of ways; without them, writing this or any other book would have been unthinkable. I need to acknowledge as well Faye Angelozzi, Alice Conklin, Charles Gillispie, Jim Goodman, Michael Jimenez, Phil Katz, Mike Mahoney, Toni Malet, Laura Mason, Andrew Mendelsohn, JoAnne Morse, Larry Owens, Peggy Reilly, Erik Sageng, Pam Selwyn, Richard Sorenson, Geoff Sutton, Rachel Weil, and Norton Wise.

During my research stint in London, the members of the Wellcome Institute for the History of Medicine—including Janet Browne, Bill Bynum, Stephen Jacyna, Chris Lawrence, Roy Porter, Katharina Rowold, Sonu Shamdasani, and Molly Sutphen—generously allowed me to join their community and to learn much about the history of medicine from them. At Cornell, the Department of Science and Technology Studies proved to be a similarly welcoming home. In addition to my "boss," Sheila Jasanoff, I thank Eileen Crist, Michael Dennis, Steve Hilgartner, Christine Leuenberger, Tracie Matysik, Trevor Pinch, Neelam Sethi, Charis Thompson, Frank Trocco, and Deb van Galder, as well as the other faculty members and graduate students in the department; most of my training in STS derives from this remarkable group.

Finally, for the last seven years Ann Arbor has provided as wonderful and sustaining an intellectual and personal community as I could have asked for. In addition to acknowledging the members of the University of Michigan History Department and Program in Science, Technology, and Society collectively, I would like particularly to thank Paul Anderson, Richard Cándida Smith, Sueann Caulfield, C. S. Chang, Fred Cooper, Laura Downs, Geoff Eley, Dario Gaggio, Dena Goodman, David Hancock, Joel Howell, Nancy Hunt, Kali Israel, Sue Juster, Mary Kelley, Kim Leary, Rudi Lindner, Jonathan Metzl, Michele Mitchell, Maria Montoya, Gina Morantz-Sanchez, Marty Pernick, Mary Piontek, Stephanie Platz, Helmut Puff, Sonya Rose, Julius Scott, Rebecca Scott, Parna Sengupta, Stefanie Siegmund, Alexandra Stern, Mills Thornton, Maris Vinovskis, and Liz Wingrove. Throughout my career, members of the History of Science Society, the Forum for the History of the Human Sciences, and Cheiron–The International Society for the History of Behavioral and Social Sciences have been unstinting in their encouragement. These include Ken Alder, Mitchell Ash, Betty Bayer, John Brooks, John Burnham, Jim Capshew, Kurt Danziger, Lynn Gorchov, Ben Harris, Sarah Igo, Stephanie Kenen, Henry Minton, Jill Morawski, Ted Porter, Dorothy Ross, Irina Sirotkina, Jeff Sklansky, Roger Smith, Mike Sokal, Mary Terrall, Richard von Mayrhauser, Andrew Winston, and Leila Zenderland.

Like any work of scholarship, this book has been made stronger by the research and insights of any number of scholars. For comments on or aid with individual chapters, I thank Ken Alder, Wendy Allanbrook, Mitchell Ash, Jordanna Bailkin, David Bien, Richard Cándida Smith, Jim Capshew, Rita Chin, Fred Cooper and the UM Department of History Executive Committee, Matthew Countryman, Peter Dear, Eric Daniels, Greg Dowd, Frances Ferguson, Dena Goodman, Lisa Jane Graham, Ray Grew, David Hollinger, Sarah Igo, Mary Kelley, Tomoko Masuzawa, Erin Mays, Katharine Norris, Hannah Rosen, Andrea Rusnock, Charlotte Sussman, Rachel Toor, Richard Turits, Rachel Weil, Liz Wingrove, and Caroline Winterer. I received extensive, beyond-all-calls-of-duty help from Susie Blumenthal, Tom Broman, Jay Cook, Ellen Herman, Sheila Jasanoff, Martha Jones, Ted Porter, and Dan Rodgers, whom I cannot thank profoundly enough. The book is certainly far better because of their acute readings; the flaws, alas, remain only my own.

Finally, without certain individuals this project simply never would have been completed. My mom, dad, sister, brother-in-law, and nephews have continued to provide love and support, even without always understanding what I was up to. Tom, Betsy, Andrea, Paul, Ken, Marybeth, Emily, Elspeth, Ellen, Jonathan, Susie, Martha, Jay, Matthew, Hannah—

each of you helped to keep me going when I really did want to quit, and I appreciate deeply the steadfastness of your friendship and support. Even after this long study, I can't say that I know for sure what intelligence and merit really are. However, I can say with complete certainty that each of you combines a keen mind with a generous heart, and thus embodies merit however it may be defined. I thank you more than you can know for being willing to extend those virtues to me.

THE MEASURE OF MERIT

Introduction

> We hold these truths to be self-evident, that all men are created equal, that they are endowed by their Creator with certain unalienable rights
> —*Declaration of Independence* (1776)

> Men are born and remain free and equal in rights. Social distinctions may be founded only upon the general good.
> —*Declaration of the Rights of Man and Citizen* (1789)

AMERICANS PROCLAIMED it so boldly, the fundamental equality of all human beings, that even their more radical co-revolutionists in France drew back a bit, conceding an equality in rights but asserting at the same time the possibility of social distinctions, of difference and inequality. Not that most Americans failed to come to the same conclusion. Few on either side of the Atlantic advocated a thoroughgoing social leveling. Human differences seemed real enough even in the late eighteenth century, Enlightenment proclamations of human perfectibility notwithstanding, and the labors of various scientists investigating human nature during the nineteenth and twentieth centuries would largely serve only to confirm this fact. Physically, mentally, perhaps even morally, by race and region and class, along lines of gender and age, differences were manifest everywhere. But the fact that not everyone was exactly the same did not mean necessarily that some were better than others, and the visibility of all these differences gave little insight into which ones should matter for what purposes. What was clear, however, was that the social systems of the old world and the ancien régime, founded on notions of inherited status and hereditary aristocracy, had been rendered suspect, if not unworkable. Thus, if social distinctions were to be reclaimed and elites legitimated, they would have to be justified along new lines, ones that could accord with republican celebrations of equality and the sovereignty of the people.

The Measure of Merit tells the story of how the American and French republics turned to the sciences of human nature to help make sense of the meaning of human inequality. These sciences' exploration of the status and character of human differences, particularly as related to mental ability, it contends, provided a range of political theorists, social commentators, and practical politicians with seemingly objective grounds for interrogating the limits of human equality and developing what could be represented as a justifiable basis for social distinctions. Indeed, from the earliest ruminations about human beings in a state of nature, discussions of the implications of human differences for the establishment of a social order promoting equality but also consonant with nature's truths were central. Jean-Jacques Rousseau and Thomas Jefferson puzzled over the natural inequalities in the eighteenth century; the naturalist Louis Agassiz and abolitionist Frederick Douglass battled over the white race's presumed innate mental superiority in the nineteenth century; and the psychologists Alfred Binet and Lewis Terman worried over the meaning of intelligence differences for democracy in the twentieth century. In each instance, questions about the correct way to understand inequalities in human abilities also became questions about the appropriate way to organize society, and vice versa. In general, mental philosophers and political theorists on both sides of the Atlantic argued that if the "false" distinctions of wealth or family background or beauty or any of the other accidents of birth could be eliminated, then the "true" ones, those reflecting fundamental aspects of a person's nature, could come to the fore. Almost all believed that social differences would not disappear; rather, they would be placed on a new footing—merit—and made to seem legitimate expressions of how individuals manifested those abilities.[1]

Two issues persistently arose for those seeking to understand social inequality in terms of ability and merit. First, there was the question of the nature of the differences themselves. Which were the ones that mattered, what was their origin, and how easily could they be altered or improved? Did nature ensure that some individuals were better than others at certain tasks, or was it all a function of education and experience? Second, once ability was acceptably defined and understood, what should happen next? Should all be trained to the same level in all things; or should those with ability be identified and receive special instruction; or should talents be allowed to develop as they would, neither promoted nor hindered by the state? And what about those deemed decidedly weak in abilities? At first, educators, mental philosophers, and political leaders in America and France responded to these questions similarly. Sharing Enlightenment commitments to the primacy of reason and the need to reconcile social structure with the dictates of nature, and sure that education could improve the populace and ready it for citizenship, they imagined social worlds in which individual differences were many, open to training, and easily harnessed to the benefit of state and society. Whereas virtue was to constitute one foundation of these new polities, mental attributes, understood as the vaguely defined term "talents," was to be the other. Thus both nations promoted broad-based education as a means of making opportunity available to all, and both emphasized individual differences in the plural—talent*s*, and facultie*s*, and abilitie*s*—whether understood as products of nature or nurture or both.

Over the course of the nineteenth and twentieth centuries, however, the specific ways in which each society responded to the evolving sciences of human nature diverged sharply as these nations addressed the problem of balancing equality and difference. Four distinctions stand out. First, in America, a political culture celebrating liberal market-based approaches, wary of placing power in the hands of the state, and deeply skeptical about claims to expertise dominated. Consonant with this outlook, throughout much of the nineteenth century both liberal and conservative writers firmly believed that human talents were multiple and diverse, and argued that a proper social order was one where the free play of talents among self-determining individuals allowed the most meritorious to rise to the top. In postrevolutionary France, by contrast, belief in the state as guarantor of equality and individual rights was strong, worry about the unrestrained market pervasive, and faith in the power of experts to act in the public interest high. Under these conditions, most favored some form of state-centered solution to the problem of equality and difference, usually one featuring a universal educational system that would identify and nurture individual talents, at least among bourgeois males.

Second, although both America and France strongly favored basic education for all classes, the kinds of educational systems that each nation developed, and thus their responses to the problem of difference, were strikingly dissimilar. Until the second half of the nineteenth century, American education was almost entirely a local affair. Primary-level training was broadly available, and often provided by local government; beyond that, however, education at the secondary or collegiate levels was principally in private, often sectarian hands, and available mostly to a small elite destined for one of the professions. Major changes occurred only after the Civil War, when federal and local governments began to invest more heavily in mass secondary and university education. In contradistinction, the French from the start adopted an approach to education that was national, universal, and comprehensive. Pyramidal in design, the French system eventually established primary schools in every region open virtually to all, with the most successful students continuing to secondary school and then elite institutions for the most advanced training. Rigorous examinations determined who could move up, with the goal of ensuring that the most talented received the best education and became the core of the nation's technocratic elite. Throughout the vicissitudes in their nation's political structure, the French continued to use the educational system to identify and train an elite who would represent the triumph of merit in service to the nation. Until the twentieth century, Americans, on the other hand, placed much more weight on personal attributes than on formal education as a means of social advancement or distinction. Once educational credentials did become more essential, however, Americans too embraced more systematic approaches to identifying and promoting the most talented, though ones adapted to its decentralized educational system, and consonant with the desire to employ objective methods of selection and to disaggregate masses of students quickly and efficiently to meet the needs of rapidly expanding urban school districts.

Third, because of the centrality of race in American culture, explorations of group-level differences had much more resonance there than in France. During the nineteenth century, anthropologists and biologists in both nations—including Samuel Morton and the American school of anthropology and later Paul Broca and the Société d'Anthropologie de Paris—constructed a language of mental capacity in the singular, based on the concept of intelligence, to describe and analyze human beings at the level of groups. Created by transforming reason from an absolute into a characteristic manifested in degrees, intelligence and its synonyms justified the arrangement of animals and humans in a simple linear order based on mental power. The result was a scientific explanation of two largely uncontested "truths" of the period: that humans, specifically white male Europeans, held pride of place in the animal kingdom, and that European civilization was distinctly superior to all others. The growing authority of scientific rationales for racism in the late nineteenth century synthesized these "truths," pushing intelligence to the forefront of explanations of the hierarchical ordering of the races. Notoriously, in America anthropological determinations of levels of intelligence by race were used to "prove" the inherent and unalterable inferiority of nonwhite peoples, and quickly became part of the nationwide debate about the place of African Americans in society. French anthropologists were no less certain about the existence of an intelligence-based racial hierarchy with whites on top, but their claims had much less resonance in a nation that saw itself as racially homogeneous and superior on the basis of culture alone to the nonwhite societies it was colonizing. Thus segments of the American public became accustomed by the latter half of the nineteenth century in ways that their French brethren simply did not to using the language of intelligence to debate whether certain groups deserved access to opportunities denied others.

Fourth, and finally, the extraordinary social changes of the late nineteenth century—political upheavals in France and social/cultural transformations in the United States—opened space for new methods of understanding and evaluating humans and their behavior. In both countries mental scientists unhappy with previous approaches to human nature began to push for more "scientific" alternatives, ones that led psychologists to recast many of their fundamental conceptions about the mind, including the notion of intelligence.[2] American psychologists turned to methods of quantification and measurement associated with the experimental laboratory to try and create an exact science of the mind, one where every mental attribute was accessible to measurement and perhaps statistical characterization. For them, intelligence as understood by anthropologists such as Morton or Broca beckoned as a biologically based, unitary, quantifiable entity that might not only usefully distinguish races, but rank individuals within a given group as well. Although French psychologists, too, were impressed by laboratory science's power, their approach emphasized clinical observation, where intensively investigated individual pathological cases were used to understand the mind's normal features. To be sure, the inventor of the modern intelligence test—the Binet-Simon intelligence scale—was a French psychologist, Alfred Binet; nonetheless, the test became an American sensation rather than a French one. French psychologists and administrators were ambivalent about the nature and intrinsic significance of intelligence and preferred to assess individuals on the basis of methods reliant on expert judgment. While few French psychologists rejected outright the intelligence test and the knowledge it could produce, most favored understanding intelligence as a complex multivalent phenomenon useful for shedding light on the abilities of the elite or diagnosing the deficient, and thus best approached through clinical/observational modes of analysis.

By the 1920s and 1930s, the combination of these factors had produced distinct ways of understanding differences in mental abilities and using them to explain who got access to what opportunities. In America, intelligence proved to be an attractive concept with which to unify the democratic and meritocratic, to help regulate the increasing demand for limited educational resources and occupational opportunities in ways that could appear objective and fair even to those least successful in garnering rewards from the system. The U.S. Army employed intelligence testing on an unprecedented scale during World War I, when more than 1.75 million soldiers were examined and sorted; such testing then underwent an enormous postwar boom. Administrators in education and industry looked to intelligence as a means of classifying their charges on the basis of a socially sanctioned criterion; professionals and experts invoked intelligence to justify their privileged status while maintaining allegiance to the ideal of equal opportunity; and the growing cadre of white-collar office workers and bureaucrats used intelligence to distinguish their mental labor from what they saw as the inferior hand labor of the factory and farm. Because intelligence signified a measurable biological construct, it could readily be represented as transcending class lines and thus as an inherently egalitarian and objective criterion. However, because members of privileged socioeconomic groups generally scored well on intelligence tests, the concept in its twentieth-century guise also offered a way to maintain the overall stability of the American social hierarchy while keeping it open to exceptional members of historically excluded groups.

In France, by contrast, the educational system continued to serve as the primary gatekeeper for entrance into the technocratic elite. Through the 1930s, intelligence and its tests were associated with identifying and classifying the mentally deficient rather than the skilled. French psychology's fascination with representing individuals in multiple registers meant that intelligence as such was rarely seen as either singular in nature or a unique determinant of an individual's future. While neither French psychologists nor the French public dismissed the importance of assessments of individual intelligence, the institutional and cultural roles of such determinations were primarily diagnostic, ways for pathology to be identified or personal mental characteristics to be known. Moreover, as Theodore Porter has demonstrated, French technocratic culture was confident of its own expert authority, and wary at best of reliance on simple quantitative determinations.[3] Ironically, therefore, it was in America, for all its individualism, that mental difference was collapsed down to a single register—intelligence as something unitary—and relied on especially in education and industry to establish a justifiable basis for differentiating the masses, while in France that intelligence was regarded as multiple and most relevant to individual self-understanding.

The Measure of Merit thus tells the story of divergent conceptualizations of intelligence and their relation to merit, showing that scientific objects such as intelligence must be seen as "product[s] of history, not of nature."[4] This is not to deny the reality of intelligence or talents or other aspects of human nature discussed in this study, but rather to show that such terms are multivalent, constantly shifting in meaning and significance as they are deployed to solve problems of social order and accomplish other essential cultural work. Adopting an approach that highlights the inseparability of the ways we understand the world and the ways we live in it—what STS scholar Sheila Jasanoff has termed "co-production"—*The Measure of Merit* demonstrates how entities such as talents and intelligence and the mechanisms that made them seem real have become constituents of the societies in which they were produced and adopted, continually shaping and being shaped by these cultures' particularities.[5] The efforts of psychologists, anthropologists, mental philosophers, and other scientists to comprehend the source and significance of human inequality, it thus contends, helped define the terms in which the new American and French republics sought to fashion and legitimate their systems of merit. Yet this intermixing of the scientific with the political, *The Measure of Merit* insists, did not result in the wholesale triumph of empirical methods for understanding intelligence and assessing merit. Rather, the negotiations between these fields of knowledge and practice generated complicated settlements, in which those appropriating knowledge about human nature—be they administrators, educators, business people, or members of the general public—were also transforming this knowledge, such that it was deemed both authoritative and yet subject to dispute. In the process of reckoning with natural inequalities, therefore, these various actors proved unable to entirely domesticate or stabilize such concepts as intelligence and merit, which remained always contestable terms in the recurrent debates about the social and political implications of inequality for a modern democracy.

Part I

MENTAL ABILITIES AND REPUBLICAN CULTURES

One

“The most precious gift of nature”

NATURAL ARISTOCRACY, REPUBLICAN POLITIES, AND THE MEANINGS OF TALENT

> For I agree with you that there is a natural aristocracy among men. The grounds of this are virtue and talents. . . . There is also an artificial aristocracy founded on wealth and birth, without either virtue or talents. . . . The natural aristocracy I consider as the most precious gift of nature for the instruction, the trusts, and government of society. . . . May we not even say that that form of government is the best which provides the most effectually for a pure selection of these natural *aristoi* into the offices of government?
>
> —Thomas Jefferson (1813)

So THOMAS JEFFERSON wrote his old friend and rival John Adams on October 28, 1813, provoked by Adams's more cynical contention that “Birth and Wealth together have prevailed over Virtue and Talents in all ages.” Adams responded to Jefferson's lofty sentiments swiftly, making clear his own rather different views on aristocracy:

> [B]oth artificial Aristocracy, and Monarchy, and civil, military, political and hierarchical Despotism, have all grown out of the natural Aristocracy of “Virtues and Talents.” We, to be sure, are far remote from this. Many hundred years must roll away before We shall be corrupted. Our pure, virtuous, public spirited federative Republick will last for ever, govern the Globe and introduce the perfection of Man, his perfectability being already proved by Price Priestly, Condorcet Rousseau Diderot and Godwin.[1]

Adams's sarcasm notwithstanding, Jefferson's conception of rule by natural aristocracy struck a nerve, reminding Adams vividly of the doctrines of all those other high priests of Enlightenment political speculation he had spent a career disputing.[2] By 1813 questions of a natural aristocracy would have already seemed rather fusty in a country moving headlong toward dramatic expansions of the suffrage and dismantlement of some of the previous century's intricate codes of deference. But both ex-presidents were looking as much backward as forward, continuing debates that had been pressed since the mid-eighteenth century in France, Britain, and America about the nature of society and governance in a republic.

Three aspects of their exchange highlight important features of this transatlantic conversation.[3] First, it is striking that the author of the Declaration of Independence, with its seemingly grand commitment to human equality, could so blithely assert that some individuals were by nature superior to others and thus better suited to rule.[4] Jefferson was not simply contradicting himself, or only engaging in the doublespeak that slaveholders were already employing to justify owning other human beings.[5] Rather, like many puzzling over a republican society's character, Jefferson distinguished between fundamental political rights, which most white men would enjoy, and opportunities for political leadership, which he argued should result from an individual's particular virtues and talents. In this Jefferson was by no means alone. The language of universal natural rights, Daniel T. Rodgers has demonstrated, took on enormous importance in America and other parts of the Atlantic world during the eighteenth century.[6] At the same time, many political theorists sought to wed commitment to a universal right, such as equality of opportunity, with an ideology that emphasized selecting persons for positions of power and authority on the basis of individual ability. Thus lower-level nobles argued for determining officerships in the French army by competence, not social rank, and natural philosophers advocated open competitions for positions and prizes such as those sponsored by the Parisian Academy of Sciences.[7] Indeed, for many in America and France one attraction of republican government was its promise to make opportunity broadly available, so that the ablest could reach positions of power.

Second, both Jefferson and Adams acknowledged, albeit from very different perspectives, that social stratification was inevitable, even within the most privileged class of white, propertied males. "Was there, or will there ever be," Adams wondered in 1787, "a nation, whose individuals were all equal, in natural and acquired qualities, in virtues, talents, and riches? The answer of all mankind must be in the negative."[8] While more optimistic about the potential of education and abundant land to produce rough equality (at least for adult, white males) within a republic of yeoman farmers, Jefferson nonetheless also conceded that society would remain divided. He hoped, however, to ensure that civil distinctions and political rule would be tied to individual virtues and talents rather than social rank.[9] Other, much more radical writers concurred: even those who launched ferocious attacks on the privileges of birth and blood generally assumed that a range of hierarchies—be it men over women, adults over children, or Europeans over other peoples—would persist, whether the product of nature or custom. And, as many American historians have noted, post-Revolutionary elites, especially those who would form the Federalist Party, often employed the ideology of natural aristocracy to justify their claims to leadership, because it could accord with the representation of the emergent republic's social order as democratic and egalitarian.[10]

Third, Adams no less than Jefferson agreed that nature was a major impediment to achieving complete equality, even in an ideal republic: some people were simply born with more talents or virtues than others, and these superior abilities either entitled them to (in Jefferson's eyes) or helped them attain (in Adams's view) positions of influence in government and society. Jefferson celebrated this fact about human nature; in his view, maintenance of the nation's republican legacy depended on ensuring that the true *aristoi* attain leadership positions. Adams was more wary, finding neither natural nor artificial aristocracy to his liking. But he, too, had to concede that natural differences between individuals would of necessity produce distinctions and stratifications, as some used their superior endowments, education, or opportunities to accrue greater power and wealth than others.[11]

Central to this discourse was the notion of "nature," a concept whose rhetorical power for Enlightenment authors can scarcely be overemphasized. A touchstone for both natural and moral philosophers, and a point of conjuncture for their often separate endeavors, nature was a key means of legitimating claims to knowledge and plans for reform, be it the nature of Isaac Newton's *Principia* (1687) or Adam Smith's *Wealth of Nations* (1776).[12] This chapter examines what it meant for Jefferson, Adams, and their peers to invoke nature, and thereby link social or political hierarchies to certain "natural" facts about human beings. In so doing, it introduces the book's central question: How have modern republics mitigated the democratic and equalizing tendencies that their establishment seemed to promise? The democracies have managed this balancing act in any number of material and structural ways. Here we explore an ideological one: the recourse to nature—specifically human nature—as a means of arguing for both equality and hierarchy.

To investigate how ideas about human nature and republican governance came to be linked, this first chapter analyzes some of the critical contributions to Enlightenment social theory and their connections to eighteenth-century sciences of the mind. The Atlantic community's republic of letters conducted this conversation mostly through the production of weighty treatises pondering the kind of society that was consonant with human nature and reason. Not surprisingly, most major participants—including Adams's entire gallery of rogues—populated the liberal and radical wings of the eighteenth-century intellectual elite. Their attempts to envision a new social order, while presented in often abstruse philosophical tomes, nonetheless helped set the terms for all subsequent discussions of the nature of an enlightened republic.

Using one of the defining articulations of republican governance, *The Declaration of the Rights of Man and Citizen* (1789), to set the stage, the chapter begins by examining the place and meaning of natural human differences—especially those suggesting scales of superiority and inferiority—in Enlightenment political and mental philosophy. It then turns to the midcentury social theorizing that made human differences a crucial political question: Jean-Jacques Rousseau's renowned exploration of the origins of social inequality and Claude-Adrien Helvétius's provocative responses celebrating the fundamental equality of all human beings. It concludes with the marquis de Condorcet, who exemplifies how late-century writers intertwined the sciences of human nature with theories of republican governance. The development of a language of talents is central to the story, which focuses on the ways Enlightenment writers—while articulating new understandings of society and human nature—employed this language to envision a social order both egalitarian and stratified, and human talents as products of nature and experience.[13] This connection between conceptions of human nature and society became so well established, it suggests, that even radical denunciations of the old order often acknowledged as well individual talents and the rights/opportunities their possession could merit. By tracing how the language of talents developed and drew together theories of mind and governance, the chapter examines the influence natural philosophical notions exerted on the discourse of republicanism, and vice versa, particularly around the issue of limiting the egalitarian possibilities of a republican society founded on virtue.[14] In the conceptions of political and social order that emerged, openness to the reasoned opinion of all was tempered by the conviction that some, through education, nature, or both, possessed talents that privileged their voices in the cacophony of public opinion.

Human Nature and a Language of Talents

Although few Enlightenment authors rejected outright the highly stratified societies they knew so intimately, a group of natural philosophers, moral philosophers, and what might anachronistically be called political theorists, ranging from Thomas Hobbes to Thomas Paine (and including all the figures about whom Adams was so chary), sought nonetheless to define new foundations for civil and political society. They looked to what they perceived as nature's realities and reason's dictates, not to mention the experiences of everyday economic life, as guides for their analyses, rejecting in principle the conventions of custom and the precepts of Christian religion.[15] Because they saw hereditary monarchy and its attendant aristocracy as underlying the structure of the ancien régime, many such writers targeted inherited political status in their attempts to imagine a society reconstituted on "rational" principles. In general, they contended that hereditary rule violated reason because of its susceptibility to instability and corruption, the natural result of its inability to ensure that the most talented individuals reached positions of power.[16] In place of this "irrational" approach, some writers went so far as to argue for a form of democratic republic, one in which "the people," however construed, would be the foundation of the state, with such social hierarchy as remained established on the rational basis of merit rather than the arbitrary accidents of birth.[17] Turning to analyses of human nature as a guide, these authors typically anchored their systems of merit in those attributes that reason deemed valuable—an individual's virtues and talents.

Article VI of *The Declaration of the Rights of Man and Citizen* (1789) provides one of the most striking articulations of this vision of a republican social order founded on democracy and merit. Helping to define the relations between the people and government for the new French republic, Article VI proclaimed that the citizenry would henceforth be equal as to rights, and demarcated solely according to individual virtues and talents:

> VI. Law is the expression of the general will; all citizens have the right to concur personally, or through their representatives, in its formation; it must be the same for all, whether it protects or punishes. All citizens, being equal before it, are equally admissible to all public offices, positions, and employments, *according to their capacity, and without any other distinction than that of virtues and talents.*[18]

As a summation of the tangle of often antithetical ideas lying at the center of late eighteenth-century political discourse on both sides of the Atlantic, Article VI had few rivals. In one short paragraph, it suggested that the people were the source of all law, that representative democracy was the preferred form of government, that law should apply equally to all citizens, and that merit provided the only legitimate grounds for limiting opportunity. Tellingly, distinction was not abolished, but simply placed on a new footing. In this regard, Article VI was characteristic of the era's political writing. Desiring a state founded on rational principles, believing in an original social compact among free (white male) individuals without distinctions of rank, hostile to the traditions of the past, and convinced that power and social roles must be distributed by some rational mechanism, Enlightenment authors seized on merit as the rational counterweight to demands for universal rights.[19]

In a world where social place was still determined largely by birth, the idea that political leadership and social preeminence should derive from an individual's talents and virtues was in many ways revolutionary. Nevertheless, demands to create a thoroughly egalitarian society or democratic government, perhaps by eliminating private property or selecting rulers by lot, were few.[20] Instead, eighteenth-century debate about civil society and political governance focused principally on balancing the claims of rights and the privileges of merit, and thus the competing pulls of equality and difference.[21] It was this debate that pushed questions concerning the nature of virtues and talents to center stage.

While Enlightenment authors made many attempts to define precisely what they meant by "virtues," they were much more circumspect about the term "talents."[22] Almost always used in the plural, "talents" was often connected with other words—"abilities," "capacities," or "faculties"—that together signified attributes that different individuals possess to different degrees.[23] Pronounced disagreement existed, however, concerning whether these differences were many or few, resulted from education or nature, and, in the latter case, whether they were products of heredity or chance. In the main, specific variations mattered less than the term's general signification: an avowedly *natural* criterion for delineating and discussing human differences. "Talents" was one way of speaking like a democrat and yet still being able to justify inequalities.

One consequence of this attempt to anchor the language of merit in such natural attributes was to link the emerging sciences exploring human nature with republican speculations about state, society, and economy. For Enlightenment political thinkers, contemporary investigations into the nature of talents and other human characteristics served a critical role, helping establish the horizon of possibilities within which to theorize their new societies. The widespread preoccupation with nature and reason, along with the development of a class of people defined by their cultural, scientific, and economic accomplishments, rather than birth, rendered speculations about human nature central to the republican project, and oriented the emerging human sciences toward social formations consonant with the developing notions of the republican citizen, the enlightened society, and the self-interested economic actor.

With the significant early exception of John Locke, few of the period's most important political writers themselves undertook serious natural philosophical investigations into entities such as talents. Nonetheless, to Rousseau, Helvétius, William Godwin, James Madison, and many others, a proper understanding of human beings seemed essential, the only means of ensuring that their speculations accorded with the truths of nature.[24] For some, what beckoned was the ancient Aristotelian language of the "faculties," for centuries the primary way of conceiving of the mind. A growing minority of authors, however, looked instead to two new forms of experiential psychology derived from Locke's *An Essay Concerning Human Understanding* (1690, and especially the fourth edition in 1700): the sensation-based psychologies, associated in England with David Hartley and in France with Etienne Bonnot de Condillac, and Scottish Common Sense Realism.[25] Whichever approach they adopted, by making human nature central to their analyses, political theorists linked conceptions of the sociopolitical order with the findings of mental philosophy. At the same time, their varied understandings of human nature revealed the new psychology's ultimate inability to prescribe any one reading of the essence of an individual's talents.[26]

The starting point for all the new Enlightenment psychologies was Locke's *Essay*, particularly his famous dictum that the mind begins as a tabula rasa, devoid of ideas until experience provides sensory perceptions.[27] Locke opened by clearing the philosophical landscape of the possibility of innate ideas, using the examples of children and idiots to demonstrate that no idea is necessarily present in every human being. In the remainder of the *Essay* he then examined the nature of human ideas, establishing that all are derived from experience or the operations of the mind on experience. Locke's goal was not so much to explain the workings of the mind—indeed, in the *Essay* Locke abjured interest in questions of brain physiology—as to argue that all knowledge must be considered empirical and to ascertain to what degree such experientially derived ideas could yield certainty.

For all of Locke's emphasis on epistemology, the *Essay* was read, almost from the start, as a significant contribution to human psychology, particularly in three respects. First, Locke singled out the special importance of investigating the understanding: "it is the *understanding* that sets man above the rest of sensible beings." While this was by no means a new sentiment—since Aristotle, reason had been used to distinguish humans from other animals—Locke's decision to continue to privilege the intellect over the other traditional divisions of the human mind, the emotions and the will, was significant. Second, although Locke had no truck with the notion of innate ideas, his work did accommodate the concept of innate powers. Indeed, Locke's theory required that the mind have an inborn *potential* to act on ideas once acquired through experience, so that it could, via reason, produce new ideas and new combinations of existing ideas.[28] Third, Locke exhibited little interest in explaining how the mind varied in different individuals. Although on occasion he referred to those of lesser understanding, such as idiots and children, their role in his argument was to illuminate the normal mind as abstract and universal, not to underscore differences between minds.[29]

For a long, dense, philosophical work, the *Essay* achieved extraordinary popularity, going through four editions, an abridgment, and translations into French and Latin within ten years, and meriting more than forty reprintings as the century progressed.[30] Widely influential, the *Essay* nonetheless never came to define a psychological school, though three emerged that drew variously on Locke's principles: English associationism, French sensationism, and the Scottish Common Sense school. The first, associationism, was codified by David Hartley in his 1749 treatise *Observations on Man. His Frame, His Duty, and His Expectation.*[31] Although Hartley's theory was derived from Locke's, he saw the mind as passive instead of as the active agent that Locke supposed acquired knowledge by subjecting experience to reason. In Hartley's system, perception induced vibrations, represented to the mind as simple ideas, which either were or could be connected with other ideas by the two forms of association, experiencing sensations simultaneously or sequentially. Although Hartley did not deny that the mind had intrinsic faculties, he believed that the faculties acted within, and arose from, the interplay of vibrations and associations alone.[32]

The second school, sensationism, derived in France principally from the work of Etienne-Bonnot de Condillac. In 1746, the abbé de Condillac published his first major work, *Essai sur l'origine des connoissances humaines,* which articulated elements of his theory of sensationism, a philosophical approach that sought to ground all human knowledge on sensations and the associations of ideas produced by sensations. Like Hartley, Condillac followed Locke in rejecting innate ideas, but denied Locke's contention that reflection could generate ideas not ultimately grounded in experience or in combinations of ideas generated from experience. In Condillac's view, the so-called independent faculties—consciousness, attention, and memory—were different phases of a single power, perception. Condillac detailed sensationist theory most completely, and to greatest effect, six years later in his *Traité des sensations* (1752), where he asked his readers to imagine a statue constructed like a human being, but with a mind devoid of ideas.[33] He then sequentially endowed it with each of the five senses and showed how all human faculties, ideas, and abilities would arise solely from experiences acquired via sense perceptions and their associations.[34]

For Condillac, the brain was a container, with experience providing the sum of the contents—or so he argued explicitly.[35] Although Condillac claimed throughout the *Traité* that "sensation contains within it all the faculties of the soul," in reality the agency of the mind was constantly, and necessarily, present in his theory. To render the statue sentient, Condillac had to assume that from the outset it possessed at least the intrinsic abilities to remember and to manipulate ideas. *With* these powers he could derive all of the higher mental functions; without them, his statue could not have exhibited even the most rudimentary aspects of reason. Certainly there could be no doubt about what Condillac *wanted* his readers to conclude; the final sentences of the *Traité* were unequivocal: "The statue is therefore nothing but the sum of all it has acquired. May not this be the same with man?"[36] And those in France who would push sensationism the furthest, most notably Helvétius, adopted this premise as the cornerstone for their investigations.

If the ambiguous philosophical legacies of Hartley and Condillac suggested that all human beings were endowed at least potentially with equal mental capacities, such was even more the case for those theories' main rival, the Scottish Common Sense school of Thomas Reid, Francis Hutcheson, and Dugald Stewart.[37] The Scottish philosophical tradition, with roots that can also be traced to Locke's *Essay* and David Hume's response to it, united Locke's empiricism with a more dynamic view of the mind, strongly qualifying Locke's notion of the mind as a tabula rasa.[38] In its stead they posited a mind full of active powers, the faculties, each of which could act independently on ideas derived from external sensations and thereby add new elements to them.[39] While the initial set of faculties was innate and universal, most argued that a faculty's power could be strengthened through education and use. As a group, the Common Sense philosophers defined their approach to the mind around two issues: skepticism and morality. They sought to combat Hume's skepticism by accepting the common ways in which humans understood the world—hence the term "common sense"—and to ensure a firm foundation for morality by positing the existence of a moral faculty present in all people.[40]

The philosophical legacies of the Common Sense school, on the one hand, and Hartley and Condillac, on the other, left many questions unresolved. Were all people born with equal mental endowments, or did there exist from birth profound differences in the strength and vigor of the faculties? Could experience (including education) alone shape an individual's talents, and if not, exactly what effect could it have? If most people did possess the same basic faculties, what explained why some excelled where others did not? It was left to other, more politically minded Enlightenment intellectuals, to explore these questions by merging the equivocal visions of the mind articulated in these philosophical treatises with a political discourse in which human nature was of central concern. Rousseau raised the problem in 1755 with his *Discourse on Inequality,* and for many it was Helvétius, just three years later, who articulated the most radical vision of what the new sciences of the mind both promised and threatened.

Rousseau's Excursion: Siting the Natural Inequalities

No eighteenth-century writer formulated a more influential account of the relations between republican societies and natural human characteristics than Jean-Jacques Rousseau. Rousseau's fame, or notoriety, spread quickly throughout the republic of letters upon his entrance onto the Parisian social/intellectual scene, accomplished through a series of dazzling early works that set the agenda for much later eighteenth-century social theorizing. In his 1755 essay *Discours sur l'origine et les fondements de l'inégalité parmi les hommes (Discourse on the Origin and Foundations of Inequality Among Men),* Rousseau sought to diagnose the problems underlying contemporary society.[41] Like Hobbes, Locke, and others before him, Rousseau took his readers on an expedition, a journey to the state of nature to discover humanity in its pristine "original condition" and then back to civilization to describe the evolution of the state of nature into the state of society.[42] His main point was deceptively simple: in a state of nature, outside the "corrupting" influences of civilization, the natural condition (of adult males) was to be free, independent, and equal.

Rousseau's very title suggests his quest: to trace the extreme and extensive inequalities manifest around him. The *Discourse on Inequality* argued that civilization itself produced these differences and that a properly organized society could eliminate many, if not most, of them. Seven years later, he would return to the theme of a new civil order based on the free association of equals in *Du contrat social (On the Social Contract)*.[43] In both works, however, Rousseau betrayed a certain uneasiness about one significant exception in his accounts. His exploration of the state of nature convinced him that inequality had not one source, but two—civilization, and nature itself:

> I conceive of two kinds of inequality among the human species, one I call natural or physical, because it is established by nature and consists of differences in age, health, physical strength, and qualities of mind [*esprit*] or soul; the other may be called moral or political inequality, because it depends upon a kind of agreement and is established or at least authorized, by the consent of men. The latter consists of the different privileges that some enjoy to the detriment of others.[44]

This second type of inequality could not be found in the state of nature. But the first, which included natural inequalities of mind, could. Rousseau did argue that human beings in a state of nature were essentially equal, concluding that however great the differences might be between one person and another, they mattered little outside of civilization. Equality prevailed in nature, he believed, because there was scant opportunity for differences to manifest themselves in the daily struggle to survive, because the solitude characteristic of nature prevented differences from being noticed, and because without civilization there was no chance for small differences to be further developed and passed along.[45] The arrival of even the most rudimentary form of society, however, changed everything. Able to compare themselves to each other, individuals discovered that their talents differed, that some were better at one thing, others at another. Not just difference but, in Rousseau's view, inequality became manifest, and with this inequality pressures increased for the development of civilization and the stratification of the social order:

> Things in this state might have remained equal, if talents had been equal. . . . [However,] the strongest did more work; the most skillful turned his [work] to better advantage; the most ingenious found ways to curtail his work; the farmer needed more iron, or the blacksmith more wheat; and, by working equally, one earned a great deal, while the other had barely enough to live on. Thus, natural inequality spreads imperceptibly along with contrived inequality, and the differences among men, developed by differences in circumstances, make themselves more obvious, more permanent in their effects, and begin, in the same proportion, to influence the fate of individuals.[46]

Two important elements stand out in Rousseau's juxtaposition of the state of nature with that of society. First, he posited an irreducible naturalness about variations in mental and physical characteristics. Unlike the inequalities of wealth, family, and power, differences in physical constitution and quality of mind could persist outside civilization. And this was true even though, like so many of the philosophes, Rousseau embraced the theory that ideas were derived from sensations.[47] Second, differences in "natural" talents could have profound social consequences, undergirding the inequalities and hierarchies that Rousseau found most characteristic of, and deplorable about, civilization.

Rousseau's analysis of the origins of human differences typified Enlightenment discussions of human nature.[48] From such midcentury philosophers as Hartley and Condillac to late-century political writers William Godwin, Mary Wollstonecraft, and Thomas Paine, not to mention Jefferson, liberal and radical authors tended to depict mental and physical characteristics as more fundamental, more real, and more natural than the "accidents" of birth, wealth, or class. Despite intense and important disagreement over exactly what these essential human attributes were, whence they derived, and whether some were more important than others, there was little dispute that such characteristics were anchored in nature and that this connection had political and social significance. Like Rousseau in his investigation into human inequality, many concluded that conceptions of civil society and of human nature must be inextricably linked.[49]

Secondly, Rousseau's "discovery" that even in the state of nature humans would manifest significant differences in talents suggested that those who dreamed of a society shorn of all marks of distinction and difference were misguided. While the social contract might well be fashioned out of independent agents coming together for mutual good, the result might not be the simple equality of the republican *agora*.[50] However equal political rights were among the citizenry, some people would remain naturally better at certain tasks, be they farming, blacksmithing, governing, or reasoning itself. These superior abilities, once revealed in civilization, might be readily translated into political and social advantage, even within the participatory system Rousseau envisioned.[51] Indeed, Rousseau and other philosophes—by and large members of the middling orders or lower—may well have found this conception of human nature attractive precisely because it confirmed their most cherished belief, that there should be rewards for having minds superior to the common run.[52] As Rousseau explained in *Emile* (1762), in a proper society the superiority of the naturally talented would be self-evident to the multitude, who would then look to such individuals for guidance (if only by example).[53]

Having articulated the problem of how to manage the tension between equality and difference, Rousseau provided little in the way of a solution in *Discourse on Inequality* and indeed was still wrestling with its implications in *Social Contract* and *Emile*. Others inspired by Rousseau were equally engaged, confronted by the twin puzzles of first understanding what and how natural the talents were, and then trying to imagine a society that would accommodate these fundamental characteristics of human nature while answering demands for equality. Among the first to weigh in was the tax-farmer turned philosopher, Claude-Adrien Helvétius.

Helvétius, the Egalitarian Challenge, and the Enlightenment Legacy

Renowned champion of equality and sometime protégé of Voltaire, Helvétius was also one of the French Enlightenment's most notorious figures. His two major philosophical works, *De l'esprit* (1758) and *De l'homme* (1772–73), pushed arguments for the natural equality of human beings to new limits, and came to represent for some the most extreme possibilities of Enlightenment thought, influencing—if only negatively—almost all who later considered such topics.[54] Enormously controversial, both works generated criticism from almost every point on the political and intellectual spectrum.

On the surface, there was seemingly little novel about Helvétius's theory of mind. Bypassing all of the equivocations and hesitancies apparent in Condillac's formulation, Helvétius accepted without reservation Condillac's insistence on the primacy of experience, made experience the cardinal principle of his own philosophy, and boldly deduced the consequences.[55] Experience alone, Helvétius maintained in both *De l'esprit* and *De l'homme,* determined every facet of the mind.[56] Innate ideas, innate faculties—both were banished from Helvétius's account, which posited instead a mind composed solely of sensations and the associations of sensations. Helvétius then concluded, in a striking departure from Condillac, that if nothing but sensations produced the mind, then all minds must intrinsically be the same and all differences, even those between geniuses and ordinary people, must arise from variations in experience.[57] "Genius is not the gift of nature . . . a man of genius spends his time in study and application."[58]

Almost any person, therefore, could become a genius. But if that were so, Helvétius wondered, why in fact was genius so rare? The final sentence of *De l'esprit* suggested his answer—politics:

> It is certain that the great men that are now produced by a fortuitous concourse of circumstances, will become the work of the legislature, and that, by leaving it less in the power of chance, an excellent education may infinitely multiply the abilities and virtues of the citizens in great empires.[59]

In *De l'homme,* Helvétius proved much less sanguine about empires and more enamored of republics as the way to ensure the education of the public.[60] Most salient here, however, is not which political system Helvétius favored, but that the conclusion to an essay on mind would be a call for civic reform. Other Enlightenment writers merged philosophy and politics, but Helvétius's linking of the potential equality of human minds with their actual inequality underscored the need for social change. Pushing republican visions of equality and sensationist visions of human malleability to their limits, Helvétius argued that nothing intrinsically prevented the entire population from reaching the same intellectual level. All that was necessary to increase the overall mental abilities of the populace, Helvétius contended, was the desire to do so and the willingness of the state to promote social change, particularly by establishing a comprehensive educational system.[61]

Helvétius devoted little attention to the practicalities of achieving such a transformation, nor did he demonstrate an empirical connection between training and genius. Nevertheless, rather than being dismissed as idle philosophical speculation, both of Helvétius's books—but especially *De l'esprit*—generated extraordinary public controversy. As David W. Smith has noted in his excellent study of the reaction to *De l'esprit:*

> No book during the whole of the eighteenth century, except perhaps Rousseau's *Emile,* evoked such an outcry from the religious and civil authorities or such universal public interest. Denigrated as the epitome of all the dangerous philosophic trends of the age, condemned as atheistic, materialistic, sacrilegious, immoral, and subversive, it enjoyed an immense *succès de scandale.*[62]

While Helvétius's pronouncements about equality helped fuel the fire, the *scandale* was provoked mainly by his theory of morality and seeming willingness, as Smith argues, to resurrect the dreaded materialism of Julien Offray de La Mettrie's *L'Homme machine* (1748). Like La Mettrie, Helvétius was read as denying both the human soul and free will by advocating a strictly mechanical determinism driven by a sensibility potentially present in all matter.[63] As late as 1797, Jean-François LaHarpe condemned Helvétius for having "materialized the mind *(l'esprit)*" and thereby having "systematically attacked all the foundations of morality."[64] Nonetheless, the dramatic events surrounding *De l'esprit's* publication—its quick loss of the *privilège* allowing its publication in France, its condemnation by Parisian religious and civil bodies, and its author's three retractions (twice before the Jesuits and once before the Parlement as he tried to forestall further censure)—made its arguments widely known and much discussed.[65]

Broad dissemination, however, did not produce many converts. For all of the egalitarian rhetoric of the day, few were willing to follow Helvétius in espousing full, natural human equality, nor to sanction his perceived blatant irreligion.[66] Adams, writing to Jefferson in 1813, summed up in his no-nonsense manner the common reaction to Helvétius: "I have never read Reasoning more absurd, Sophistry more gross, in proof of the Athanasian Creed, or Transubstantiation, than the subtle labours of Helvetius and Rousseau to demonstrate the natural Equality of Mankind."[67] Adams did not get Rousseau quite right, as Rousseau was little more sympathetic to Helvétius than was Adams. But Adams did capture the incredulity with which many contemporary and subsequent writers reacted to Helvétius's theories.[68]

Most prominent among the immediate responses was Denis Diderot's, who between 1773 and 1776 wrestled for the second time with the implications of Helvétius's philosophy. Like Adams's offhand remarks, Diderot's long refutation of *De l'homme* emphasized the perceived absurdity of Helvétius's doctrine of human equality. "Why does it seem to [Helvétius] to be proven," Diderot asked rhetorically, "that every man is equally fitted for everything and that his dull-witted janitor has as much intelligence as he himself, potentially at least, when such an assertion seems to me to be the most palpable of absurdities?" Diderot justified his position on the basis of an understanding of human nature that combined sensation psychology and human physiology. In Diderot's view, Helvétius's theory was untenable not just because it violated common experience, but also because it ignored differences in the physical nature of brains. If structure explained the difference between humans and beasts, Diderot argued, then it must explain as well "the varying degrees of intelligence, sagacity, and cleverness between one man and another."[69]

Diderot did not deny the role of experience. Following Condillac and Helvétius himself, Diderot assumed that ideas were generated from sensations and that the mind's primary function was to combine those sensations into ever more complex ideas. Where he diverged was in the primacy to be accorded experience. For Diderot, physiology mediated experience: each individual had a particular mental structure that influenced how and which sensations were combined into ideas. Diderot never developed a full-blown physiological theory of the mind; he was more concerned with critiquing Helvétius's ideas than advancing mental science. Nonetheless, his antagonism to Helvétius's philosophy and his turn to an alternative account of mental function illustrate the connections developing between mental philosophy and political theory.

The visceral response to Helvétius is illuminating. Helvétius struck a nerve, raising questions about the very nature of the republics that Enlightenment theorists were so assiduously attempting to imagine. The problem was simple: What space existed for difference in Helvétius's theory? If Helvétius were right, then it ought to be possible for the state, by educating all equally, to produce similar levels of talents and virtues in every citizen. Fine in theory, but what would it have meant for an actual social and political order if distinctions could not be justified rationally? The result would have to be either an absolutely egalitarian society—a prospect with few proponents—or recourse to arbitrary distinctions, a solution difficult to square with condemnations of hereditary aristocracy.

By advocating arguments based on a not easily dismissed theory of the mind, Helvétius threatened to bring the whole developing edifice of merit tumbling down. While some theorists, such as Diderot, responded by uniting experiential psychology with notions of physiological difference, these claims were at best open to dispute. The safest and most common approach to Helvétius was simply to dismiss his ideas as chimerical, and then move on to less troubling matters. Adams did so in his letter to Jefferson. Helvétius's erstwhile patron Voltaire was even more pointed in a 1773 letter: "No one will convince me that all minds are equally suitable to science, and that they differ only in regard to education. Nothing is more false: nothing is demonstrated more false by experience."[70] After the American and French revolutions, few would have disagreed. Even radical republicans such as Condorcet, Paine, and Wollstonecraft formulated their approaches to remaking the social order on decidedly different grounds.

"Fashion has introduced an indeterminate Use of the Word 'Talents'"

By the late eighteenth century, the newfound proximity of the languages of mental philosophy and social theory had taken on considerable importance for many authors, shaping their sense of the possibilities for organizing republican societies and structuring republican governance. William Godwin, for example—high on Adams's list of suspect political authors—in 1793 boiled down his organizational principle for society and government to its essence: "The thing really to be desired, is the removing as much as possible arbitrary distinctions, and leaving to talents and virtue the field of exertion unimpaired."[71] Like Jefferson and many other late-century republican theorists, Godwin was clear that merit should be preeminent and that talents were integral to any conception of merit. But complexity and ambiguity reigned when it came to defining one of these central concepts, "talents." While late eighteenth-century authors often invoked the phrase "virtues and talents," they rarely attempted a precise definition. Instead, "talents" functioned mainly as a placeholder, an open signifier suggesting a multitude of positive characteristics without, like "virtue," being restricted to particular denotations. In *The Rights of Man* (1791–92), for example, radical pamphleteer Thomas Paine littered his text with references to talents and how successful government required a variety of them.[72] But he provided few concrete examples and never a specific definition. He simply assumed that his readers would know what he meant.[73]

For our purposes, "talents" must be defined more precisely, given the centrality of talents to the project of determining the boundaries that human nature set on republican government and society. Paine did provide one important insight in this regard: he often suggested in *Rights of Man* that talents were intimately connected with powers or faculties of the mind and were critical to the possibility of good leadership. "We have heard the Rights of Man called a levelling system," he declared while emphasizing the importance of intellectual talents for rulers and decrying the inability of hereditary mechanisms to ensure such abilities,

> but the only system to which the word levelling is truly applicable is the hereditary monarchical system. It is a system of mental levelling. It indiscriminately admits every species of character to the same authority. Vice and virtue, igno-

> rance and wisdom, in short, every quality, good or bad, is put on the same level. Kings succeed each other, not as rationals, but as animals. It signifies not what their mental or moral characters are.[74]

Contemporary dictionaries corroborate Paine's use of the word. Samuel Johnson, for example, defined "talent" in 1755 as "faculty; power; [or] gift of nature," a meaning little different from that provided by *Le Grand vocabulaire françois* (1773) and the *Dictionnaire de l'Académie françoise, nouvelle édition* (1786), both of which defined "talent" as a "gift of nature, natural disposition or aptitude for certain things, capacity, [or] ability."[75] In both languages, "talent" was taken to refer to a potential for superior achievement, with the suggestion that it was inborn rather than developed.[76] The French dictionaries implied as well that "talent" referred to specific, externally manifested characteristics, especially those facilitating the accomplishment of particular tasks.[77]

In eighteenth-century dictionaries, therefore, "talents" denoted attributes of mind or body, perhaps present from birth, with a hint in French that it referred to external accomplishments rather than internal potentials. Left at this level of generality, "talents" could signify almost any operation that an individual could perform. Indeed, Adams exploited just this openness when responding to Jefferson's missive on natural aristocracy:

> We are now explicitly agreed, in one important point, vizt. That "there is a natural Aristocracy among men; the grounds of which are Virtue and Talents." . . . But tho' We have agreed in one point, in Words, it is not yet certain that We are perfectly agreed in Sense. Fashion has introduced an indeterminate Use of the Word "Talents." Education, Wealth, Strength, Beauty, Stature, Birth, Marriage, graceful Attitudes and Motions, Gait, Air, Complexion, Physiognomy, are Talents, as well as Genius and Science and learning.[78]

Adams, of course, was having a bit of fun at Jefferson's expense, twitting Jefferson's faith in the common citizen's judgment and his idealized vision of democratic politics. Nonetheless, Adams's remarks are revelatory in at least two respects. First, his acknowledgment of the political implications of talents made clear the centrality that aspects of human nature had assumed in republican discourse. And second, Adams's cynicism highlighted a critical feature of Jefferson's conception of "talents": Jefferson restricted it to those attributes he deemed essential to a democratic republic's continued flourishing—genius, science, and learning. Nor was Jefferson alone: if "science" were dropped or broadened to denote any kind of systematic thinking, and if reason were included, then Jefferson's version of "talents" would approach what was routinely assumed about politics and society by all of the Enlightenment theorizers Adams held in such contempt. Each considered himself/herself the final judge of what constituted true talent, and most sought to legitimate their own authority by establishing the value of such talents as theirs to a republic.

Mary Wollstonecraft, a leading light in Godwin's circle of radical Enlightenment authors, used "talents" in precisely this way.[79] Wollstonecraft was concerned throughout *A Vindication of the Rights of Woman* (1792) with delineating which attributes made a good republican citizen and which did not. Style, cunning, coquettishness, sensualism, and dissimulation—skills commonly associated with both women and aristocrats—had no merit in republics.[80] Rather, Wollstonecraft valorized virtues and talents that revealed independent minds well-stocked with knowledge; reasoning faculties finely sharpened; passions and interests dominated by reason; and politeness, modesty, and concern for the good of the whole.[81] Characteristics that accorded well with Jefferson's ideal of the independent yeoman farmer and Robespierre's of the virtuous citizen, they were mostly "male," associated in particular with hardworking men of the middling sort whose talents had fostered success in public or economic life.[82] "Abilities and virtues," Wollstonecraft noted, "are absolutely necessary to raise men from the middle rank of life into notice, and the natural consequence is notorious—the middle rank contains most virtue and abilities."[83]

In a democratic republic, therefore, as even Adams would have agreed, not all talents were meritorious; the trick was to determine which would contribute to an individual's wisdom or virtue, and whether and how those talents could be developed. For many philosophers theorizing the republican project, the general answer to the first question was clear: the most critical talents involved the ability to acquire and generalize from information, to rise above local prejudices, and to channel passions and desires according to reason. Paine lauded dispassionate wisdom; Wollstonecraft declared that "[t]he power of generalizing ideas, of drawing comprehensive conclusions from individual observations, is the only acquirement, for an immortal being, that really deserves the name of knowledge"; and even Rousseau proclaimed that "the faculty of self-improvement" differentiated humans from beasts.[84] Such characteristics were also, as many feminist scholars have demonstrated, routinely gendered masculine, thus providing justification for excluding women from active membership in the polity while remaining committed to universalist principles.[85]

The question of where talents came from, why some people had more than others, and why it mattered—crucial to exploring how far beyond rights equality a republic could theoretically go—was more complicated and generated little consensus. Some authors argued for the primacy of native endowment, others education; some saw heredity as essential, others thought it inconsequential; some distinguished between individuals and groups, others did not. The differences were enormous, indicating the high stakes riding on the outcome and the evidence's equivocal nature.

Probably the most important distinction worried over was that between talents as gifts of nature and as products of education. As we have seen, contemporary dictionaries suggested that talents were present from birth. It underlay both Jefferson's ideal of rule by a natural aristocracy and Paine's celebration of the potential of republican democracy to bring forward the talented, whatever their social origin.[86] Even in *Emile,* Rousseau repeatedly returned to the distinction between natural and moral inequalities, arguing that methods of education and levels of expectation must be calibrated to individual natural abilities. "Each advances more or less according to his genius, his taste, his needs, his talents, his zeal, and the occasions he has to devote himself to them," Rousseau observed early in his discussion of Emile's education.[87]

Rousseau, of course, dedicated most of *Emile* to illustrating the power of education to shape an individual's talents, and in this he was by no means alone. Virtually all writers conceded that external influences could have a significant effect on human talents. In *Rights of Woman,* for example, Wollstonecraft contended that women lacked many of the abilities required of republican citizens because they had been denied the necessary training. Proper education, she asserted, would demonstrate that women had the same kinds of faculties as men, though whether to the same degree she left an open question, at least rhetorically. Only once women could exercise their mental faculties to the fullest, she argued, would it be possible to determine their appropriate political rights and social roles. What she did not deny, however, was that individuals, whether male or female, naturally differed in the talents with which they were endowed and the degree to which development was possible.[88]

Other writers—including Godwin, Helvétius, Antoine Louis Claude Destutt de Tracy, Joseph Priestley, and Benjamin Rush—pushed the power of education much further, according it a definitive role. For these authors the mind was a tabula rasa, empty of all ideas and potentials before being filled by experiences. Godwin's attacks on physiological explanations of human abilities in his *Enquiry Concerning Political Justice* (1793) illustrate well this "environmentalist" position. First, he maintained that while individuals might differ in their original endowments, those differences were insignificant. "But, though the original differences of man and man be arithmetically speaking something, speaking in the way of a general and comprehensive estimate they may be said to be almost nothing."[89] Second, he reversed the notion that physiological characteristics, such as skull size, might cause differences in abilities by suggesting that human physiognomy was largely shaped by individual action.[90] And third, Godwin rejected the notion that capabilities were heritable. In place of native endowment, Godwin emphasized the powers of education: "whether the pupil shall be a man of perseverance and enterprise or a stupid and inanimate dolt, depends upon the powers of those under whose direction he is placed. . . . [T]here are no obstacles to our improvement, which do not yield to the powers of industry."[91] In thus denying the importance of original endowments and maintaining that experience created the most significant distinctions between individuals, Godwin summed up nicely the fundamental precept underlying the arguments of those most committed to explaining human talents on the basis of education.[92]

Reflecting their environmentalist orientation, Godwin and others like him for the most part discounted heredity as a determining factor in individual abilities. But even some who adopted a more physiological approach also gave short shrift to hereditarian explanations of natural capacities. Paine, for example, denounced hereditary monarchy on the grounds that mental and moral abilities did not follow any simple law of heredity. Good kings, he proclaimed, rarely produced talented heirs:

> When we see that nature acts as if she disowned and sported with the hereditary system; that the mental characters of successors, in all countries, are below the average of human understanding; that one is a tyrant, another an idiot, a third insane, and some all three together, it is impossible to attach confidence to it, when reason in man has power to act.

Rather, in Paine's view the talents necessary to good government were many and varied, and were scattered anew throughout all classes with each generation. Only a society open to developing abilities wherever they lay, he concluded, would operate in concert with the dictates of reason and nature:

> Experience, in all ages, and in all countries, has demonstrated, that it is impossible to control Nature in her distribution of mental powers. She gives them as she pleases. . . . [Wisdom] is like a seedless plant; it may be reared when it appears, but it cannot be voluntarily produced. There is always a sufficiency somewhere in the general mass of society for all purposes; but with respect to the parts of society, it is continually changing its place.[93]

Others, however, were not so sure. Jefferson and Adams, for example, were certain that talents could be bred, if only humans would marry with the same attention to lineage as was employed in mating sheep.[94] Physician Benjamin Rush, though a vociferous advocate of the transformative power of education, also adopted something of a hereditarian stance, speculating that heredity might prove to be the most significant factor in the apportionment of talent. "The time may come," Rush speculated, "when we shall be able to predict, with certainty, the intellectual character of children by knowing the . . . different intellectual faculties of their parents."[95]

What is most striking about late Enlightenment discussions of heredity is that there was no orthodox position. As much intellectual respectability lay in denying the power of heredity as in strictly endorsing it, and a vast array of intermediate positions—involving belief in the inheritance of acquired characteristics and/or in environmental forces acting on the fetus—were also advanced.[96] Even among eighteenth-century French physicians, Carlos López-Beltrán has shown, the clear reality of familial resemblances and transmission patterns of ancestral pathologies yielded little certainty about what was or was not hereditary.[97]

The consequences of such an open-ended understanding of heredity for notions of republican governance were twofold. On the one hand, individual republican theorists could not escape the need to decide what they believed about the heritability of talents, and what those beliefs implied for a functioning republic. Paine's vision of an open, liberal, and nonaristocratic society stratified by talent but founded on equal opportunity for all, for example, depended on his conviction that human minds were replete with faculties, that these faculties were largely a consequence of physiology, that powers of mind were rarely inherited, and that abilities were distributed unpredictably throughout a population. Different "facts" about heredity might have resulted in a much different vision of the sociopolitical order.

On the other hand, the variety of credible approaches to the heritability of talents meant that different visions of republican governance could accord with the truths of nature. Science might dictate the terms, but much depended on which scientific rendering of human nature an author credited. For radically egalitarian thinkers such as Godwin, for example, the benefits of conceiving of the mind as malleable were obvious: every putative difference between individuals or groups that could justify social hierarchy was vitiated if all differences were the result of education.[98] The same theory, however, was used by Adam Smith, among others, to endorse a much different social agenda: the development of a differentiated and stratified workforce. Belief in the plasticity of talents did nothing to mitigate Smith's conclusion that an unequal social structure was essential to a nation's prosperity and a society's happiness.[99]

Similarly, while reference to human mental characteristics as the products of native endowment could readily sustain arguments for the status quo, neither Paine nor Diderot, for example, had any difficulty in reconciling such a position with an interest in the radical transformation of society.[100] Convinced that talents, while gifts of nature, were spread equally throughout all social classes and that republics required a diversity of talents to prosper, both concluded that a republican society must open opportunity to all.[101] Those who adopted a physiological orientation, however, had more difficulty conceiving of a society in which hierarchy of every sort was deemed artificial and thus eliminable. For them, the reality of talents made not only distinctions but stratifications seem both inevitable and justified. Moreover, because the meaning of "talent" carried some sense of comparative superiority, and the kinds of talents that they deemed relevant to republican societies were so circumscribed, the slippage between superiority in one domain and superiority overall occurred easily, as Jefferson's conception of the natural aristocracy vividly demonstrates. Thus, even when denouncing traditional aristocracy in the name of equality, authors with a physiological understanding of the origins of talents tended to reinscribe inequality into the heart of their approaches.[102]

The number of possible ways in which conceptions of merit, understandings of talents, and theories of society and governance could be rendered and combined by republican theorists was thus quite large. Even liberal and radical writers substantially disagreed over just how perfectible (in Adams's word) human beings were, as well as over issues of the heritability of talents, their nature and number, and their relative importance in defining the kind of republican state and society that seemed feasible.[103] To explore these connections in more depth, the chapter concludes by examining how one such contributor to the radical Enlightenment attack on traditional society, the marquis de Condorcet, navigated these complexities and allowed his understanding of talents to help shape his conception of the nature and limits of a republican polity.

"Nature has set no term to the perfection of human faculties"

The writer perhaps closest in outlook to Helvétius and the author of one of the late Enlightenment's greatest paean's to human perfectibility—*Esquisse d'un tableau historique des progrès de l'esprit humain* (1795)—was Jean-Antoine-Nicolas de Caritat, the marquis de Condorcet. Mathematician and protégé of *encyclopédiste* Jean d'Alembert, permanent secretary of the Parisian Académie des Sciences, champion of human equality, and tireless advocate of social reform during the first stages of the French Revolution, Condorcet was also one of the Revolution's most prominent martyrs.[104] A liberal aristocrat, Condorcet, like Helvétius, became a zealous convert to republican ideals and a staunch supporter of many of the Revolution's early goals. Nonetheless, his vote against executing the king, support of universal suffrage, and public pronouncements in favor of the peaceful resolution of differences through reason guided by moral science eventually put him at odds with the increasingly powerful and violent Jacobin faction. Denounced for his moderation, Condorcet went into hiding from July 1793 until he was captured and then died in prison in March 1794, using those months to compose the *Sketch for a Historical Picture of the Progress of the Human Mind.*

In this and other works from the 1790s, Condorcet imagined a future that in many ways realized Helvétius's most exuberant hopes.[105] Starting from principles in agreement with sensationist conceptions of human nature, Condorcet argued for, and celebrated, the unlimited possibilities for human perfection. Like Helvétius, Condorcet considered education critical: with comprehensive public instruction for females and males, poor as well as rich, all would be able to embrace their basic natural rights, participate fully in public functions, and contribute to the endless improvement of humanity. Nor did Condorcet limit his vision of human perfectibility to Europeans. All peoples, even the "savages" of Africa and Asia, Condorcet suggested, were fundamentally the same in their ability to progress. With proper education, that provided by European "teachers," he explained, any nation could pass through the stages of civilization described in his *Sketch* and reach the exalted heights so far attained only by the French and their brethren.[106] Notwithstanding the intense Francocentrism, Condorcet's conception of human nature was strikingly universalistic: for all, education was the royal road, and perfection the common destiny.

Despite this profound commitment to unlimited human improvability, Condorcet did not follow Helvétius in denying the reality or importance of individual differences. Rather, he conceded that different individuals endowed by nature with different talents would contribute each in their own way to attaining the final goal of human perfection.[107] Condorcet provided no account of how the sensationist theory of mind he adopted explained these natural differences in talent. Like Rousseau, he simply assumed that education and environment, as powerful as he thought they were, failed to account for all human difference and theorized about what implications those natural inequalities might hold for the construction of a rational society. "All individuals are not born with equal capacities," Condorcet observed in "The Nature and Purpose of Public Instruction" (1791), "nor, taught by the same methods for the same number of years, will they learn the same things." Condorcet declared that the purpose of education in a republic was not to give every person the same mental endowment, but only to raise all individuals to a level sufficient to exercise their political rights. Education, he insisted, should be calibrated to the individual's mental capacities:

> It is therefore incumbent upon society to offer each and every man the means of acquiring an education commensurate with his mental capacities and the time he can devote to his instruction. No doubt, a greater difference will result in favor of those endowed with greater talent and those to whom an independent fortune allows the liberty to devote a larger number of years to study. But if this inequality does not subject one man to another, if it strengthens the weakest citizen without giving him a master, it is neither wrong nor unjust.[108]

Condorcet sought to temper the elitism of his educational plan in two ways. First, like Paine, Condorcet believed that natural talents were distributed at random throughout the population, and thus that the system of education must encompass the entire citizenry. Liberal that he was, Condorcet accepted that one of the fruits of wealth would continue to be access to a superior education. However, alongside these privileges of wealth he placed the opportunities that should be accorded on the basis of talent. Arguing that abilities needed to be nurtured wherever they were found, Condorcet called for the establishment of a comprehensive system of primary education.[109] Everyone, he argued, deserved instruction in the basic skills necessary to be a good republican citizen. From this starting point, however, only some—those identified as having the requisite talents—would continue to the advanced training that would produce the most important social distinctions. These students would receive a purely scientific education, one that would allow them to generate new knowledge for the benefit of humankind.[110]

In addition to arguing that the elite should be open to all ranks of society, Condorcet maintained that the role of that elite must be well circumscribed. An advocate of a minimalist state in which individual liberty and freedom were maximized, Condorcet desired that as many public duties as possible be open to the abilities of all.[111] As he observed in his essay on education: "The freest country is that in which the greatest number of public functions can be exercised by individuals who have received only an elementary education. Thus it is essential that the laws seek to simplify the exercise of these functions."[112] The most talented could help accomplish this not by ruling in the name of the people, but by devising those laws that would allow citizens of average ability to operate the levers of state power with skill.

Indeed, at the center of Condorcet's vision of a properly organized republic lay this tension between his desire for broad access to political leadership and his acknowledgment that certain people were naturally more able than others. His solution was to try to limit the role for talent to creating a machinery that would extend the powers of all, and to insist that in most areas of public life the average citizen's abilities must be sufficient and popular sovereignty paramount. As he explained in his essay "On the Admission of Women to the Rights of Citizenship" (1790):

> with the exception of a limited number of exceptionally enlightened men, there is absolute equality between women and the remainder of the men. . . . Since it would be completely absurd to restrict to this superior class the rights of citizenship and the eligibility for public functions, why should women be excluded from them any more than those men who are inferior to a great number of women?[113]

Condorcet imagined one final way in which talents need not undermine republican government. Employing an understanding of heredity that gave great space to the inheritance of acquired characteristics, Condorcet concluded his *Sketch* with a prophesied future in which the gains in talent produced by education would be passed on to the next generation as elements of their physical constitutions. However disparate natural abilities might now be, Condorcet assured his readers, future generations would eventually approach perfection together and all would become members of the elite.[114]

Condorcet's optimism brought him to a position similar to Helvétius's. In the long run, Condorcet concluded, education would help render all people socially as well as politically equal, and would also ensure the progress of humankind. Nonetheless, echoes of Rousseau's worries about the implications that differences in natural talents would have for social equality haunted Condorcet's account. In Condorcet's own age, and in the near future, human difference as well as human equality would shape the social order. Condorcet worked assiduously to limit the effects of those differences on his imagined republic, but he could never quite escape the implications of what he took to be this fundamental truth about human nature. Always, the natural elite loomed behind his speculation, constituting both the engine of universal progress and the possibility that some might remain more equal than others.[115] While progress might mitigate this natural aristocracy's power, would it ever (save in some impossibly distant future) eradicate the special role that Condorcet, with whatever reluctance, had had to assign the naturally talented? Could governance derived from the people ever become governance completely by the people? And in the end, how egalitarian would human nature allow Condorcet's imagined republic to become? These questions troubled not just Condorcet's vision of a republic founded on reason, but the conceptions developed, as we have seen, by most other Enlightenment republican theorists as well.

Conclusion

The responses to Helvétius's "provocation" illustrate the enormous number of fault lines that radiated around the concept of "talents" and its place in the Enlightenment vision of a republican society. The ambiguities surrounding "talents" gave the term enormous power and resonance. It could be used simultaneously to legitimate social and occupational distinctions and to validate broad-based educational systems designed to foster talent across those distinctions' very lines. It provided a language in which to argue for greater political and social power as well as a means to exclude whole groups from all but the most basic rights. It helped to consolidate a system that seemed to offer opportunity to all and yet justified restricting those opportunities. In other ways, however, the ambiguities surrounding "talents" presented real problems. By leaving the concept ill defined, the boundary demarcating where universal rights ended and the privileges of talent began was open to continual renegotiation. According mental characteristics such a prominent place in the imagined republic also left open the possibility that changes in theories about the mind could have significant ramifications for how the sociopolitical order was understood and structured. As long as the system of governance remained justified in terms of claims about human nature and the illuminations of reason, it was imbricated in the knowledge systems that gave such entities definition and meaning.

The precariousness of this arrangement is particularly visible in Wollstonecraft's arguments for the rights of women. As we have seen, the core of Wollstonecraft's argument was that women—and any other group—deserved just that rank in society consonant with the mental capacities with which nature had endowed them and that education had developed.[116] Unquestionably Wollstonecraft believed women to be men's equal in terms of native intellectual abilities. Indeed, her essay is replete with references to prominent women whose manifest abilities established incontrovertibly their possession of faculties at the level of any male's.[117] But by couching her argument in terms of native endowments and faculties that could differ by degree, she left open the possibility that new truths about the actual potentials of various intellects could have significant political and social repercussions. If women's faculties were proven physiologically inferior, for example, there would be strong grounds, according to Wollstonecraft's logic, to limit women's roles in political and civil life. Given her own middle position on the origins of talents, she thus left open to question how much education could remedy inequities in the distribution of talents and thus how egalitarian a republican society could be.

In articulating a prominent role for talent in the new republic, therefore, Jefferson and Adams looked forward as well as back. Their vision of an aristocracy of intellect captured an element in republican culture that would continue to attract the energies of political thinkers and the interest of the emerging human sciences, not to mention the passions of reformers of all stripes. Jefferson and Adams identified, if nothing else, a language through which outsiders could claim inclusion and experts positions of authority within a democratic system of governance. What is more, their sense that human differences were relevant solely to the degree that they were derived from nature (because nature alone was seen as standing outside the partisan conflict of personal interests) meant that those theorists able persuasively to interpret nature might attain a significant public voice. By the early nineteenth century, two separate lines of discourse were emerging around this interaction of political theorizing and understandings of human nature. The first, explored in chapter 2, followed directly from these Enlightenment interrogations of human nature and republican societies and focused on how individuals both could and should be demarcated one from the other. The second, present more as a shadow hovering over Enlightenment analyses such as Wollstonecraft's, concerned races and groups. As we will see in chapter 3, this discourse, much more physiological in orientation, functioned principally to explain why the differences between at least certain groups could not be readily altered.

Two

Mental Capacities and Orthodox Minds

MENTAL SCIENCE, EDUCATION, AND THE POLITICS OF INDIVIDUAL DIFFERENCE

> When hereditary wealth, the privileges of rank, and the prerogatives of birth have ceased to be and when every man derives his strength from himself alone, it becomes evident that the chief cause of disparity between the fortunes of men is the mind. Whatever tends to invigorate, to extend, or to adorn the mind rises instantly to a high value.
>
> Alexis de Tocqueville, *Democracy in America* (1840)

As THE NINETEENTH century opened, both the United States and France faced crossroads. With the fervor of revolution fading, each republic confronted the sobering prospect of establishing a stable political order out of the theoretical pronouncements and specific grievances that had characterized their revolutionary discourse. On the surface, their trajectories during the next seventy-odd years could scarcely have diverged more. With the adoption of the Constitution in 1789, the Americans created a political structure robust enough to withstand not only a second war with the British Empire, but also regional antagonism and civil war. France, meanwhile, careened from one form of government to another, exchanging republic for empire for monarchy for liberal monarchy for republic for empire and once again for republic.

Nonetheless, both the United States and France wrestled with a similar political problem: how to fuse elements of democratic, republican, and meritocratic theory into a practical political structure, one that could also sustain market capitalism. Neither country's revolutionary legacy presupposed exactly how these pieces would be combined. Both, however, assumed that the ideals of equality and realities of social difference would remain and would need to be reconciled. Although by 1800 few in France or America doubted that republican citizens had replaced monarchical subjects as the fundamental political units, fewer still could imagine a social world void of distinctions. Even the most radical—Jacobins in France and anti-Federalists in the United States—assumed that fundamental, perhaps permanent, differences in abilities distinguished one individual from another, and that such differences could have significant political repercussions. For most, therefore, Helvétius's thoroughgoing egalitarianism seemed largely irrelevant, even though traditional formulations for managing difference—especially those reliant on hereditary rank—had become suspect. Instead, political thinkers turned to different categories—including nature, character, virtue, and talents—to explain the persistence of difference despite the equality of rights in their new societies. In their view, the solution was to eliminate all *artificial* distinctions, so that an individual's *natural* talents could flourish, benefiting both individual and nation.

This chapter focuses on how nineteenth-century writers used "talent" to integrate meritocracy with democracy (in America) or with bureaucratic centralization (in France) so as to explain difference and justify unequal allocation of political or social goods within the population. It explores why many Americans and their French cousins felt compelled to refashion their political and cultural languages in light of new understandings of human nature and the notion of talents. In particular, the chapter investigates the growing influence of philosophies representing the mind as a collection of faculties whose powers arose from both native propensity and acquired facility, and how these notions intersected with and helped justify social class and occupational stratification, the gendering of separate spheres, and restrictions on access to higher education.

A vigorous historiography exists, especially for the United States, on the roles virtue and character played in linking human characteristics to nineteenth-century political and social issues.[1] Eighteenth-century "virtue," historians agree, underwent something of a gender split in the early nineteenth century, as the vitality of republican ideology waned.[2] Virtue proper became primarily a female attribute, tied to moral purity, sociability, and protection of the domestic realm, and something that could exist in greater or lesser degrees.[3] For men, the nineteenth-century term of greatest potency was "character," an inner-directed sense of self supposed to guide individuals through the increasingly competitive and chaotic public worlds of politics and the market economy.[4] Both were treated primarily as unitary, acquirable characteristics, things one either did or did not possess and ideals toward which bourgeois women and men should strive.

Talents, however, were much different. Heterogeneous and largely inborn, though open to development, the lure of talents, especially to the American middle classes, lay primarily in their ability to capture and naturalize individual differences. Talents provided a language with which to frame distinctions between people and legitimate the social and gender inequalities that resulted. Because of its enormous flexibility, however, the language of talents was attractive to commentators of various political persuasions.[5] Liberals and radicals often employed the term to challenge hierarchies in the name of equal opportunity for "all," regardless of class origin (but rarely age, race, etc.).[6] More conservative commentators were equally enamored with talents, using them to justify social hierarchies on grounds seemingly consonant with nature and democratic culture. The malleability and occasional instability of the concept and the politics it underwrote, however, became problematic for many commentators when confronting group-level distinctions, most notably of gender and race. As we will see in chapter 3, the very difficulties in pinning down the term—which made talents so attractive when debating how to organize the educational system, for example—became liabilities when attempting to demarcate permanently one group from another. In response, a new language was fashioned to account for those differentiations, one centered on a term that also foregrounded mental ability, but with very different connotations: "intelligence."

Languages of Politics and Visions of Difference in Antebellum America

Predictably, perhaps, many Americans greeted the dawn of the new century with a zeal for retrospection, seeking to divine their new republic's future from its tumultuous past. Throughout the nation, century sermons were delivered on the last Sunday of the eighteenth century or the first day of the nineteenth, few better known than the one preached by Samuel Miller, a prominent New York City Presbyterian minister, evangelical Calvinist, and Jeffersonian Republican.[7] Elaborated and extended into the two-volume work *A Brief Retrospect of the Eighteenth Century* (1803), Miller intended his observations "to attempt a review of the preceding age, and to deduce from the prominent features of that period such moral and religious reflections as might be suited to the occasion."[8] For the most part, Miller simply synopsized the major intellectual trends of the eighteenth century. At least twice, however, he went further, roused to angry denunciation of positions he felt flouted morality and orthodoxy.

The first—surprisingly for an ardent Jeffersonian—he associated with the names of Helvétius, Condorcet, and Godwin: belief in the "perfectibility of man." Miller argued that their philosophy embraced a materialism and faith in education's power that contradicted both experience and the dictates of reason and revelation. Believing that human nature could be perfected, Miller declared, was empirically false and morally dangerous, as it substituted human action for God's redemption and obviated Christian religion and individual salvation. Nonetheless, Miller did *not* reject Helvétius's account of the power of education to form individual talents: "It will be readily granted, indeed, to the advocates of this delusive system, that education is extremely powerful; that much of the difference we observe in the talents and dispositions of men is to be ascribed to its efficacy." Rather, what Miller objected to most strenuously was the notion that as individuals or societies increased their knowledge, moral character would necessarily improve.[9] Education might explain difference, but it could not, in Miller's view, eradicate sin.

Miller's ire flared again when considering another notorious freethinker, Mary Wollstonecraft. Upset by the *Vindication of the Rights of Woman*'s wide circulation, Miller devoted his chapter on education largely to a lengthy critique of Wollstonecraft's position. As with Helvétius et al., Miller criticized little about Wollstonecraft's understanding of the mind or her conclusions about women's capacities. He happily conceded that with equal education women *could* evince mental talents equivalent to men's:

> the idea of an *original* difference between the mental characters and powers of the two sexes has been pushed greatly too far. . . . Females, if it were practicable or proper to give them, in all respects, the same education as that bestowed on men, would probably discover nearly equal talents.

Miller argued, rather, that women *should* not receive the same education as men, because nature and God destined women for functions distinct from men's, and because morality and decency required separate educations for men and women. In Miller's view, women were "destined for different pursuits," which called for different educations and would give rise to different talents.[10] The sexes were distinct, according to Miller, not because nature had endowed them with unique talents, but because society and morality dictated separate, morally equivalent spheres for each. Miller thus accepted some of the most radical late eighteenth-century contentions about the origin of difference, the nature of equality, and the effects of education in shaping talents; by subsuming the issue of difference to morality and religious orthodoxy, however, he drew strikingly different conclusions. Where Enlightenment writers had used such principles to assault artificial distinctions in the name of talent, Miller reinscribed difference, merging the potentials of talent with the prescriptions of convention and social order.

Miller was not alone. Numerous Americans—anxious about economic and social changes, desiring political stability, and fearful that the French Revolution's excesses were a direct consequence of radical Enlightenment thought—demanded that order be reestablished in their troubled republic, combining calls for a renewed commitment to moral values and religious orthodoxy with celebrations of merit-based hierarchy.[11] Articulated most explicitly by Federalists and Whigs, including John Quincy Adams,

Lyman Beecher, Catherine Beecher, and Daniel Webster, and spread nationally during the Second Great Awakening, this ideology of perfectionism and cultural conservatism became a dominant way for nineteenth-century middle-class Americans to make sense of the world. At its core, the view promoted by leading Whigs emphasized the need to harmonize the population's diverse interests and talents in order to ensure a stable social order. Celebrating the importance of character, propriety, respectability, and deference, they touted as well economic expansion and the agency of government, both personal and social, to accomplish their agenda.[12]

The Federalists/Whigs were not alone on the political stage, however. Heirs to the Enlightenment radical tradition continued to push for comprehensive, egalitarian solutions to the problem of governance and human nature. In 1802, for example, the Irish-born Philadelphian physician James Reynolds, a staunch advocate of workingmen's rights, published "Equality, A Political Romance," an eight-part article in *The Temple of Reason*. Reprinted in book form as *Equality; A History of Lithconia,* Reynolds's tract imagined a utopian society, Lithconia, founded on the principle of complete equality for all. Property was abolished, men and women were treated absolutely equally, and governance, to the slight degree necessary, was participated in by everyone:

> Every man and woman, if they live long enough, will succeed in their turn to the duties of administration. . . . Elections for the purpose of choosing men of great abilities, or men best acquainted with the interests of nations, or who are most conversant in the constitution and laws of the state; or, those who have much at stake in the country, and are supposed on that account to have the greatest interest in its prosperity and welfare—are unnecessary here. Every man's stake in the country is equal.

A kind of reworking of Rousseau's tale of the move from the state of nature to civilization, Reynolds's story added a twist. In his telling, a return to human nature, guided by "the united reason of man," could overcome civilization's inequalities.[13] Reynolds's Lithconia evolved out of an initial condition little different from that of the new American republic. Working within and then beyond the nation's constitution and laws, however, "men of superior talents" slowly guided the development of Lithconia into an egalitarian, communally organized, producers' republic where individual human nature flourished and society achieved perfection.

Despite his thoroughgoing commitment to fundamental human equality, Reynolds no more than Rousseau or Condorcet could entirely escape belief in nonartificial inequalities, the natural distinctions. In part he attempted to finesse the issue by arguing that the most important distinctions were those of age.[14] When describing the Lithconian educational system, the special rewards for mechanical inventiveness, and especially the "men of intelligence" who had guided Lithconia's development, however, Reynolds went further, conceding that important, permanent human differences existed that rightly resulted in favors, opportunities, or authority not accorded to all. How exactly such rewards for the talented meshed with his radical egalitarianism, Reynolds never addressed. But their persistence suggests that even the most extreme Jeffersonians could not imagine a democracy completely stripped of difference.

More moderate Jeffersonian/Jacksonian Democrats addressed the issue of equality, democracy, and difference by emphasizing economic and social laissez-faire mixed with political egalitarianism, at least for white males.[15] Targeting the lower and middling classes, those who used this second major language for antebellum Americans celebrated "the people," unfettered competition, and a dynamic social order in which individual talents and skills alone should determine one's status and possibilities.[16] The Jacksonians, too, emphasized character, but used it to denote the attributes and determination by which individuals (white males) made their own ways in the world, without the help or interference of the powerful.[17] President Andrew Jackson summarized this ideology well in his famous 1832 bank charter veto:

> Distinctions in society will always exist under every just government. Equality of talents, of education, or of wealth cannot be produced by human institutions. In the full enjoyment of the gifts of Heaven and the fruits of superior industry, economy, and virtue, every man is equally entitled to protection by law; but when the laws undertake to add to these natural and just advantages *artificial* distinctions . . . to make the rich richer and the potent more powerful, the humble members of society . . . have a right to complain of the injustice of their Government.[18]

Jackson granted nature a larger role in the origins of an individual's talents than had Miller. But he shared with Miller and his rival Whigs the sense both that society required diverse abilities and that the language of merit could coexist with a language of egalitarianism through the intermediary of talents. For Jackson, a just government did not ruthlessly impose material equality, but rather allowed "superior industry, economy, and virtue" to flourish, with social place determined by the free play of an individual's talents and virtues.

Jackson certainly was not relying on a formal philosophical theory to anchor his arguments about talents, character, and merit. But, like the pronouncements of Miller, Reynolds, or even his Whig opponents, Jackson's rhetoric does signal that the Enlightenment project of founding society and government on human nature had not been completely superseded. Both Jackson's call for laissez-faire and Miller's for moral governance derived, in part, from their different understandings of how talents originated, for Miller as products of human effort and character and for Jackson as gifts of nature. In the wake of postrevolutionary America's many upheavals—including the Constitutional debates, first and second party systems, and persistent demands of the disenfranchised (workingmen, women, blacks, etc.) for a more inclusive political system—questions such as Miller's and, later, Jackson's about human nature and its possibilities abounded. The more educated addressed such issues largely by turning to the approach to moral philosophy increasingly dominant in higher education, the American Common Sense school. Drawing on Scottish Common Sense philosophy, American educators and moral philosophers developed a theory of the mind that could account for individual talents without sacrificing commitment to fundamental human equality. It was a view that both Whigs and Jacksonians found attractive.

An American Common Sense

As discussed in chapter 1, Scottish Common Sense realism understood the mind as composed of a set of faculties present in every individual from birth. Well institutionalized in Scottish universities, the Common Sense philosophy emigrated to America in the mid-eighteenth century, along with Scottish Presbyterianism.[19] It found an institutional home first at Princeton under John Witherspoon, who arrived from Scotland to assume the presidency of the College of New Jersey in 1768 bearing Thomas Reid's *Inquiry into the Human Mind on the Principles of Common Sense* (1764). The philosophy soon spread, carried along by the waves of religious enthusiasm sweeping America during the early nineteenth century, and by the rage, especially among Presbyterians, for founding colleges. By 1851, as Rand Evans has noted, "two-thirds of the colleges in the United States were directly or indirectly under the control of the Presbyterian Church. . . . With the founding of those new colleges on the frontier went the ideas of the Scottish philosophy."[20] Indeed, by the mid-1820s, Evans observed, Yale and even Harvard had succumbed, supplementing or at times supplanting Locke with the Common Sense school.[21]

Common Sense philosophy succeeded more completely than simple dissemination suggests. Most American colleges gave the philosophy pride of place in the curriculum, assigning the college president to teach it in a baccalaureate's final year, in a course on moral and/or intellectual philosophy whose purpose was to explore and reinforce the values expected of a Christian gentleman.[22] Positing the universality of an independent, God-given, moral faculty, Common Sense not only provided space for a Christian God, but suggested that middle-class morality (its strictures themselves just being consolidated) was anchored in the fundamentals of human nature.[23] Moreover, and most significantly, the philosophical system placed a premium on individual responsibility by promoting a conception of the mind in which character—moral and intellectual—emanated primarily from an individual's choices and self-discipline. The linchpin connecting choice and self-discipline to character development was the notion of a mental faculty and how it was shaped. Common Sense philosophers envisioned the mind as replete with faculties that would engage with phenomena once experienced as sensations. They further believed the mind could not act before acquiring sensations and that there was no a priori difficulty in gaining knowledge of the external world. Nonetheless they also gave great scope to the mind's powers to shape experience and, in the moral realm, to intuit from experience principles of just behavior.[24]

Two characteristics are particularly salient in Scottish Common Sense notions of the mind's faculties. First is their sheer variety. Although most Common Sense philosophers affirmed that mental activity was unitary, they nonetheless divided the mind into two or three broad powers: most often intellect, sensibilities or emotions, and will; or intellectual and moral powers. Under these gross divisions they then grouped almost any number of specific faculties, depending solely on the ingenuity and parsimony of the author.[25] Thus Dugald Stewart in 1793 described the intellect as encompassing nine faculties, and over half a century later, the American Common Sense philosopher Francis Wayland identified eight: perception, consciousness, original suggestion, abstraction, memory, reason, imagination, and taste.[26]

Second, a central Common Sense precept was that the strength of an individual's faculties derived largely from the effort devoted to developing them. Few proponents of Common Sense doubted that there might be individual natural differences in the power of mental faculties, but they typically consigned such differences to the margins of their analyses, concentrating instead on the effects of habit, education, and environment.[27] This orientation would prove particularly attractive to those developing the American versions of Common Sense.

The leading figures in American Common Sense were Thomas Upham and Francis Wayland, along with Joseph Haven, Mark Hopkins, Noah Porter, and Leicester Sawyer. By the 1830s they were producing textbooks for American colleges that sought to adapt the Scottish philosophy to American undergraduates.[28] In the process, Upham and Wayland in particular created eclectic versions of Common Sense by uniting ideas derived from Reid and Stewart with aspects of Locke and of those Continental idealists whose thought they found especially attractive, most notably Immanuel Kant and Victor Cousin.[29] In general, American Common Sense philosophers stressed the malleability of the mental powers and the responsibility of parents, teachers, and ultimately individuals themselves to develop their faculties appropriately. As Wayland succinctly declared in 1854, "every man is thus made the architect of his own fortune."[30]

All agreed that each individual, with the possible exception of the idiot, was born with a relatively complete set of faculties, though there was much uncertainty as to the relative importance of education and native endowment in determining their power. Thus George Peck noted in his 1841 review of Upham's *Mental Philosophy* that "there is diversity in original constitution—and diversity as the result of training," and Wayland observed that "though [all are] endowed with the same faculties, we perceive that these faculties are bestowed in different degrees."[31] Nonetheless, both Upham and Wayland argued strenuously that a significant factor determining the strength of the faculties in adulthood was the education and amount of disciplined training those faculties received. "A weak memory may be rendered strong," Wayland observed, "by resolutely laboring to improve it. The remedy, however, resides in ourselves, and it is the same for all."[32]

Neither Upham nor Wayland, nor any of the Scots who fashioned Common Sense, rigorously explained why developing the faculties required exercise. In part they relied on an implicit or explicit analogy with corporeal development: physical powers grew stronger through exercise, and so mental powers must respond likewise.[33] They also marshaled empirical evidence to buttress their claims, particularly of the difficulties formerly deaf or blind persons experienced in using their restored senses.[34] Generally, however, they relied on the evidence of children: clearly babies' mental powers were not well developed, but as children matured, a vast array of mental abilities began to appear.[35]

The basic picture of the human mind was thus established: Common Sense philosophers held that every individual possessed diverse mental faculties, with training strongly influencing the faculties' development. Because these authors were convinced that the faculties were largely independent of one another, they argued that variations in endowment and education could produce, even in close relatives, wildly different characters and talents.[36] What accounted for the proliferation of these variations? According to Upham, a mixture of differences in native endowment, knowledge possessed, power of attention and memory, ability to discover relations, and most of all, "Habit."[37]

The wide variety of individual abilities and personalities apparent in contemporary society could thus be readily explained.[38] From the perspective of the Common Sense philosophy drilled into American college graduates, individual differences resulted inevitably from small variations in native mental endowments, magnified by each person's unique experiences and personal efforts toward education and self-betterment. A strong emphasis on personal responsibility thus underlay Common Sense formulations. "The great truth," George Peck observed about Upham's work, "that *all the intellectual, sentient, and voluntary powers are susceptible of cultivation,* is clearly brought out; and the necessity of this cultivation to the perfection of the mind is strongly impressed."[39] Thus what talents one possessed and what one became for better or worse were represented by American Common Sense philosophers as dependent on early education and personal choices: one's mental attributes were the product of will, not destiny.

The Nature of the Individual: Common Sense Meanings of Talent

The connection Common Sense philosophy posited between individual abilities and personal effort resonated widely in antebellum America. To the project of inculcating and reinforcing moral beliefs, associated with evangelical Christianity and Whig ideology, the value of such a view was clear. It explained both how there might be moral principles applicable to all and why each individual could be held responsible for his or her own character. Indeed, one of the most pervasive arguments for teaching the Common Sense philosophy and relying on its analyses of human nature was precisely its concordance with Christian orthodoxy and middle-class morality. Upham proclaimed this straightforwardly in his first textbook: "We are taught by this science to revere the wisdom of our Creator."[40] Wayland, Haven, Sawyer, Hopkins, and Porter all followed suit, insisting that Common Sense should form the curriculum's capstone precisely because it inculcated the values appropriate to a Christian gentleman.[41] Noah Porter, for example, in a generally critical review of a number of mid-century psychology texts, singled out precisely this aspect of Wayland's *The Elements of Intellectual Philosophy* for praise: "We take no exception to the principles of this work on religious grounds. . . . The philosophical principles which he advances are eminently safe and sound."[42]

For middle-class Americans, especially conservative ones, Porter's relief that Common Sense was sufficiently "safe and sound" spoke volumes about their own anxieties and the benefits Common Sense might offer. Working-class militancy; uncertain social restraints in such a vast, sparsely populated country; rumblings of mass democracy; echoes from abroad of Jacobinism and other social unrest; tensions over Irish immigration; clashes over slavery; and evangelical fervor for individual and world reform—these all underscored for many the importance of assuring one another that the nation could indeed become an orderly Christian republic. Spurred by the upheavals of the post-Revolutionary period, many middle-class Americans turned with a new urgency to notions of morality, Christianity, and especially character to define how Americans should conduct their affairs. Visible in the language of revivals, reform movements, and middle-class domesticity, the concern with character and conscience, often mediated through the commonplaces of Common Sense, was integral to the way the bourgeois, especially Whigs, fashioned their cultural identities and the politics of what Lee Benson has called "the positive liberal state."[43] Evangelical ministers such as Lyman Beecher—one of the period's most famous preachers—relied on conceptions of human nature consonant with Common Sense formulations to praise industry, condemn idleness, and promote "temperance," a word that itself suggested the restraint at the philosophy's core.[44] Beecher's daughter Catherine also turned to Common Sense to derive connections between human nature and moral duty, making clear the power of this approach in her 1831 work *The Elements of Mental and Moral Philosophy, Founded upon Experience, Reason, and the Bible.*[45]

For the more political, the Common Sense vision of the mind as a set of faculties requiring training, vigilance, and balanced development resonated strongly with metaphors used to understand society as a whole. Whig social thinkers were deeply concerned with organizing and sustaining a diversified, expansive economy in which there would be a "harmony of interests," as Henry C. Carey dubbed it, among its constituent members.[46] Prominent Whig writers such as Calvin Colton, allied with Henry Clay, and Carey, a political economist, employed Common Sense notions of faculties and talents to justify this view and the stratification of society it implicitly sanctioned. Thus Colton argued that a person rose or fell in America "according to his talents, prudence, and personal exertions," and Carey provided a detailed explanation of the prosperity attendant on diverse talents:[47]

> The greater the *power* of association—the greater the diversity of the demands upon the human intellect—the greater, as we have seen, must be the development of the peculiar faculties—or individuality—of each member of the society; and the greater the *capacity* for association. With the latter comes increase of power over nature and over himself; and the more perfect his capacity for self-government, the more rapid must be the motion of society—the greater the tendency towards further progress—the more rapid the growth of wealth.[48]

Carey's integration of the celebration of diversity with the need for moral oversight—"self-government"—and the careful balancing of the economy's parts, all in the name of progress and by means of talents, characterized much antebellum Whig rhetoric. The stratification implicit in this outlook, and the ways in which a Common Sense–style understanding of human nature could justify the resulting development of hierarchy, also underlay the Whig conception of the social order, as Alexis de Tocqueville noted in *Democracy in America* when describing the susceptibility of democracies to capitalist aristocracies:

> While the workman concentrates his faculties more and more upon the study of a single detail, the master surveys an extensive whole, and the mind of the latter is enlarged in proportion as that of the former is narrowed. In a short time the one will require nothing but physical strength without intelligence; the other stands in need of science, and almost of genius, to ensure success. This man resembles more and more the administrator of a vast empire; that man, a brute.
>
> The master and the workman have then here no similarity, and their differences increase every day. . . . Each of them fills the station which is made for him, and which he does not leave; the one is continually, closely, and necessarily dependent upon the other and seems as much born to obey as that other is to command. What is this but aristocracy?[49]

To the Whigs' main rivals, the Jacksonian Democracy, aristocracy was exactly what the Whig approach to mind and society promised and what they feared.[50] And yet Jacksonians, too, drew on a Common Sense–style understanding of human nature in constructing an ideology for mass democracy. Evoked by Andrew Jackson in his bank veto message, the Democrats' language stressed the rough-and-tumble competition of American life, depicting the nation as constituted through the untrammeled development of its citizens' diverse talents. It was the language of party politics, small-producer capitalism, and frontier expansion, the language in which Jacksonians celebrated the intelligence of "the people" as against the machinations of the privileged, and championed the notion of the negative liberal state. The introduction to the first issue of the *United States Magazine and Democratic Review,* a principal organ of Jacksonian Democracy, said it forthrightly: "We believe, then, in the principle of *democratic republicanism,* in its strongest and purist sense. We have an abiding confidence in the virtue, intelligence, and full capacity for self-government, of the great mass of our people, our industrious, honest, manly, intelligent millions of freemen." The author went on to conclude, in a classic statement of Jacksonian ideology, that "the best government is that which governs least."[51]

For Jacksonians, an approach to human nature similar to that of Common Sense was doubly valuable. First, it helped to warrant reliance on the opinions of the "common citizen" and thus to justify both the radical extension of suffrage and the opening of public office to all voting citizens. In this vein, George Bancroft argued in 1835 that if "the gifts of mind and heart are universally diffused," then it followed that "common judgment" was "the nearest possible approach to an infallible decision."[52] And second, the notion of the diversity of talents provided one rationale for the Jacksonian belief, as William Leggett observed in 1834, that that society was best in which all members' "talents and their virtues shape their fortunes," with no impediments or advantages to some rather than others.[53] Like their Whig rivals, most Jacksonians realized such competition might yield a differentiated, indeed stratified, society; they rationalized this outcome, however, by arguing that such a development would be just insofar as it resulted purely from individual merit, the unrestrained expression of self-realized talents.[54] As the editors of *U.S. Magazine and Democratic Review* argued:

> As far as superior knowledge and talent confer on their possessor a natural charter of privilege to control his associates and exert an influence on the direction of the general affairs of the community, the free and natural action of that privilege is best secured by a perfectly free democratic system, which will abolish all artificial distinctions, and, preventing the accumulation of any social obstacles to advancement, will permit the free development of every germ of talent, wherever it may chance to exist.[55]

Echoing their intellectual forebearer Thomas Paine, these Jacksonian editors suggested that the key to a republican democracy was allowing individual talents, wherever they were found, free rein. However, the consequences of this move, which thereby left open the possibility of merit-based stratification, for establishing a truly republican society haunted many Jacksonians. Some, such as the Ohio author of an 1849 essay entitled "The Absolute Equality of Mind," strove diligently to reconcile belief in "the manifest differences in the mind's instrument" with "the absolute equality of mind." Assuming the homogeneity and unlimited power of human comprehension, the author argued that differences in the endowment and developed strength of the faculties were nothing as compared to the mind's "immortal nature and *infinite* power." As a result, the Ohioan concluded that belief in "equal mental power" was not simply "a demagogical *fiction* to found the humbug of a democratic government upon," but that the "great democratic principle—the radical equality of man" could actually be proved.[56] On the eve of the Civil War, J. C. Hope still maintained this position, arguing in *DeBow's Review* for the overwhelming importance of diligence and "industry" in determining a person's mental powers and worldly accomplishments.[57]

Others, however, even in the Jacksonian camp, dissented from this emphasis on the power of self-fashioning. In a footnote to an 1834 article on "Genius" in the *Southern Literary Messenger*, for example, the journal editor demurred emphatically from the author's conclusion that while "some difference in natural faculties no doubt exists," it "is probably small." Undoubtedly worried about such a position's implications for the defense of slavery, the editor noted that arguments such as those put forward by the author "lag far behind the spirit of the age."[58] The editor was unquestionably correct: notions of "absolute equality" were not only contested in the antebellum era, but seemed increasingly like the quaint vestiges of an eighteenth-century optimism no longer relevant to the highly variegated American republic, especially one with a class of permanent slaves. Core constituents of Jacksonian Democracy—white workers, small farmers, artisans, and the like—while overwhelmingly committed to political equality, based arguments for broad expansion of their rights principally on equality of moral worth, the scattering of natural endowments across classes, and the value of all labor and talents, not on the fungibility of all citizens however talented.[59]

Clearly, one attraction of the Common Sense philosophy was that it provided intellectual support for this Jacksonian social ideology, by upholding the moral equality of all human beings while explaining why, on the basis of nature or effort, individuals differed sharply in their talents. Common Sense philosophers thus had succeeded in constructing a portrait of the mind amenable to both visions of antebellum America. For reformers, advocates of middle-class respectability and character, and believers in government power, Common Sense provided a human nature in which individual responsibility was paramount, a universal moral standard could be applied to all, and the balanced development and interrelation of the faculties was central. For proponents of laissez-faire economics and participatory politics, Common Sense philosophers' emphasis on the diversity of talents fit with a system envisioned as most just when rewarding those who, through talent and virtue, merited prosperity, one in which the polity was organized less around deference than the creative combination of multitudes of distinct interests. And the close link that Common Sense philosophers posited between talent and character was one way in which these two parties could speak, ironically, with something of a common voice.

This shared vocabulary was far from yielding a shared political program. Nonetheless, the profound interest evinced by both Whigs and Jacksonians in the diversity of individual talents suggests that they faced a common problem: how to combine commitments to equality with the prerogatives of merit. How was the triumph of talent, rather than of privilege or class solidarity, to be ensured? Many of the period's most important political disputes—whether over national banks, restrictive tariffs, or the availability of cheap land—attempted to answer this question or responded to the tensions it highlighted. Nowhere was this more apparent than in discussions surrounding basic and advanced education, where the problem of the meaning of talents and the institutionalization of their rewards was paramount. Whigs no less than Jacksonians pushed for a broadly inclusive system of basic education, though they also argued for advanced levels of training, about which most Jacksonians were more skeptical. However, whereas Jacksonians praised education as a means of promoting talents wherever they were found and of strengthening mental faculties to place governance in the hands of the people, Whigs emphasized instruction as a means of disciplining the faculties and morals of the lower orders and ensuring balanced minds among the college-educated elite. Nonetheless, in part by using the language of talents and Common Sense, proponents of both positions managed their disagreements sufficiently to collaborate on the creation of educational institutions and on placing talent at the heart of their understanding of individual difference.

"To convert men into republican machines": Orthodox Minds and Republican Citizens

In 1802, Nassau Hall, the heart of the College of New Jersey (later Princeton University), was in flames. Students had already rioted twice in 1800, predisposing college president Samuel Stanhope Smith to assume that this too was a deliberate act of rebellion. The fire, Smith later lamented, was "one effect of those irreligious and demoralizing principles which are tearing the bands of society asunder, and threatening in the end to overturn our country."[60] According to their elders, materialism, atheism, immorality, and irreligion were rampant among American youth at the new century's outset, beliefs fueled by the seeming decline in religion and ubiquity of riots and other challenges to authority.[61] College officials advanced many explanations for their charges' unruliness, but the most common focused, as had Smith, on the dangerous heritage of the eighteenth century. The trustee committee investigating the Nassau Hall fire, for example, concluded that Jacobinism underlay the torching and that firmer discipline was the solution.[62]

Whether or not it inspired American undergraduates, the glow of the French Revolution filled many of the older generation with dread. Symbolizing the triumph of mob and guillotine, the Revolution and associated philosophies—especially those of Helvétius and Condorcet—appeared to undermine all social order. The election of Thomas Jefferson, a Francophile democrat and well-known deist, as president in 1800 simply underscored the sense of impending doom.[63] Over the next four decades, college presidents and trustees, mostly Federalists or Whigs-in-the-making, would attempt to reimpose order on their troubled campuses. Most visibly, they clamped down on misbehavior, resorting even to mass expulsions to underscore their resolve; more subtly, they sought to structure a curriculum that would enhance the development of disciplined minds and upright Christian characters.[64] The problem of talents—what they were and how to encourage the "right" ones—loomed large, for most believed higher education aimed to develop superior abilities and morals within the next generation of political and civic leaders.[65]

The goal of molding a disciplined citizenry also weighed heavily at the other end of the educational spectrum, where Federalist/Whig reformers sought to institute at local and state levels systems of broadly inclusive common schools.[66] They too worried about how to develop the faculties necessary for a republican citizenry entitled to choose its own leaders. Deeply concerned with promoting proper mental development, both sets of reformers relied on presumptions about human nature—including the malleability of the individual, the multiplicity of human talents, and the preeminence of character—reinforced by mainstream American psychological thought.

The interweaving of education, republicanism, and talents resulted, in many senses, directly from the Constitutional settlement and the contradictory forces it unleashed. Joseph Perkins laid out the issues clearly in 1797 in his *Oration upon Genius:*

> A republican constitution, by rendering every meritorious citizen eligible to the most dignified and lucrative offices of state, must of necessity have a very happy and efficacious influence on the cultivation of talents. So long as real abilities shall constitute an indispensable qualification in every candidate for public honor and confidence, and continue to meet their deserved reward, Columbian genius cannot want a very powerful incentive to the most strenuous exertion.

Perkins's challenge, derived from the Constitution and its proclamation of government of, by, and for the people, was simple: if the only acceptable criterion for choosing leaders was merit, and merit depended on talent, then how was the development of talent to be widely encouraged? Perkins suggested that competition for state offices would spur citizens to develop their talents fully. Most other commentators, however, assumed that formal education would also be crucial, as Perkins himself acknowledged at other points.[67] Perkins's contemporary, the eminent physician Benjamin Rush, for example, specifically addressed this issue in his own treatise on education, "Of the Mode of Education Proper in a Republic" (1798), arguing that the new republic's very survival depended on the general education of the populace. "I consider it is possible to *convert men into republican machines,*" Rush observed. "This must be done if we expect them to perform their parts properly, in the great machine of the government of the state."[68]

Numerous Federalists and Whigs would follow Rush's lead, arguing that the people's intellectual and moral development should not be left to chance, because a republic required citizens willing to sacrifice self-interest for the common good and to choose as leaders those of superior talents. "While our supreme magistrates make the union of eminent talents and inflexible integrity their invariable object in the nominations to offices of responsibility and honor," Perkins declared, "our independent freemen should ever be studious to regulate their elections by the same excellent and infallible standard."[69] Education, Perkins continued, was the only way to ensure that such excellent judgment by the common citizen would be realized.

This sense that a broad public education was necessary to maintain the republic was widely shared in the antebellum period, although the specific justifications for such a project were varied. Jacksonians certainly celebrated the common school, but less so that the masses could choose as leaders the "natural aristocracy" than that artificial distinctions could be eradicated through the development of the talents of all.[70] E. H. Chapin argued in an address to the Richmond lyceum in 1839 that every free citizen deserved an education at least "sufficient to qualify him for all the duties which it will be incumbent on him to perform in after life, as a man and an American."[71] And James Carter made the point even more forcefully in his *Essays upon Popular Education* (1826):

> While the best schools in the land are *free* all the classes of society are blended. The rich and the poor meet and are educated together. And if educated together, nature is so even handed in the distribution of her favors that no fear need be entertained, that a monopoly of talent, of industry and consequently of acquirements will follow a monopoly of property.[72]

The value of the common school, Carter argued, lay not in its ability to inculcate order and deference, as the Federalists/Whigs would have it, but rather in discovering and nurturing talents wherever they might arise. The free play of talents, he and other Jacksonians concluded, was crucial to the maintenance of a true republic. Education, properly organized, would thus promote the interests of the people as a whole, by developing the talents of each individual. "The great purpose of life," Thomas Upham declared, was "to render all qualified to fulfil [*sic*], in the best manner possible, the duties of men, of citizens, of members of families, and above all of Christians."[73]

How were young people to be readied for such republican and Christian citizenship? Drawing on the Common Sense tradition, the overwhelming consensus was that a balanced approach to the development and strengthening of the faculties was essential. The principal fear was of the unbalanced mind, strong in some areas but left weak in others. "I do not put much trust in particular genius," America's preeminent antebellum scientist Joseph Henry observed. "Give me a mind of general powers not deficient in any one faculty and we have the elements of a great mind."[74]

When considering curricula, whether collegiate or common-school, therefore, most reformers favored some version of a liberal arts education, which they believed best able to strengthen the whole range of an individual's faculties. Although the curriculum's specific content was hotly debated, especially around the relative roles of classical languages versus modern languages and sciences, all agreed on the overall goal. "One part of education is the storing of the mind with new ideas," Upham noted in 1827, "another, and not a less important one, is the giving to all the mental powers a suitable discipline; exercising those, that are strong; strengthening those powers, which are weak; maintain among all of them a suitable balance."[75] From the Common Sense perspective, such concerns were perfectly reasonable. Common Sense philosophers made questions of training the diversity of faculties paramount, and suggested that exercising and thereby developing a full and balanced complement of faculties was critical to a properly structured education.[76]

Whig educational reformers were particularly obsessed with the balanced development of the mind. Fearful of the effects of the rise of a new class of businessmen schooled in moneymaking but not gentlemanly conduct, such authoritative voices of the old New England order as Yale president Jeremiah Day and Professor James L. Kingsley offered in their widely influential "Yale Report" (1828) a spirited defense of the traditional course of study and the values it could impart.[77] Conceding that in a democratic republic government positions were theoretically open, as they put it, to "*merchants, manufacturers,* and *farmers,* as well as professional gentlemen," they nonetheless dismissed the notion that a common-school education alone was therefore suitable for all.[78] Rather, they insisted that broadening the pool of potential leaders meant that more classes needed access to higher education, where their talents could be appropriately developed and they could be prepared for their future social roles.[79] "The great object of a collegiate education," Day and Kingsley observed, "preparatory to the study of a profession, is to give that expansion and balance of the mental powers, those liberal and comprehensive views, and those fine proportions of character, which are not to be found in him whose ideas are always confined to one particular channel."[80]

From both Whig and Jacksonian perspectives, therefore, education was a critical arena for working out the interplay of character and talent. However different individuals might be, the common features of mind and the faculties' susceptibility to improvement gave hope to many that the nation might endure as a vast republican experiment. The question was how education could help negotiate the tension between demands for unity and celebrations of individuality, between the citizen as republican machine and as purveyor of talents to the nation at large. As Francis Wayland astutely observed in 1842:

> There will always be produced native talent, vast power of influencing mankind, united with restless, aspiring and insatiate ambition. And this talent will be unfolded in greater proportion as common education is more generally diffused. The question, then, is not whether such talent shall or shall not exist. The only practical question is, whether these rare endowments shall be cultivated and disciplined and cautioned and directed by the lessons of past wisdom, or whether they shall be allowed to grow up in reckless and headstrong arrogance. . . . It is merely a question whether the extraordinary talent bestowed upon society by our Creator, shall be a blessing or a curse to us and to our children.[81]

Broad-based education might or might not be able to create talents, Wayland and other antebellum Americans believed, but it could certainly develop them. Education was thus critical to determining whether talents would prove "a blessing or a curse" to the republic. A source of authority for both natural aristocracy and popular democracy, talents helped to define what a republican culture should strive to develop in its citizenry and to explain why different individuals must play different roles in a diversified economy and polity. They were also used, albeit in more limited ways, to explain differences between whole groups of people, such as men and women, while still sustaining a commitment to equality, by suggesting that what differentiated one kind of person from another was their specific complement of abilities. This "separate but equal" approach to talent, as it might be called, was most thoroughly elaborated in the attempts to define gender-specific domains of action and authority for middle-class men and women on the basis, in part, of the particular talents it was contended each sex should manifest.

Separate but Equal? Mental Faculties and Gendered Minds

The reimagining of the divisions between private and public, household and market, masculine and feminine that characterized the emergence of the middle-class culture of character and respectability during the late eighteenth and early nineteenth centuries was predicated on the transformation of the social world into a gender-segregated union of a private, female-dominated domestic domain and a public, male-dominated political/economic one.[82] While these formulations might or might not characterize women as secondary to men, the salient point was less women's inferiority than their difference: women were represented as equal in value and rights to men, but assigned to their own distinctive social tasks, a differentiation based on the possession of abilities and deficiencies complementary to men's. "The true attitude to be assumed by women," explained Catherine Beecher, one of the architects of the antebellum discourses structuring domesticity,

> is that of an intelligent, immortal being, whose interests and rights are *every way* equal in value to that of the other sex. . . . Instead of rushing into the political arena to join in the scramble for office, or attempting to wedge into the over-crowded learned professions of man, let woman raise and dignify her own profession, and endow posts of honor and emolument in it, that are suited to the character and duties of her sex.[83]

For most male writers, female duties derived principally from the attributes of sentiment and moral character that they believed constituted female nature and thus served as the primary demarcators separating the male world from the female. Turning on its head the Puritan equation of morality and maleness, these authors argued that women, by virtue of either endowment or education, possessed faculties especially suited to the realms of emotion and proper conduct. Thus A. B. Muzzey, no friend of women's rights, asserted in 1840: "there is one realm where woman reigns in undisputed supremacy; it is the realm of Moral power."[84] Even a champion of the cause of women, Thomas Branagan, had expressed similar sentiments in 1808: "I do, indeed, exalt the female character higher than the male, in those qualifications which ennoble human nature, and make it almost angelic; and those are benevolence, sympathy, commiseration." Branagan continued, however, in a vein that Muzzey specifically repudiated, by praising women's intellectual capabilities: "and, as it respects every other acquirement which men have, or ever will attain, I contend that the natural genius of women can, if improved, make, on an equal par, the same attainments."[85]

Branagan here echoed a view of the female intellect advanced by women across the political spectrum, including not only such zealous advocates of female equality as Judith Sargent Murray or Sarah Grimké, but also staunch promoters of the complementary nature of men's and women's roles, including Catherine Beecher and Sarah Edgerton. In an unpublished essay, "The Education of Women," Grimké expressed her position emphatically: "Women are gifted with the same powers and are as susceptible of cultivation as men. Why then should they not have the same facilities and the same inducements for improving their faculties?"[86] Beecher, for all of her commitment to a gender hierarchy and gendered realms of activity, resolutely contended throughout her life that women possessed valuable intellectual capabilities requiring development. "The success of democratic institutions," Beecher explained,

> as is conceded by all, depends upon the intellectual and moral character of the mass of people. . . . It is equally conceded, that the formation of the moral and intellectual character of the young is committed mainly to the female hand. . . . Let the women of a country be made virtuous and intelligent, and the men will certainly be the same. The proper education of a man decides the welfare of an individual; but educate a woman, and the interests of a whole family are secured.[87]

Her point, reiterated by many, was not that women were born with unique intellectual faculties, but that society would be best served if women developed their faculties differently, toward the performance of distinct, though still socially useful, tasks in a separate arena.[88]

For Alexis de Tocqueville, such an approach to relations between the sexes was a Smithian "division of labor," as he explained in his chapter "How Americans Understand the Equality of the Sexes," which Beecher cited approvingly in her own *A Treatise on Domestic Economy:* "The Americans have applied to the sexes," he observed, "the great principle of political economy which governs the manufacturers of our age, by carefully dividing the duties of man from those of woman in order that the great work of society may be the better carried on."[89] What justified this division of labor and corresponding linkage of difference and equality? According to Tocqueville, the diversity of the faculties. Human talents varied, and thus some individuals or types of individuals were or became better suited to certain kinds of endeavors. Although Tocqueville granted physical differences some role in such distinctions, he and other commentators turned mainly to mental and moral characteristics to bear the brunt of the argument. Common Sense–type conceptions of the mind proved particularly valuable in this regard, because they could readily be used to validate the notion of a differentiated social space, in which particular roles went to those with specific abilities.

While the language of mental faculties was thus useful in arguing for the difference in gender roles, its success in justifying such distinctions was more problematic. Although different abilities could be used to warrant different social roles, the question remained: What produced variations in abilities? If nature, the rationale behind establishing gendered realms was clear-cut: men and women were born with different faculties, and it would make no more sense to encourage one to perform the other's duties than to ask a blind person to act as a lookout. And certainly many such propositions—that men and women had "natural" propensities to develop certain faculties—had been and continued to be advanced.[90] Nevertheless, this was not the dominant form in which the argument was couched.

Recall that Samuel Miller challenged only Wollstonecraft's notion that the sexes *should* receive the same education. Similarly, Hannah Mather, descendant of the illustrious Puritan stock that had produced Cotton Mather, argued in 1818 that "the wise Author of nature has endowed the female mind with equal powers and faculties, and given them the same right of judging and acting for themselves, as he gave to the male sex."[91] Christian religion, Mather continued, could allow nothing less, because all human beings, male and female, must in moral terms be considered equally open to God's salvation.[92] Even Catherine Beecher acknowledged the possibility and, indeed, desirability of training in women many mental faculties developed in men. "*A well balanced mind,*" Beecher argued in 1829, "is the greatest and best preparation for her varied and complicated duties. Woman, in her sphere of usefulness, has an almost equal need of all the several faculties."[93]

Given the widely accepted belief, consonant with Common Sense, that human abilities derived largely from their possessors' efforts, the fundamental issue in justifying the development of separate spheres became not whether women *could* manifest the same abilities as men, but whether they *should*. Those who articulated the ideology therefore concentrated enormous effort on what was socially and morally appropriate and on policing those who violated what was separated, in a sense, only by convention. Hannah Mather explained the position clearly:

> There can be no doubt, that, in most cases, their [women's] judgment may be equal with the other sex . . . but it would be morally improper, and physically very incorrect, for the female character to claim the statesman's birth [*sic*], or ascend the rostrum to gain the loud applause of men, although their powers of mind may be equal to the task.[94]

Propriety and Christian morality, not nature, demanded that women seek occupations distinct from those of men, and that they develop talents particular to the private, domestic needs of their roles as wives, mothers, and guardians of virtue. Women's domestic endeavors were varied enough and their definitions of them supple enough, that this rubric justified training most of the faculties.[95] Nonetheless, conventional opinion held that women ought to differ from men and that domestic duties embodied and helped to produce those differences. Evidence of good character demanded nothing less; otherwise, women were likely to be branded with one of the period's most unflattering epithets, "unwomanly" or "unsexed," and thereby written entirely out of the social world most were raised to inhabit.[96]

The language of faculties was thus a double-edged sword. While the multivalency of mind helped to make separate spheres intellectually plausible, the faculties' plasticity ensured the inherent instability of separations of human beings on the basis of their propensities, as there could be no "natural" reason for parsing social roles any particular way. Social arrangements were thus left, intellectually, in a state of flux: it could not be assumed that women would by nature become domestic, any more than that the children of workers or farmers would by nature become what their parents had been. While a dynamic, industrializing economy required just this occupational flexibility, many middle-class Americans found its social implications unsettling. Without the conviction that society was based on a natural order, partisans of stability had to revert to some combination of persuasion and coercion, with all the attendant danger that not every citizen would heed such messages.

As a result, attempts to either contain or exploit this sense of the malleability of human nature reverberated not only throughout the discourse on gendered domains, but also many other movements characteristic of antebellum culture. Hannah Mather's earnest talk about women's duty to devote themselves to the private rather than the public realm relied on the same basic conception of the mental faculties as her contemporary Judith Sargent Murray when arguing that women, properly trained, could contribute as much to the political and professional worlds as men. The same could be said of those antebellum adversaries Catherine Beecher and Sarah Grimké. Similar positions were advanced within the abolitionist movement and among advocates of universal public education. Neither defenders of the status quo nor advocates of reform were willing simply to reject the notion that most (white) human beings *could* be the mental and moral equals of adult, white, propertied males; the question was rather to what degree the nation was better served by maintaining a highly differentiated citizenry.

Most authors who turned to human nature to ground their social visions thus incorporated the universality, diversity, and malleability of the mental faculties into their meritocratic versions of republican democracy, seeking simultaneously to satisfy demands for egalitarianism and calls for a differentiated social order. As we have seen, this was as true of Whigs as Jacksonians. By so doing, however, the systems of distinctions and/or stratifications that they sought to confirm by means of the language of talents were never fully naturalized, and thus remained open to debate and alteration. As we will see in chapter 3, this flexibility in the notion of talents encouraged some Americans confronting that most freighted of all antebellum demarcations—race—to develop a new, more deterministic language of difference, one in which multiple, malleable "talents" were replaced with singular, inborn "intelligence." While the influence of this developing science of race remained limited in antebellum America, in the post–Civil War decades it became one of the dominant ways in which human beings and their capacities were understood.

Talent, Merit, Equality: French Education and the Revolutionary Legacy

Before considering this alternative discourse of difference, it is necessary to step back from the American picture and examine notions of talent in America's sometime sister republic. France's republican revolution, too, pushed questions of equality, merit, democracy, and difference to the fore. Although the vicissitudes of Revolutionary/post-Revolutionary France undercut much of the Revolution's emphasis on democracy and equality, the elitist and (potentially) meritocratic structures put in place during Napoleon's reign came to define important aspects of French state and society for the rest of the nineteenth century, and, in many respects, up until today. As a result, the place of talent in the educational and philosophical discourses of nineteenth-century France was, paradoxically, both more secure and less central than in America.

On the one hand, the system of broad-based basic education, *concours* (competitive examinations), and *grandes écoles* (elite schools) that marked the early nineteenth-century refashioning of French education created a structure designed, among other tasks, to identify, or create, talent within the (male) citizenry and to enlist it in service to the state. Combining the hallmark of the ancien régime's educational approach—*émulation,* or excellence through competition, honors, and the desire for glory—with the Revolution's distrust of self-perpetuating elites, the Bonapartist system celebrated (however disingenuously) the discovery and development of talent wherever it might be found, and the consequent merit-based awarding of honors and positions.[97] On the other hand, the predominant philosophical approach to mind in the academy, Victor Cousin's eclecticism, was much less concerned with human talent and differences than with moral character and the universal attributes of the self in its most abstract, refined sense. Even in biomedical/physiological circles, where disdain for eclecticism ran high and interest in individual minds and their variations flourished, the focus centered on materialistic investigations of the brain or its pathological deviations. "Talents," to be sure, did not disappear entirely from the lexicon employed by early- to mid-nineteenth century French liberals to describe their social world or reimagine its structure. However, the term's importance receded and its meaning underwent a subtle but significant shift, symbolized by the use of "talent" in the singular, rather than the plural form typical earlier.

In certain respects there was no clearer example of the legacy of Enlightenment concerns with equality, talents, and merit than the debates over the nature of the post-Revolutionary educational system. Like their American counterparts, virtually all parties in France regarded public instruction as key to maintaining or (re)shaping the Revolution's heritage. "The most important thing, that on which all others depend, that alone which can insure among us the maintenance of liberty and equality, that which can, in a word, regenerate the culture [*les mœurs*] and the nation with it," C. L. Masuyer, a deputy to the National Convention, observed in December 1792, "*is public instruction common to all citizens.*"[98] During the Revolution and Directory—before Napoleon finally ended the seemingly endless discussion—Jacobins and more moderate republicans fiercely debated the character of education: Should it aim to produce a broadly educated citizenry or a highly trained elite? [99] Talents figured centrally in the arguments, because all conceded the Republic's need for virtuous citizens with talents at the ready for the nation's disposal. Condorcet summarized the issue in his *Rapport sur l'instruction publique* (April 1792), the document that, with Charles-Maurice de Talleyrand-Périgord's September 1791 report, constituted the starting point for most revolutionary-era discussions about organizing public education:[100]

> To assure each one the facility of perfecting his skill, of rendering himself capable of the social functions to which he has a right to be called, of developing to the fullest extent those talents with which Nature has endowed him; and thereby to establish among all citizens an actual equality, thus rendering real the political equality recognized by the law.[101]

As in America, no revolutionary party doubted that the new educational system's first aim must be to strengthen political equality by making more manifest the natural equality of all. But virtually every commentator also conceded that education could go only so far, because talents originated largely from native endowment.[102] "Nature has made strong men and weak men," Masuyer declared, though denouncing Condorcet's proposal, "[and] it is the same for the powers of the soul or the intellectual faculties."[103] Condorcet, Talleyrand, François Lanthenas, and eventually Pierre-Claude-François Daunou, among others, concluded that because of these natural inequalities education should have not one function, but two: creating an educated republican citizenry, and training the talented as the next generation of republican leaders. Indeed, the features of the Condorcet and Talleyrand plans that generated the most hostility—even from many moderate republicans, as R. R. Palmer has shown—was precisely their advocacy of multitier education and focus on advanced grades of training.[104]

To accommodate this natural inequality, Talleyrand proposed a system with four levels of institutions, Condorcet one with five. While both were careful to provide for students from all social classes, in reality only primary education was open to all (including girls), though mandatory for none; the upper levels were limited to the sons of wealthy families plus a few *pensionnaires* (scholarship students).[105] Neither democratic nor completely meritocratic, the higher degrees of instruction advocated by Talleyrand and Condorcet promised a bourgeois republican elite, whose talents would be ensured by incorporation of a signal characteristic of ancien régime education, émulation.[106] "Talent can have no other prompting than *émulation,*" Pierre-Louis Lacretelle suggested in 1791, revealing the continuing faith in émulation's ability to spur individuals to ever greater accomplishments through competitions, prizes, and public acclamations of success.[107] Having admitted that very few could ever attend the proposed National Institute standing at his educational hierarchy's apex, for example, Talleyrand strove to reassure the National Assembly's Constitutional Committee that admittance into the higher levels would be strictly merit based, by means of competitions and unparalleled success at the lower grades.[108] "Do not shudder, that your brows will be encircled for a moment with crowns," president of the Department of Paris, Louis-Pierre Dufourny, advised prize-winning University of Paris students in August 1793, "they are not the crowns of pride, nor the crowns of tyranny; they are the crowns awarded for *émulation,* for the talents by which republics have been established and are distinguished."[109] Even Robespierre's spokesperson, Michel Le Peletier, conceded the need for advanced grades open to those who showed exceptional talent in the levels below.[110]

High among the benefits to the Republic of this multitiered system, according to its proponents, was its ability to identify and prepare a cadre whose exceptional talents, as Condorcet observed, could "fulfill those public functions which require the highest degree of enlightenment."[111] Talents, in other words, not only secured republican equality, but also justified unequal access to positions of power. It was around this point, not surprisingly, that those who favored mass education founded on the ideal of complete social equality coalesced. "Education is a monstrosity when it is unequal," François-Noël Babeuf announced. "[S]uperiority of talents and industry is only a chimera or a decoy that has already unduly served the plots of conspirators against equality."[112] To Masuyer, Condorcet's plan promised not equality but simply a new social hierarchy, now founded on intellectual difference rather than birth: "this system [of Condorcet's] is antithetical [*éversif*] to every principle of liberty and equality, for it would have no other effect than to create two classes of men, *those who think and reason, and those who believe and obey.*"[113] Even those more open to training talents at various levels, such as Daunou, worried that establishing a corporation of teachers and elite educators selected by talent, as envisioned by Condorcet, might create a new aristocracy that could unduly influence public opinion and policy.[114]

In response critics insisted—as had their kindred spirits in the United States—that any new instructional system should above all raise the citizenry's talents so that all had the skills necessary for a republican democracy. "All citizens *are destined to exercise public office* [*les magistratures populaires*]," Masuyer argued, "*thus if you place* the knowledge necessary for public office, for administration of public goods, *outside the common comprehension of the citizenry,* you will violate the first law of equality."[115] Everyone conceded that raising the talents of the multitude was critical to the republic's survival; neither Talleyrand nor Condorcet denied that.[116] But Le Peletier's report—as presented to the National Convention by Robespierre in July 1793 after Le Peletier's assassination in January—went much further, proposing a system of common education rigorously equal for all:

> I demand that you decree that, from the age of five up to twelve for boys, and up to eleven for girls, all children without distinction or exception be educated in common, at the expense of the Republic; and that all, under the holy law of equality, receive the same clothing, the same nourishment, the same instruction, [and] the same care.[117]

Addressing directly the inequalities of wealth and power rampant under the monarchy, Le Peletier proposed the Revolutionary era's one truly comprehensive, universal plan. He argued that only by removing children from their families and placing them in *maisons d'égalité,* enforcing strict equality of treatment, and instructing children in republican principles—controlling "the totality of the child's existence"—could a new generation of virtuous, talented citizens be produced for whom equality was not "a specious theory, but a constantly effective practice."[118] Even among Le Peletier's allies, such as the chemist Antoine Fourcroy, few would go quite so far in the search for republican equality.[119] Most believed that removing children from their parents was impractical, if not simply wrongheaded, and many worried about the project's enormous expense. Moreover, some concluded that talents simply *were* distributed unequally, and thus that national instruction should not try to make all intellectually equal, but rather to find and train superior talent for the nation.

In the end, the only significant legislation that emerged from the Revolutionary era's welter of proposals and years of debate was the law of 3 brumaire An IV (October 24, 1795). Authored by Daunou, it decreed the establishment of primary schools in every canton, and provided for up to one quarter of students in any school to attend free of charge.[120] With attendance optional, private schools legalized, a free marketplace for education established, and fees for most students, it was a far cry even from Condorcet's proposal, much less Le Peletier's inclusive, national system.[121] Still, by adhering to the principle of basic education for all, the law of 3 brumaire did not completely retreat from commitments to republican democracy, although it did foreshadow the compromises between principle and expediency that Napoleon would institutionalize in the law of 11 floréal An X (May 1, 1802) and the decree organizing the Imperial University (March 17, 1808).

These two pieces of legislation established the basic Bonapartist structure of public instruction. The law of 11 floréal mandated a multilevel system of voluntary primary education, whether public or private, open to all (although only one-fifth of places were reserved for the indigent); and secondary and advanced training (including the lycées that would dominate the system) reserved for the few, those from middling to wealthy families plus a small group (6,400) of *élèves nationaux* funded by the state. Even of these latter, only 4,000 were to be chosen strictly by merit—success on competitive examinations and in the lower grades—while the remaining spots would be allotted partly on the basis of family loyalty and service to the regime.[122] The decree of March 17 sought to consolidate nationwide imperial control of education by establishing the new Imperial University as the sole educational authority and by requiring all educators to have obtained their degrees from one of the University's faculties.[123]

Although numerous private, mostly Catholic, primary and secondary schools remained, and public schools were grossly underfunded, the tendency throughout the first half of the century as a result of these Napoleonic reforms was toward increasing national control and the domination of the system by a few elite postsecondary institutions.[124] Even as late as the Falloux law of March 15, 1850, primary education was organized principally on the basis of a free market, albeit with some provision for the poor, and the upper levels of education, whether dominated by state or church, were open primarily to those able to pay.[125] The Imperial University retained oversight of the education system through its authority to grant *brevets de capacité* (teaching licenses), though its grip was loosened somewhat during the Restoration and Empire when certain religious orders received exemptions.

Overall, the secondary institutions—especially the lycées and grandes écoles—did most to shape French educational culture. Particularly important were the Ecole Polytechnique, founded in 1793 to produce military engineers, and the Ecole Normale Supérieure, founded first in 1795 and then refounded in 1810 to train lycée professors.[126] Located at the apex of the system of elite instruction, the Ecole Polytechnique and the Ecole Normale exerted a powerful influence throughout the century, training most government bureaucrats and the upper echelons of industry. Ambroise Rendu, former inspector-general and University council member under Napoleon, explained in 1816 the necessity of such Napoleonic/republican institutions even in Restoration France: "society must offer special instruction, suitable to form permanently a small number of men, called by their tastes and their talents, from which society itself immediately derives great benefits."[127]

To ensure that education's upper levels would be supplied with this "small number" exhibiting appropriate tastes and talents, French administrators turned again to émulation, institutionalized in the *concours*.[128] Examinations, mostly national in scope, were established at almost every stage in the French system to determine which students qualified for advancement. To enter the Ecole Polytechnique or the Ecole Normale, for example, a candidate had to place near the top of rigorous qualifying examinations, widely conceded to be the most difficult academic exercises in France.[129] Similarly, an examination was required to enter many prestigious lycées and to obtain the *baccalauréat* showing completion of secondary studies. Even the appointment procedure for University faculty chairs was shaped by the ideology underlying the concours system. As François Guizot proclaimed,

> faculty chairs are not at all awarded, as in the colleges, by the head of the teaching corps; they are obtained by competitive examinations, where the candidates can exhibit all the newest and most sublime knowledge; no one is restricted from the examination; no idea is prohibited or prescribed, and the public is at the same time judge of [the candidates'] talents and guarantor of their true independence.[130]

For officials such as Guizot, the concours symbolized not only a commitment to merit, but also to what might anachronistically be called transparency and accountability. At least theoretically, the public could see why certain candidates succeeded and others failed, and could be assured that strictly objective criteria, and not the background of the candidates, governed the proceedings. Given the vital social role of the concours, it is not surprising that throughout the nineteenth century the various registers that governed public instruction were filled with decrees or rules like that of December 7, 1850, for admission to the Ecole Normale:

> ARTICLE 1. — Student places at the Ecole Normale are determined by the set of examinations that take place each year
>
> ART. 14. — . . . The members of each Commission, after having compared the results of the written examination and the oral examination with the various pieces of information gathered on the candidates, draw up and propose to the Minister the list of those who should definitively be admitted.[131]

Examinations and ever more advanced training did not mean, of course, that all and only the most talented reached the top of the educational hierarchy. Secondary education was expensive, scholarships were few, women were largely excluded, and moral character and political reliability were explicit factors in candidate selection at almost all levels.[132] It did mean, however, that those who managed to graduate from the grandes écoles or other elite institutions were seen to be among the nation's most talented, and generally treated as such.[133] Rewarded with plum civil or military positions, called *polytechniciens, normaliens,* and the like to emphasize their separate and corporate identities, those who successfully scaled the educational hierarchy constituted the body of technocratic experts consulted for advice and leadership throughout the vicissitudes of nineteenth-century French politics.[134] Even during the ultraroyalist phase of the Bourbon restoration, when distrust of the traditionally liberal republican graduates of the grandes écoles flourished, an attempt to suppress the Ecole Normale proved short-lived and reliance especially on the technical expertise of graduates continued apace.[135] The elaborate and minutely detailed procedure for ranking Ecole Normale candidates, enacted on July 12, 1820, reveals how thoroughly and early selection through examination became embedded within French administrative culture:

> ART. 10. — The examiners . . . will communicate their judgment, by indicating the order in which the students should be ranked respectively in each of the tests of the examination: they will draw up a report at the end of each session [*séance*].
>
> ART. 11. — The director of the Ecole Normale will draw up a table of all the students competing in letters and another table for all the students in science. These tables will be divided in columns. The director will indicate there the grades obtained by each student. Similarly, he will inscribe the rank that each competitor obtained, on the basis of the judgment of the examiners, in each of the tests of the examination; and finally, he will add in their observations on the character of each student, his habits, and the conduct he exhibited during the entire time of his residence in the school.[136]

The Jacobins were thus not far wrong when they worried that educational reforms along meritocratic lines might contribute more to creating a new elite than to sustaining radical democracy.[137] Indeed, their fears might have been accentuated had they anticipated just how the form of selection and the meanings attached to it would develop. The upper reaches of the French system quickly homogenized, in terms not only of class and gender, but intellect as well. Where republican theoreticians had spoken primarily of multiple talents, the educational system focused on only a few and on ranking students incessantly according to their relative success in developing them.

The examinations were integral to this process. At the secondary level, and for admission to the Ecole Polytechnique and the Ecole Normale, questions were concentrated in a few domains: classics, mathematics, and at times philosophy. All candidates took the same examination, and all were ranked according to their performance. Once in lycée or one of the écoles, students followed a standardized curriculum, and instructors were supposed to rate each pupil daily on a composite of academic performance and behavior, with final rankings published at year's end.[138] Thus, both the internal workings of the system and its external structure were designed to ensure a single standard of performance for measuring all, and the rise to the top of those who best satisfied it.[139] In place of "talents," French education substituted "the talented," and suggested that the endowments and training of this elite group readied them to succeed at virtually any endeavor.[140] Madame de Staël had thus imagined part, but only part, of what would transpire when she observed in 1800 that "the principle of a republic where political equality is sacred, must be to establish the more marked distinctions among men, according to their talents and virtues."[141] Only the pluralization of these terms would prove illusory.

Victor Cousin and the *Via Media* of Mind

One of the earliest great successes of the post-Revolutionary system of talent, not to mention a significant architect of its later phases, was Victor Cousin, founder of the philosophical system, eclecticism, that predominated in mid-nineteenth century France.[142] The son of a Parisian watchmaker, Cousin was created, as Alan Spitzer has shown, by the post-Revolutionary concours system: Cousin's "unrivaled series of triumphs in scholastic prize competitions," Spitzer noted, "made him the first great athlete of the meritocracy, star of the first *promotion* of the Ecole Normale in 1810."[143] Adopted as a symbol of the educational system's openness to talent, Cousin commenced a glittering academic career almost immediately upon graduation from the Ecole Normale. In 1815 he was taken under the wing of Pierre Paul Royer-Collard, spiritualist philosopher, favorite of Napoleon, and Sorbonne professor.[144] Asked to substitute for Royer-Collard in his lecture series at the University, Cousin proved immensely popular. From 1815 until 1820, when his lectures were suppressed after the ultramonarchists gained power, Cousin drew enormous crowds and established himself as a cult figure, especially among the new educational system's young products.[145] With the final, more liberal phase of the Restoration, Cousin regained his post as lecturer in 1828, but achieved his greatest influence from 1830 to 1848 during the July Monarchy, when, as director of the Ecole Normale and member of the Council of Public Instruction, he virtually ruled the French philosophical world, establishing his philosophical psychology as a required course in the lycées and training almost the entire next generation of French philosophers.[146]

Like the Scottish Common Sense realists, from whom he borrowed much, Cousin developed his philosophical system largely in reaction to the empiricism and sensationism of Locke, Hartley, and especially Condillac.[147] Versed in the work of both the Scots—Reid and Stewart—and the German idealists, particularly Kant, Cousin sought to fashion an approach to the nature of the mind at once empiricist and spiritualist, while avoiding the pitfalls of each.[148] On the one hand, Cousin was little attracted to ultraconservative Restoration philosophies such as orthodox Catholicism and extreme monarchicalism, as exemplified by the teachings of Joseph Marie de Maistre and Louis Gabriel Ambroise de Bonald.[149] On the other hand, Cousin distinguished his philosophical system from approaches to mind and matter inspired by physiology. Typically starting from Condillac's sensationism, physiologically oriented investigators ranged from the moral and social philosophers known as the *idéologues*—Antoine Louis Claude, Comte de Destutt de Tracy, Pierre-Jean-George Cabanis, François Magendie, and Joseph Victor Broussais, among others—to mental pathologists such as physicians Philippe Pinel and Jean Etienne Dominique Esquirol, to followers of Franz Joseph Gall advocating organology or phrenology.[150] Varied in method and orientation, all privileged material explanations for mental phenomena and were particularly attentive to individual differences in behavior and intellect.[151] Because the idéologues and phrenologists were unabashedly committed to explaining mental activity in terms of the brain's material organization, both approaches quickly became associated with such politically suspect positions as materialism and atheism, thus alarming moderates such as Cousin.[152]

Too committed to liberal constitutional monarchy to look to the right for guidance, Cousin was also too convinced of the importance of traditional morality and religion to sympathize with the idéologues. Instead, Cousin situated himself in the middle, the *juste milieu,* and via eclecticism attempted to appropriate the best from all philosophies.[153] "The sole means of escaping error," Cousin observed after being restored to his post in 1828, "is by discovering and embracing all truths no matter what the systematic and defective forms in which they are clothed."[154] Central to his philosophical doctrine was the notion that humans were both spiritual and material beings, and thus that philosophy must provide space for morality and God:

> Our true doctrine, or true flag is spiritualism. . . . It teaches the spirituality of the soul, the liberty and responsibility of human actions, moral obligation, disinterested virtue, the dignity of justice, the beauty of charity; and beyond the limits of this world it shows a God. . . . It teaches all men to respect and value

> themselves, and, little by little, it conducts human societies to the true republic, that dream of all generous souls which in our times can be realized in Europe only by constitutional monarchy.[155]

To justify this vision of bourgeois rectitude in an orderly republic as the proper organization for society, Cousin turned to an analysis of the nature of the mind. Not completely hostile to Locke and Condillac, Cousin conceded to empirical investigation a fundamental role in exploring human nature. He dissented, however, on what constituted the empirical. Reviving the term "*psychologie*" to distinguish his approach from the suspect *idéologie,* Cousin was less concerned with amassing facts about the mind and its variety than with the search for features characterizing all minds.[156] Cousin did not dispute that human beings differ and that these differences could have important social consequences. Equality, except as it concerned moral rights and responsibilities, little interested him. The product of a hierarchical, merit-based system, Cousin believed God had ordained social stratification, by means of unequal mental endowments: "It is not true that men have the right to be equally rich, beautiful, robust," he declared. "God has made us with powers unequal in regard to all these things. Here equality is against nature and eternal order; for diversity and difference, as well as harmony, are the law of creation."[157] However, for Cousin, psychology's task was to explore, especially in the moral realm, similarities among human beings, not their differences. Like the Scottish Common Sense school, Cousin believed in a universal moral sense, present in every individual regardless of age, gender, or race. "It is a fact," he noted in his examination of Locke's philosophy, "that in the presence of certain actions, reason qualifies them as good or bad, just or unjust. . . . There is not a man, ignorant or instructed, civilized or savage, provided he be a rational and moral being, who does not exercise the same judgment."[158]

Cousin gave little attention to the varieties of actual individual minds. "The only difference between one man and another," he explained, "is the greater or lesser clarity in the manner in which they are aware of these elements [multiplicity, unity, and the relations between them]."[159] While, as Jan Goldstein has argued, Cousin left himself room to marginalize women and the popular classes because of their presumed unwillingness or inability to analyze their ideas exactingly, he nonetheless erected no absolute differences between categories of people.[160] God endowed all human beings, in his view, with the same basic capacities. Some might develop certain faculties more fully than others, but these "great men," as he called them, still represented everyone else. Indeed, Cousin went so far as to argue that no true differences in reason could exist: "there is nothing less individual than reason: if it were individual, we would control it as we control our determination and our desires; every minute we would change our acts, that is to say our concepts."[161]

As a consequence, and because of Cousin's unique institutional position, the dominant philosophical approach to human nature well into the Third Republic evinced little concern with explaining why talents varied and what those differences might mean.[162] Except for rather formulaic suggestions that God ordained such differences or, as Cousin's chief disciple Théodore Jouffroy argued, that individuals applied their wills differently to disciplining their capacities, the eclectic school said little about talents.[163] One of Cousin's followers, L. E. Bautain, for example, suggested in his 1839 elementary psychology textbook, *Psychologie expérimentale,* that distinctions in children's intelligence arose because of the ways children were spoken to: most children, hearing only of worldly things, developed an intelligence "little more lively than at birth"; those few, however, who were exposed to speech that "came from the soul" and that "proclaimed with faith the sacred name of God," were able to experience true intellectual development.[164]

Via the eclectic philosophy, Cousin and his followers could preach belief in moderation, in harmonizing psychology with Christianity and state power, and in fundamental equality tempered by the elite's preeminent role. Politically savvy, Cousin lost few opportunities to demonstrate the close fit between eclecticism and the liberal state, praising the state, for example, as the true guardian of liberty and equality: "The State thus does not limit liberty, as some say; [rather] it develops and insures it. . . . Before the State . . . all are equal." As director of the Ecole Normale, Cousin preached this position to all who attended France's most elite educational institution, and as formulator of the policy mandating the teaching of philosophy in the lycées, he extended his empire across French secondary education. Official French culture was thus saturated with eclecticism, a philosophy that emphasized the characteristics common throughout humanity but also celebrated hierarchy, and one ideally suited, according to Cousin, to meet the needs of the nation and the historical moment.[165]

The eclectics' dominant position did not mean that their rise to power went unchallenged. As has been suggested, during periods of close church-state connection—especially in the 1820s and the 1850s—more avowedly Catholic philosophies than eclecticism received official sanction. The contributions of de Maistre and Bonald notwithstanding, W. Jay Reedy has shown, contemporary philosophical approaches of this kind were actually short-lived. In the end, Catholic theologians felt more secure with philosophies derived from Aquinas than with modern speculation, and thus the influence of traditionalist Catholic philosophies tended to be limited primarily to the highly orthodox and to strictly theological issues.[166] The most direct challenge to eclecticism came from the idéologues, self-avowed heirs of Condillac who dominated the Second Class of the new *Institut national des sciences et arts,* the *Classe des sciences morales et politiques,* from 1795 to 1803 and propounded their program from that position.[167] Their interest in human difference and sensationist and physiological approaches to mental phenomena was clear as early as volume 1 of Destutt de Tracy's textbook on idéologie, *Projet d'éléments d'idéologie* (An IX [1801]). Following Condillac, Destutt de Tracy argued that the mental faculties—sensibility, memory, judgment, and desire—arose from the interaction of the brain's physical organization with different types (*espèces*) of sensation, and thus were shaped largely by the individual's particular experiences and activities. One of Destutt de Tracy's closest colleagues among the idéologues, Cabanis, went even further, and fashioned a theory of the mind fully rooted in physiology.[168]

Like so many eighteenth-century writers investigating the mind's operations, Cabanis began with Locke. Cabanis departed from Locke, and from Locke's successors Condillac and Helvétius, however, by tying sensationist psychology directly to human biology. Physiology, he argued at length in *Rapports du physique et du moral de l'homme* (1802), teaches that human beings naturally vary, that "the different organs or systems of organs do not have the same degree of force or effect in different subjects. Each person has his strong and his weak organs." Cabanis treated the brain simply as another organ, no different—in an analogy soon to become notorious in French intellectual circles—than the stomach or the intestines. Like any other organ, to operate properly he believed that it must be in balance, both internally—in regard to its various faculties—and externally, as part of the body's system of organs. Because brains naturally varied, that balance would differ from individual to individual. And because the brain was, for Cabanis, a real physical entity, that balance was liable to disturbances or alterations.[169] Cabanis—and most of the other idéologues—thus saw the brain's physical nature and the forces acting thereupon as vital elements shaping individual human nature and talents.

For all of the rigor and clarity of their work, the idéologues' cultural prominence was brief. With the suppression of the Second Class of the National Institute in 1803 after Napoleon's concordat with the church, idéologie fell into political disfavor because of its materialist and antispiritualist implications, a status intensified with the Bourbon restoration and then the rise of ultraorthodox monarchicalism.[170] Interest in mental physiology did not die out, but it was manifested more in straightforwardly medical settings or work on obviously pathological conditions, exemplified by the nosologies of mental illness developed by Pinel, Esquirol, and their followers.[171] The one major exception, Jan Goldstein and Stephen Jacyna have insisted, came with the rise of phrenology during the 1830s.[172] Although as antithetical to certain sensationist tenets as eclecticism, phrenology's naturalistic and physiological explanations of the mind and its founder Franz Joseph Gall's liberal politics, Jacyna has cogently observed, encouraged most contemporaries to see the phrenologists as heirs of Condillac and the idéologues. The phrenological doctrine of cerebral localization of mental functions was consequently challenged on both scientific grounds, most notably by Pierre Flourens, and cultural ones, as too materialist to be politically safe. While it gained numerous devotees, especially among the popular classes, phrenology posed little threat to eclecticism's hegemony in official academic culture.

However important it was in the French medical world, the physiology of difference was thus not well positioned to influence broader discussions of the nature of the mind and individual talents. Cousin's eclecticism, while it largely ignored questions of human difference and talents, did provide sanction for conventional religious and moral beliefs in its emphasis on the integrated self, the *moi,* possessing a moral faculty common to all. Moreover, eclecticism's ability to merge its focus on the "great man" as representative type with its orientation toward investigating universal human nature, suggested how the educational system's elite products could simultaneously be selected for special attention and presumed to stand in for the people as a whole. Republican, if not democratic, eclecticism thus sustained hierarchy and difference while still maintaining the validity of such universalisms as equality of basic human rights. It also suggested that the task of French philosophers/psychologists was not to investigate why certain individuals succeeded in the system of concours and grandes écoles, but rather to understand what the French as a people ideally could or should become. Where nineteenth-century Americans, working within the discourse of democracy, had to explain restriction of certain social goods to only a few, their French counterparts, confronting an idiom emphasizing equal rights and class differences, felt less pressured to justify distinctions than to spur the nation as a whole forward via that favored ancien régime device, émulation.

Conclusion

The French approach to talent and the individual thus differed significantly from its American counterpart. While philosophers and commentators in both countries assumed that the mind was a collection of independent faculties, they diverged in the status accorded those powers' individual manifestations. Whereas Upham, Wayland et al., embraced the range of the intellect's concrete manifestations and stressed the mind's diversity and malleability, Cousin and his fellow eclectics concentrated on the mind's abstract universal features and privileged attention to individuals who fully embodied such features. Cousin's disinterest in variations in talent, so different from the American Common Sense philosophers' approach, was consonant with an educational system that focused on creating a group of homogeneously trained "talented" rather than developing a variety of abilities within a heterogeneous population. Not that the upper levels of the American system of higher education were particularly diverse, at least in terms of class, gender, and race. Indeed, both nations' educational systems combined commitments to relatively inclusive primary education—based on the belief that a republic demanded a literate citizenry—with exclusive secondary institutions, open mostly to the elite's sons and (in the United States) daughters. Nonetheless, the decentralized primary schools, private secondary academies, and sectarian colleges of nineteenth-century America bore little resemblance to the centralized, pyramidal, Paris-dominated, French system, although it too promiscuously mixed public and private elements.

These differences in the French and American notions of talent and the educational systems constructed around them, when employed to buttress each nation's claims to having established meritocratic methods of allocating social rewards, produced subtle distinctions in each society's definitions of merit. In both cultures, commentators maintained that success was talent driven and that the able would, or at least should, ultimately prevail. French writers on education and philosophy translated this position into the proposition that the basis for meritocratic decisions should be native talent winnowed through competitions open to all. American authors, less enamored of national solutions, instead believed that any individual, through hard work and determination, could manifest talent and achieve success, celebrating this possibility as the clearest indication of the meritocratic nature of American politics and society.

Regardless of the reality of these representations of French and American culture, in both cultures talents played a critical role in legitimating the social structures that actually developed. Through the language of talent, nineteenth-century French and American members of the elite and even popular classes had one means of explaining why only certain individuals received access to such valuable social goods as higher education, professional training, or coveted government positions, and also of pursuing claims that talented individuals deserved opportunities regardless of social origin or ability to pay. When the conversation shifted from individuals to species, races, or groups, however, as we shall see in the next chapter, both French and American writers turned to an alternative, and much more biologically based, language of difference. Following up, in a sense, on the idéologues' move to tie human characteristics directly to physiology, naturalists, anthropologists, and others investigating group-level differences fashioned a discourse not of malleable talents, but of nature-endowed intelligence.

Three

All *Men* Are Created Equal?

ANTHROPOLOGY, INTELLIGENCE, AND THE SCIENCE OF RACE

> Died— NEGRO TOM, the famous African Calculator, aged 80 years. He was the property of Mrs Elizabeth Cox of Alexandria. . . . This man was a prodigy. Though he could neither read nor write, he had perfectly acquired the art of enumeration. The power of recollection and the strength of memory were so complete in him, that he could multiply seven into itself, that product by 7, and the product, so produced, by seven, for seven times. . . . He drew just conclusions from facts; surprisingly so, for his opportunities. Thus died Negro Tom, this *self-taught Arithmetician,* this *untutored Scholar!*— Had his opportunity of improvement been equal to those of thousands of his fellow-men, neither the Royal Society of London, the Academy of Sciences at Paris, nor even NEWTON himself, need have been ashamed to acknowledge him a Brother in Science.
>
> *Columbian Centinel* (Boston), December 29, 1790

THOMAS FULLER, the full name of "Negro Tom," had become famous late in his life after being "discovered" in 1788 by William Hartshorne and Samuel Coates, two members of the Pennsylvania Society for the Abolition of Slavery, during a trip to Virginia. Greatly impressed by Fuller's calculating abilities, they sent a report to Dr. Benjamin Rush—secretary of the Society and perhaps America's most prominent physician—who promptly drafted an account for publication in the *American Museum* and for dissemination to American and English abolition societies. The London group, Rush explained, had requested "such accounts of mental improvement, in any of the blacks, as might fall under their notice, in order to enable them to contradict those who assert that the intellectual faculties of the negroes are not capable of improvement equal to the rest of mankind."[1] The gambit made perfect sense. Wanting to sway public opinion toward eliminating slavery, Rush and his fellow abolitionists sought to dismantle every possible support for slavery's continuation. And in the wake of the first attempts at abolition in the Americas, some of slavery's proponents on both sides of the Atlantic argued that slavery was justified because Africans were by *nature* inferior to Europeans, particularly in mental ability.[2] Opponents of necessity responded in kind, and a people's innate characteristics soon became focal points for scientific debate and ideological struggle.

In her influential essay "Ideology and Race in American History," Barbara Fields argued that race science/scientific racism developed in America as demands for emancipation increased, reflecting the emergence of a new form of bourgeois rationality dedicated to "identifying and classifying differences among people" on the basis of "scientific first principles." "Race," she concluded, "is a product of history, not of nature."[3] Fields's insistence on race as an ideology inflected by time, place, and social position, and her contention that racial science was employed to naturalize racial caste are critical insights. Nonetheless, comparison with similar scientific developments in France suggests that the production of the category "race" did not occur solely in response to American or even Western tensions over slavery and emancipation.

Eighteenth-century arguments about race couched in the language of biology were tentative, with only tepid support at best from the scientific community. But, during the nineteenth century, moves toward more naturalistic understandings of human beings occurred throughout the West. In America, this shift dovetailed with increasing anxiety over race and its cultural meanings, making recourse to natural differences, and especially the language of natural inferiority or superiority, increasingly common. The heyday of French racial science, however, came in the second half of the nineteenth century, when the French elite was concerned more with integrating the peasantry into the nation than with distinguishing among regional or social groups, and thus exhibited little interest in highlighting internal racial differences.[4] French anthropologists were implicated in France's burgeoning colonial project, it is true, and contributed variously to the nation's imperial ideology, the *mission civilisatrice*. Even so, the ameliorist aspects of that ideology and most anthropologists' commitment to liberal republicanism and neo-Lamarckian transformism meant that the biologically based racial hierarchies they fashioned had few of the cultural implications of those developed by their American counterparts. The need to stabilize notions of racial caste was simply much more acutely felt in the United States than in France.

In addition, French anthropologists, mostly centered at the Société d'Anthropologie de Paris, worked in relative isolation from the public and wrote largely for specialist audiences. Similarly engaged Americans, by contrast, lacked formal institutional structures and consequently addressed both laypersons and specialists in general-interest periodicals, creating a kind of public forum on race. Their views thus became part of a culture-wide discussion in ways that French racial science rarely did. What differed most between America and France, therefore, was not the racial science itself so much as the locations in which it was developed, its uses, and its resonance with broader cultural discussions. While race was indeed a contingent product of history, as Fields has insisted, that contingency had less to do with the vocabulary of racial difference per se than with the practices and sites through which the languages of race and difference were articulated and made vital.

When the discourse of racial science began to coalesce, anthropologists, physicians, and other natural scientists central to its formulation accorded particular prominence to a group's mental and moral characteristics. Although investigators routinely noted skin color, hair texture, and other physical features, most exponents of racial science focused on intellectual ability (viz., Thomas Fuller), particularly in the singular as the level of a race's overall intelligence. In so doing, they imported a zoological concept, one initially used to distinguish humans from other animals, but increasingly to explain the so-called scale of nature. Naturalists interested in race seized on intelligence and employed it to differentiate races and arrange them in a graded sequence (or at least to justify scientifically long-standing cultural practice). They did so by investigating some physical feature of the head—be it facial angle, cranial volume, or brain weight—readily measurable on living people or skeletal remains and plausibly tied to a group's overall level of mental power, and then collecting data with which to calculate racial averages and produce graded sequences.

In so doing, anthropologists rejected, at times almost consciously, a rival project, deeply embedded in romanticism, that privileged emotions rather than reason or intelligence as the most important features defining human nature.[5] Relegating "excessive" feeling to "lesser" groups—women, blacks, primitive societies—anthropologists on both sides of the Atlantic believed civilization derived primarily from exercising reason and disciplining the passions. In their view, the important differences between peoples were reflected, physiologically, in the average size of a group's brains, and, culturally, in artistic and scientific progress. Moreover, many suspected that the basic emotions and sentiments were both difficult to measure and common to all humans, and thus not criteria relevant to objectively distinguishing among groups.

The romantics were not alone, however, in contesting the meanings derived from the anthropologists' data, anthropologists' claims to mechanical objectivity notwithstanding. In the United States, African American intellectuals in particular vigorously debated whites' increasing use of such data to justify blacks' inferior status. And in France, where the most fiercely contested cultural battles concerned whether the nation should be empire or republic and the culture sacred or secular, few accorded findings about biological differences between groups much significance. Intelligence as a racialized and hierarchical concept thus did not completely win the day in either country. Nonetheless, in both nations, one consequence of the search for physical correlates of mental characteristics was to make intelligence seem a singular, real, measurable, physical entity, one open to appropriation by a range of scientific practitioners with a variety of agendas.

"Intelligence," Animals, and the Scale of Creation

The word "intelligence" has a long history. *Le Dictionnaire de l'Académie françoise* of 1694 provided seven meanings for the term, all relating primarily to human capacities or knowledge. Four had been present since at least 1500, including "the faculty of understanding" and the "act of or capacity for understanding"; three were newer additions, among them "knowledge or comprehension" and "great people who have extraordinary talent for government."[6] In all the definitions knowing was central, though in two different modalities: knowing as a potential or ability, and knowing as the simple possession of knowledge. Little changed through the dictionary's successive editions until An VII (1798), when the fifth edition revealed a small but significant shift: unadorned "knowledge" became "profound knowledge" and unqualified "comprehension" "clear and easy comprehension."[7] Knowledge as such was now less central; having *intelligence,* like being *intelligent(e),* required a certain type of knowledge gained in a certain way. During the nineteenth century, this shift became even more pronounced. In the Academy's 1835 sixth edition, for example, the most significant change in "*intelligence*" and "*intelligent(e)*" lay in the continued constriction of the knowledge denotations of the word.[8] By 1835, "*intelligence*" had virtually ceased to refer simply to knowledge, instead suggesting either an absolute ability shared by all, or something relative that individuals could manifest in different measures.

The story of intelligence in English is more vexed. Samuel Johnson's *Dictionary of the English Language* (1755) froze the dictionary meaning of "intelligence" well into the next century in both Britain and America.[9] Johnson's definitions—including "commerce of information, notice, mutual communication"; "commerce of acquaintance, terms on which men live with one another"; and "understanding, skill"—emphasized exchange, be it of information or mutual knowledge; only the adjectival meaning related to individual knowledge, and that primarily in the sense of an acquired skill.[10] Meanings such as the "faculty of understanding" or "understanding as a quality admitting of degree," which the *Oxford English Dictionary* now suggests were prevalent at the time, were ignored by Johnson, and indeed remained absent from English-language dictionaries until the revised second edition of Noah Webster's *American Dictionary of the English Language* (1841). Webster there did pick up on other senses of intelligence, specifically "a gift or endowment" and "the capacity for the higher functions of the intellect." Nonetheless, the term remained largely in the backwaters of English-language discourse until later in the century.[11]

Such was not the case in French. Indeed, the 1835 dictionary caught another important change in meaning, noted almost as an aside: the term's extension to animals.[12] Reflecting and made possible by the shift in the meaning of "*intelligence*" toward an ability existing in degrees, the inclusion of animals signaled important changes in ideas about the relations between humans and brutes and the natural order writ large. In part, these changes reflected the success of the taxonomic projects of eighteenth-century natural history. Swedish naturalist Carolus Linnaeus's monumental work on arranging the living world into one all-encompassing classificatory system, *Systema Naturae,* begun in 1735 and completed posthumously in 1793, inspired other naturalists to either extend his work or to create their own improved systems.[13] In each of these schemes—by Linnaeus or Georges Louis Leclerc, Comte de Buffon or Johann Friedrich Blumenbach or Jean-Baptiste Lamarck, just to name the most prominent—humans were made part of the order of nature, classified by physical and moral characteristics, and divided into subgroups.[14]

In addition, as Robert J. Richards has argued, the rise of sensationist psychologies during the eighteenth century eroded the long-presumed barrier between human and animal mentality.[15] Rejecting both the Aristotelian claim that only humans possessed reason and the Cartesian position that animals were organic machines operating on instinct, Condillac and his followers insisted that all sensate beings generated ideas, however much those ideas' power and complexity differed by species.[16] As Charles-Georges Le Roi explained in the *Encyclopédie méthodique* (An II [1795]), "this faculty [sensibility], more or less excited by needs and circumstances, produces the different degrees of intelligence that we observe, whether between species or between individuals."[17]

Sensationist pronouncements notwithstanding, throughout the nineteenth century naturalists continued to debate whether animals possessed intelligence as well as instinct and what the exact relations between the two were. Some, ranging from the great early nineteenth-century French comparative anatomist Georges Cuvier up to the American theologian and college president John Bascom in the 1870s, maintained that the distinction between brute and human mind was absolute. "There is here," Bascom observed, "a radical difference between brute and human intelligence, referable to the manner of acquisition; the one exhibiting the alertness, exactness, and limitation of the senses; the other the breadth, slowness, and uncertainty of reflection."[18] Even in Bascom's case, however, the word "intelligence" itself tellingly covered all sentient beings; different though human and brute intelligence might be, by the 1870s a common term applied to both.

For others the applicablility of "intelligence" to the animal kingdom was undisputed. Frédéric Cuvier, for example, Georges's brother as well as his colleague at the Muséum national d'Histoire Naturelle, conceded that animals exhibited intelligence, though he was unsure whether nonhumans possessed reason. Cuvier believed, Richards notes, "that the term 'intelligence' (*l'intelligence*), had its proper use in characterizing animal behavior, since it signified something less than reason."[19] Lamarck, the Cuviers' main rival at the museum, went further, conceiving of the entire animal kingdom as linked through a progressive transformism as much intellectual as physical:

> It is true that one observes a kind of gradation in intelligence of animals, as it exists in the increased perfection of their organization, and one notes that they have ideas and memories; that they think, choose, love, hate; that they are susceptible of jealousy; and that by diverse inflexions of their voice and by signs they communicate and understand one another.[20]

The pioneering American researcher into animal behavior Lewis Henry Morgan concurred. While he in no way advocated Lamarck's version of evolution, Morgan did argue in 1843 that the Creator had distributed intelligence throughout the animal world, and on a graduated scale.[21]

Such speculation was thus rife well before Charles Darwin published *Origin of Species* (1859) and especially *The Descent of Man* (1871) and made the continuities between brute and human intelligence critical to his theory. While Darwin's claim in *Descent* that "there is no fundamental difference between man and the higher mammals in their faculties" put the issue into stark relief, it did not mark a fundamental transformation in the debate.[22] Indeed, for many nineteenth-century naturalists the issue was less whether all species manifested intelligence and more how intelligence was and was not related to instinct. Although Condillac had attempted to banish instinct entirely from the lexicon, arguing that no behaviors could be innate, few researchers even among sensationists followed his lead, most being convinced that at least certain behaviors must be inborn.[23] Rather, the major source of contention was between those who argued that instinct and intelligence were different manifestations of the same phenomenon, and those who claimed that instinct and intelligence were distinct sources for behavior.

Lamarck and his followers tended strongly to the former position, explaining continuities between intelligence, habit, and instinct as the result of the gradual transformation of repeated deliberate actions into unconscious habits and then heritable instincts. "A scientific explanation or theory of instinct," the American evolutionist Joseph Le Conte declared in 1875, "must connect it with intelligence on the one hand and the lower phenomena of the nervous system on the other—must show how all these several capacities are evolved the one from the other—must bring them all under the universal law of evolution."[24] Darwin, though more open to the inheritance of acquired characteristics in the *Descent* than in *Origin,* nonetheless felt that the evidence favored the separate origin and evolution of intellectual faculties and instincts. "The more complex instincts," he observed, "seem to have originated independently of intelligence."[25]

Whatever position naturalists took on the relation of intelligence and instinct, they largely agreed on one point, that both were phenomena that could be analyzed at the level of species. Even Darwin, though resolutely focused on selection acting on individual variations, nonetheless unselfconsciously spoke of the mental characteristics of entire species:

> the difference in mind between man and the higher animals, great as it is, is certainly one of degree and not of kind. We have seen that the senses and intuitions, the various emotions and faculties, such as love, memory, attention, curiosity, imitation, reason, &c., of which man boasts, may be found in an incipient, or even sometimes in a well-developed condition, in the lower animals.[26]

In this he was not alone. Many nineteenth-century naturalists came to understand the scale of animal creation as founded on a common attribute, intelligence, understood as a singular faculty whose power varied across the animal kingdom by degrees. Moreover, although the term signified a mental characteristic, it also took on physicalistic connotations tied to the material structure of the brain. In that form, intelligence would prove to be a powerful resource for anthropologists, who took as their classificatory project not the whole of the animal kingdom, but only a small slice, *Homo sapiens,* and who saw in racial differences the key to understanding humankind.

The Language of Intelligence and the Science of Race

In 1839 Samuel G. Morton—Philadelphia physician, American Philosophical Society member, professor of anatomy at the Pennsylvania Medical College, and secretary of the Academy of Natural Sciences—published *Crania Americana, or a Comparative View of the Skulls of Various Aboriginal Nations of North and South America.*[27] *Crania Americana* was composed of six parts: a lengthy introductory essay presenting Morton's proposed division of the human race into twenty-two separate families; a detailed analysis of the cultures and crania of forty-one native American groups; an explanation of how Morton obtained his measurements; tables of anatomical and phrenological measurements on 158 skulls; an appendix by the noted phrenologist George Combe relating "the natural Talents and Dispositions of Nations, and the Developments of their Brains"; and a set of seventy-one lithographs depicting some of the skulls analyzed in the text. Morton's compendious work and subsequent publications by himself, Josiah Nott, George Gliddon, and Louis Agassiz—members of the so-called American School of Anthropology—stirred up a far-reaching controversy within American scientific, literary, and political circles during the middle decades of the nineteenth century.[28]

The scientific dispute centered on the origin of the human races. Were human beings derived from a single set of ancestors (monogenism), as naturalists believed and scripture indicated, or had there been multiple creations of distinct human species (polygenism), the races, as the American school of anthropologists would argue? The broader cultural debate focused on Negroes and members of other "nonwhite" racial groups. Were blacks, Indians, Asians, perhaps even the Irish, of equal humanity with whites (however culturally inferior some might regard them), because all represented the same species, or were they biologically distinct and, as many white Americans would see it, *inherently* inferior because they were members of different, presumably lesser, species? Beneath both sets of issues lay a still more fundamental question: What should be the ultimate authority to adjudicate such disputes, scripture or nature?

For many mid-nineteenth-century Americans, the answers remained equivocal at best. Studies by William Stanton, George Fredrickson, Mia Bay, Bruce Dain, and others have shown that while polygenism's proponents had a large influence in the scientific world until the advent of Darwinism in the 1860s altered the terms of the debate, their reception by the public was more mixed.[29] Some northern whites, and most blacks, were sufficiently antislavery to oppose a theory that so clearly supported "the peculiar institution"; a number of African Americans, in fact, crafted extensive critiques of both polygenism and its accompanying theories of African biological inferiority.[30] Other northern whites, and many southern ones, however indifferent or even positively disposed to slavery, had little stomach for an interpretation that ran directly counter to orthodox Christianity and that smacked as well of materialism and atheism.[31] As an article in the African American newspaper the *National Era* observed in 1854:

> A majority of the advocates of Slavery labor under a wholesome fear of the Devil, at any rate, if they cannot be said to love God; and they are beginning

> to see that the apostles of the new revelation are leading them into the slippery and dangerous paths of Infidelity. . . . They accordingly pass sentence of condemnation upon the infidel theory of a diversity of races.[32]

Nonetheless, polygenist arguments achieved increasing public credibility in the late 1840s and 1850s, fueled by the writings of southern physicians, naturalists, and intellectuals; by the theories of northerners such as Morton and Agassiz; and, almost perversely, by the natural rights arguments of certain abolitionists. Indeed, by 1854 the abolitionist leader and ex-slave Frederick Douglass was so concerned that he developed an address specifically refuting polygenist claims. "Let it be once granted," he eloquently argued in "The Claims of the Negro Ethnologically Considered,"

> that the human race are of a multitudinous origin, naturally different in their moral, physical, and intellectual capacities, and at once you make plausible a demand for classes, grades and conditions, for different methods of culture, different moral, political, and religious institutions, and a chance is left for slavery, as a necessary institution.[33]

Douglass highlighted the polygenists' key move: the equation of difference with scales of inferiority/superiority. Implicitly rejecting mental philosophy's model of difference, which saw faculties as many and complementary, American ethnologists turned instead to the classificatory sciences, where single, decisive differences demarcating groups were critical and interest in hierarchies was strong. They drew in particular on the vision of a hierarchy of species associated with the Great Chain of Being, and with it the chain's key criterion for distinguishing species, an organism's level of overall intelligence. In so doing, the American anthropologists helped to expand the domain of the language of intelligence, by extending it from the level of species to encompass races and groups, where it would intersect with America's growing obsession with race to create a powerful language of natural difference.

By at least the mid-eighteenth century, Winthrop Jordan has shown, the question of the origin and nature of the human types intrigued naturalists and others on both sides of the Atlantic.[34] Fueled by fascination particularly with Linnaean taxonomy and by the global reach of European trade and empire, categorizing the world's flora and fauna became a major undertaking.[35] Naturalists included humans in this classificatory gaze, most typically dividing *Homo sapiens* into a number of varieties—including "European," "African," "American," and "Asiatic"—that roughly matched the era's typical racial groupings.[36] In general, these systems relied on three criteria to distinguish the varieties: skin color and texture, hair type, and some aspect of skull conformation.[37] Although skin and hair differences constituted the most obvious markers of racial distinctness—and there was much debate over their relative permanence—virtually all parties ultimately accorded skull characteristics the greatest significance.[38] Naturalists turned to cranial features—facial angle, cranial volumetric capacity, or the cephalic index—partly because they facilitated comparisons across time by introducing skeletal characteristics often preserved for centuries. But largely, their centrality derived from two interrelated factors: the cranial measurements that gained prominence fell into a nicely graded sequence, and most practitioners assumed that physical characteristics of the head said something significant about the mind inside.[39]

The first feature, that measurements of certain cranial characteristics allowed the races to be arrayed in a distinct pattern, received its most important early articulation by the Dutch artist, naturalist, and professor of anatomy Peter Camper.[40] In *A Treatise on the Natural Differences of Features in Persons of Different Countries and Periods of Life* (1768), Camper propounded a new way of measuring human beings, the facial angle—formed by the intersection of a horizontal line from the ear to under the nose and a vertical line from brow to jaw—and applied this measure to various types of animals and human beings.[41] Much to his surprise, he suggested, he discovered that the animal kingdom could be arrayed on a single scale, with Greeks from antiquity at the top, measuring 100 degrees, followed by Europeans at about 80 degrees, Negroes at about 70 degrees, orangutans at about 58 degrees, African monkeys at about 42 degrees, and so on. Camper's findings corresponded closely to the hierarchy articulated in the chain of being and seemed to offer physical evidence in support of blacks' inferior status. Camper himself, a strict monogenist and strongly antislavery, drew no particular conclusions from this finding.[42] And one of the period's most prominent naturalists, Johann Friedrich Blumenbach, produced a devastating critique of Camper's method in 1795, arguing that individuals varied too much to obtain consistent results for any group's facial angle.[43]

Nonetheless, Camper's arguments attracted a great deal of attention, both favorable and critical, and the use of the facial angle spread widely. In and of itself, however, the facial angle said nothing about human mental attributes. That connection was supplied, Claude Blankaert has argued, by two French naturalists, Georges Cuvier and Etienne Geoffroy Saint-Hilaire, who in 1795 linked the facial angle explicitly to intellectual capacity:[44]

> it seems that the size and convexity of the skull indicates sensibility, just as the extension and thickness of the snout indicates brutality. One observes in the diverse races of man the same series of relations, as in the diverse species of animals, between the projection of the skull and the degree of intelligence.[45]

Their focus was on the orangutan, which was crucial, as it allowed them to apply a characteristic used to assess animal species—overall intelligence—to humans. Although the physiognomic teachings of Jean-Gaspard Lavater and the phrenological theories of Franz Josef Gall, Johann Spurzheim, and George Combe also linked cranial or facial features to mental characteristics, they did so based on models of the mind that stressed the multiplicity of the faculties, not some global power such as intelligence, and thus proved much less attractive to naturalists.[46]

For the next century, this putative relationship between the skull and the intellect would dominate ethnology/anthropology.[47] Although other measures soon eclipsed the facial angle, notably cranial volume and cephalic index, the investigative strategy was similar: collect a group of heads or skulls, precisely measure their features, calculate racial or group averages, array the groups numerically by mean, and then connect that ordering with a hierarchy of intelligence, generally by implying a causal link between size of the skull feature and power of the intellect.[48]

In the first half of the nineteenth century, Samuel Morton undoubtedly did more than any other scientist to establish cranial capacity as *the* preferred method for demonstrating that the human races fell into a graded series.[49] A circumspect polygenist, Morton's major interest was in the native peoples of the Americas, who he wished to show were all part of the same racial group. Morton chose the skull as his principal research material and over thirty years assembled a collection of some six hundred of them, which he analyzed to measure cranial volume and other features. His most sensational finding, presented in *Crania Americana* on the basis of his measurement of 256 skulls, was that the five races into which the human species was conventionally divided were independent in origin, character, and development and could be arranged in a hierarchy according to mean internal cranial capacity. (See table 3.1) His analyses themselves produced few surprises: Caucasians were uppermost, having a mean of 87 cubic inches, followed by Mongolians (83 cu. in.), American Indians (82 cu. in.), Malaysians (81 cu. in.), and Ethiopians or Negroes (78 cu. in.).[50] Stephen Jay Gould performed a detailed re-analysis of Morton's data in the 1970s, revealing many flaws in Morton's procedures; Morton himself came to slightly different conclusions in 1849 on the basis of 623 crania and an improved measurement technique. While Morton still found Caucasians to be on top, with Teutonic whites at the absolute summit (a mean of 92 cu. in.), he now placed the Negro group considerably above the American Indian group (by almost 4 cu. in.), and included other idiosyncrasies at the level of families within a given group.[51]

The basic message of Morton's data, however, whether from 1838 or 1849, was the same: the human races had distinct characteristics and fell

TABLE 3.1
Cranial Capacity by Race, according to Morton's *Crania Americana*

Races	*Number of Skulls*	*Mean Capacity (cu. in.)*	*Largest (cu. in.)*	*Smallest (cu. in.)*
Caucasian	52	87	109	75
Mongolian	10	83	93	69
Malay	18	81	89	64
American	144	82	100	60
Ethiopian	29	78	94	65

into a graded scale, with whites at one end and blacks at or near the other. Morton's disciple Josiah Nott, for example, glossed an illustration of three skulls—Caucasian, Mongol, and Negro—placed in order from largest to smallest as follows:

> Although I do not believe in the intellectual equality of races, and can find no ground in natural or human history for such popular credence, I belong not to those who are disposed to degrade any type of humanity to the level of brute-creation. Nevertheless, a man must be blind not to be struck by similitudes between some of the lower races of mankind, viewed as connecting links in the animal kingdom; nor can it be rationally affirmed, that the Orang-Outan and Chimpanzee are more widely separated from certain African Oceanic Negroes than are the latter from the Teutonic or Pelasgic types.[52]

Although, as Stanton points out, Morton himself was more reticent about stating unequivocally that cranial capacity equaled mental capacity, the appendix to *Crania Americana,* by phrenologist George Combe, showed no such modesty: "The aggregate natural mental power, (animal, moral and intellectual,) of the individuals composing any nation, will (other conditions being equal) be great or small in proportion to the size of their brains."[53] The richness and detail of Morton's empirical evidence was overwhelming, praised enthusiastically in all of the reviews of *Crania Americana,* even by the English-speaking world's leading anthropologist, James Cowles Prichard:

> when the care and accuracy of the observations made by its [*Crania Americana*'s] author, and the learning, ingenuity and skill manifested in his deductions from the data before him, are taken into account, together with our previous ignorance of the subject of his inquiries, few, if any, of his readers will hesitate in affirming that his book well deserves to be generally known, and to find a place in every library connected with natural science.[54]

Indeed, for most of the nineteenth century, Morton's work served as a model for investigations in physical anthropology.[55]

Matter, Mind, and Hierarchy

Unexplained in this account, however, is a fundamental question: What made this key relation of brain size to mental capacity so unproblematic, not only for advocates of polygenesis, but for their opponents as well? In part, the answer was physiology. With broad agreement by the eighteenth century's end that the brain was the physical seat of the mind, the notion that brain size might affect mental strength gained plausibility. In addition, Franz Gall's work on cerebral localization, which seemed to demonstrate that mental faculties could be identified with specific regions of the brain, and the widely accepted connection between very small brain size and idiocy, both seemed to indicate that an individual's mental power was at least partly determined by the size of the vessel housing it.[56] As Josiah Nott pithily observed, "all scientific men concede that brains below a certain size are always indicative of idiocy, and that men of distinguished mental faculties have large heads."[57] More significant, however, may have been the widespread belief that both the animal kingdom and human societies could be arranged in hierarchies, with differences in mental attributes a key characteristic. Morton articulated just such a conception when describing the ancient Peruvians: "It would be natural to suppose, that a people with heads so small and badly formed would occupy the lowest place in the scale of human intelligence."[58] In evoking the scale, Morton simultaneously drew on two discourses about hierarchy that naturalists were, by the early nineteenth century, beginning to intertwine.

The first, traceable at least to Aristotle, was the language of the chain of being.[59] According to this theory, the animal species formed a series of discrete links in a single continuous chain extending from the lowest life forms to humans—seen as unique in the animal world for the possession of reason—and then on to purely spiritual forms that occupied the chain's highest ranks. Although by the late eighteenth century the concept was no longer the sole, or perhaps even the primary, way of organizing the animal kingdom—that honor had passed to Linnaean taxonomy—the chain of being still exercised a powerful hold on the imaginations of both naturalists and laypersons. Indeed, at the turn of the century Lamarck reimagined the chain along transformist-evolutionary lines, promoting a significant alternative to the Linnaean program.[60]

Within eighteenth- and early nineteenth-century conceptions of the chain of being, reason or intelligence was increasingly understood as the trait creating the graded scale. What separated human beings from the apes, their nearest neighbors on the chain, for example, was not that humans possessed reason and apes did not, but that humans possessed reason or intelligence of a higher order than apes, who themselves possessed more of it than those lower on the chain.[61] "To the degree that we see animals rise in the progressive scale of organization," the French naturalist Julien Joseph Virey noted in 1801,

> their nervous system becomes more voluminous, their brain larger and more complicated. . . . The intelligence of animals grows in the same progression, in general; such that one arrives at man by slight differences almost successive, as it is easy to note in passing from dogs to monkeys, to the orangutan, and from there to the Hottentot Negro, and from there to the white man, the European, the most industrious and most enlightened.[62]

As natural historians were developing this biologized language of intellectual hierarchy for the animal kingdom, Enlightenment social theorists were constructing a similar discourse around notions of civilization and the hierarchy of human societies.[63] The Enlightenment period is notorious for its lack of sympathy toward cultural relativism. Although few Enlightenment writers argued that non-European peoples were of a different order of humanity, almost no one considered the "savage" nations of the world different from, but equal to, the societies of western Europe. Rather, convinced that Europeans had, in general, achieved an unparalleled degree of civilization—the word itself is of Enlightenment coinage—most Europeans who contemplated non-European nations found those societies distinctly inferior, if not irremediably so.[64]

In judging a society's degree of civilization, European commentators generally employed two criteria. The first was moral: How "barbarous" were the people? How completely were they governed by law and manners, how much by force and cruelty? And the second, related criterion, was intellectual: To what degree did genius manifest itself among the people? How advanced were their arts and sciences?[65] The scale of civilization was simultaneously a ladder of moral and mental progress, with notions of graded mental capacity underlying both. The explanation of certain races' superiority or inferiority tended, ultimately, to be couched in terms of relative mental capacity: some peoples either were endowed with, or had developed, their mental capacity much more fully than others. As Samuel Stanhope Smith observed early in the nineteenth century:

> The coarsest features, and the harshest expression of countenance, will commonly be found in the rudest states of society. And the mental capacities of men in that condition will ever be proportionally weaker than those of nations who have made any considerable progress. They become feeble through want

> of objects to employ them, and through defect of motives to call forth their exercise. . . . The Hottentots, the Laplanders, and the people of Tierra del Fuego are the most stupid of mankind for this, among other reasons . . . they approach, in these respects, the nearest of any people to the brute creation.[66]

In an era that celebrated the power of reason, it is perhaps not surprising that some notion of human intelligence would have been turned to as a critical attribute distinguishing peoples.

By so doing, writers on the science of race began to cobble together a new language for describing and exploring human differences, one in which the "talents" of mental philosophers proved of much less resonance and utility than "intelligence." For Morton and other nineteenth-century investigators, the salient "fact" to be explained was the obvious and undeniable inferiority of certain peoples. Intelligence, the naturalists' term explaining the gradation of animal species, was easily appropriated to provide a naturalistic explanation for the manifest existence of this racial hierarchy. Its connotations of global mental power, varying by degrees and related to the brain's physical nature, allowed measurable external characteristics, such as cranial capacity, to be related to an internal mental feature that could plausibly account for a people's place in the racial hierarchy.[67]

To be sure, Morton especially tended to veer between the language of intelligence and that of faculties when doing racial comparisons. "The intellectual faculties of this great family," he noted about the American Indians, "appear to be of a decidedly inferior cast when compared with those of the Caucasian or Mongolian races. They are not only averse to the restraints of education, but for the most part incapable of a continued process of reasoning on abstract subjects."[68] Even here, however, the plural "faculties" was treated effectively as a singular measure, since the comparison worked only if all powers varied in the same way. Moreover, early nineteenth-century racial anthropologists suggested that this singular entity might be fundamentally biological in origin, an inborn trait that no effort could significantly alter. "It seems to us to be mock-philanthropy and mock-philosophy," Louis Agassiz observed in 1850, "to assume that all races have the same abilities, enjoy the same powers, and show the same natural dispositions, and that in consequence of this equality they are entitled to the same position in human society."[69]

Nonetheless, during the first half of the nineteenth century such moves remained tentative. When Morton turned from comparing racial groups to describing them—that is, when he moved from comparative anthropology to ethnography—his certainty about the primacy of biology was much less clear. The bulk of *Crania Americana* comprised detailed descriptions of the customs and skulls of the forty-one Native American racial groups. Having characterized the group Native Americans as inferior intellectually to Caucasians and Mongolians, Morton might have been expected to develop this theme in his descriptions of the individual families. But confronted with the rich variations among and accomplishments of the Native American peoples, especially those he deemed Toltecans, Morton emphasized their complex characters and frequently praised their intellectual achievements. Of the Mexicans (Aztecs), for example, Morton remarked that the "state of civilisation among the Mexicans, when they were first known to the Spaniards, was much superior to that of the Spaniards themselves on their first intercourse with the Phenicians. . . . Their understandings are fitted for every kind of science." Morton went on to note Mexican accomplishments in architecture, calendrics, and science—especially arithmetic and astronomy—and even suggested that they had produced figures the equal of Descartes, Kepler, and Leibniz.[70] Not bad for a family of peoples "incapable of reasoning on abstract subjects."

Morton's shifts between denigrating and praising various Native American peoples suggest much about the status of conceptions of human mind and difference within early nineteenth-century anthropology. When viewed from a distance and searching for general characteristics with which to compare groups, human races, families, or nations could appear homogeneous, and complexities of mind could fade into a sense that some brains simply were more powerful than others. Certainly this had long been the case, even in mental philosophy, when describing the extremes of the intellectual spectrum—idiots and geniuses—or when comparing children to adults. But when the demands of quantification and comparison did not require a unidimensional scale, variety characterized the discourse, and the multiple ways that individuals within a group could develop particular talents and faculties seemed most salient.[71]

We have already seen in chapter 2 how the dominant early nineteenth-century mental philosophies—Scottish Common Sense and Cousinean eclecticism—were little inclined to represent mental power as homogeneous. The Scots emphasized the diversity of the faculties, and the Cousineans were uninterested in differential mental capacity, focusing instead on the universal characteristics of abstract mind. Around these psychologies of the *individual* mind developed a discourse of heterogeneity, of talent*s* and power*s* and capacit*ies*. In contradistinction, the language that emerged among those comparing biological or social groups referred most frequently to intelligence or mental capacity in a singular, global sense. Created by transforming reason from an absolute into a characteristic admitting of degrees, intelligence and its cognates facilitated imposing a simple linear order on the animal and human worlds. They thus explained scientifically, or at least naturalistically, two "truths" of Western thought:

that humans, specifically white Europeans, held pride of place in the animal kingdom, and that European civilization was superior to all others.

Morton's research was critical to establishing the scientific plausibility of this picture in America, and helped to codify an investigative pattern that flourished until the century's end. While the number of cranial features measured and the range of instruments employed would expand greatly, and while living heads would become as significant a source of data as skulls, the project remained the same: to analyze and order races or groups, through an investigative strategy centered on reducing cranial characteristics to a compilation of measurements that could be easily arrayed into linear hierarchies and aligned with mental attributes. This science came to its greatest technical fruition in the later nineteenth century in France, as we shall see, in the work of Paul Broca and his circle. Its greatest cultural impact, however, occurred earlier, in America. Just as Morton et alia were developing their ethnological theories, these ideas were shaping and being shaped by broader public debates over the status and nature of the Indian and especially Negro races. Many American anthropologists actively contributed to these discussions; moreover, their insistence on the status of humans as natural objects open to empirical investigation like other flora and fauna accelerated the trend of transforming what had been a moral question into a scientific one.

Debating the Natural Order

American scientific discussions of the physical and material nature of the continent's inhabitants were initiated, in a sense, by French naturalists in the mid-eighteenth century. The Comte de Buffon's assertion in 1749 that New World animals and peoples were smaller, weaker, and less prolific than their European counterparts inspired, as is well known, Thomas Jefferson to pen *Notes on the State of Virginia* (1781–85) in rebuttal.[72] Jefferson dedicated much of his text to elaborate descriptions of the health and vigor of the American landscape and its many varieties of plants and animals. In a few sections, however, Jefferson turned to the New World's human inhabitants, especially the native peoples whom Buffon had characterized as enfeebled and about whom another French commentator, Abbé Corneille de Pauw, remarked in 1768 that

> a brutish insensibility forms the basis of the character of all Americans; their indolence prevents them from being attentive to any instruction; they know no passion strong enough to move their souls, to transcend their nature. Superior to animals in the use they make of their hands and their tongue, they are nevertheless truly inferior to the lowest Europeans. Deprived of both intelligence and perfectibility, they can only obey the impulse of their instincts.[73]

Outraged, Jefferson vigorously defended the natural character and manliness of the New World's natives:

> [The Indian of North America] is neither more defective in ardor, nor more impotent with his female, than the white reduced to the same diet and exercise: that he is brave, when an enterprise depends on bravery . . . that his vivacity and activity of mind is equal to ours in the same situation.[74]

Central to both assessments was not just Native Americans' physical qualities, but their intellectual characters as well. Indeed, de Pauw's attempt to situate North American "savages" between animals and the lowest Europeans was based precisely on their presumed level of intelligence, an assertion Jefferson answered by insisting throughout the *Notes* that Indians "are formed in mind as well as in body, on the same module with the 'Homo sapiens Europaeus.'" Jefferson, however, was not so protective of Africans in America. Comparing them to Native Americans, he found blacks distinctly inferior, utterly lacking in more than the most basic reason and imagination: "never yet could I find that a black had uttered a thought above the level of plain narration; never see even an elementary trait of painting or sculpture."[75] Significantly, both Jefferson's condemnation of African Americans and defense of Native Americans relied on the perspective of natural history, employing the language of inbred physical nature, not moral worth.[76] In eighteenth- and early nineteenth-century America, as Winthrop Jordan has persuasively argued, such a position was still rare. Biblical or environmentalist explanations of human diversity predominated, whether used to argue for the fundamental equality of all human groups or their God-given differences. Antislavery activists, to be sure, did use cases like Thomas Fuller's to demonstrate the potential mental equality of Africans and Europeans, but those were by no means the primary grounds on which debates over slavery and equality were fought.[77]

By the early nineteenth century, however, the ethnological musings of Jefferson, Camper, Charles White, and Edmund Long, among others, about the African race's physical inferiority and similarity to the orangutan had achieved some publicity and respectability in the new republic. Clement C. Moore felt obligated to attack Jefferson's racial theories publicly in 1804, one indication that such positions were no longer considered marginal.[78] A second was Princeton University president Samuel Stanhope Smith's decision to extensively revise his *Essay on the Causes of the Variety of Complexion and Figure in the Human Species* (first published in 1787; republished in 1810) to combat such "dangerous" doctrines.[79] Smith, in particular, went to extraordinary lengths to press his environmentalist argument that all human variation, including skin color, was climatological in origin, and that both nature and scripture attested to the basic unity of the human species. By the first decade of the nineteenth century, however, this standard late-Enlightenment account of human differences seemed less convincing, as both white and black Americans could see that skin color, hair texture, and other "typical" African features were not undergoing dramatic transformation in the New World. Moreover, the Haitian revolution, in which blacks created a short-lived republic that many American whites thought was founded on bloodshed and corruption, seemed dramatic proof to many observers that "Negroes" were incapable of self-rule.[80] Arguments about the innate inferiority of Africans, as a consequence, became more frequent, even if almost invariably linked with scriptural accounts of them as Ham's descendants.

The central issue in these early points of controversy was the African race's potential for "civilization." Many white Americans, especially those with proslavery leanings, argued vociferously that blacks could progress only modestly toward civilization, and that only under white tutelage. "Never before has the black race of Central Africa," John C. Calhoun remarked in 1837,

> from the dawn of history to the present day, attained a condition so civilized and so improved, not only physically, but morally and intellectually. It came among us in a low, degraded, and savage condition, and in the course of a few generations it has grown up under the fostering care of our institutions, reviled as they have been, to its present comparatively civilized condition.[81]

Calhoun and others who asserted that blacks were inherently limited and always had been, however, faced a problem: modern Western civilization's acknowledged birthplace, Egypt, lay in Africa. In response, efforts to disentangle Egyptians from other Africans were rife, well exemplified by Morton's 1844 publication *Crania Aegyptiaca; or, Observations on Egyptian Ethnography, Derived from Anatomy, History and the Monuments,* in which he sought to prove that Egyptian and Negro crania were distinct and had been since antiquity.

Just as strenuously, however, critics of slavery, including most of the African American intellectual elite, argued ceaselessly for close connections between Egyptian civilization and the contiguous Black African cultures. As early as 1827, for example, the African American newspaper *Freedom's Journal* printed an article by John Russwurm denigrating attempts to split Egyptians off from the African world:

> Mankind generally allow that all nations are indebted to the Egyptians for the introduction of the arts and sciences; they are not willing to acknowledge that the Egyptians bore any resemblance to the present race of Africans. . . . All we know of Ethiopia strengthens us in the belief that it was early inhabited by a people, whose manners and customs nearly resembled those of the Egyptians.[82]

David Walker in his famous *An Appeal . . . to the Coloured Citizens of the World . . .* (1829), and Rev. Hosea Easton in his less well-known *Treatise on the Intellectual Character, and Civil and Political Condition of the Colored*

People of the U. States and the Prejudice Exercised Towards Them . . . (1837), also addressed the issue of Egypt, though they turned the tables more decisively, by assuming that ancient Egyptians were African and civilized, and then wondering about the relative savagery of contemporaneous European peoples.[83] "It is a little singular that modern philosophers," Easton observed, "the descendants of this race of savages [Europeans], should claim for their race a superiority of intellect over those [Africans] who, at that very time, were enjoying all the real benefits of civilized life."[84]

Easton identified one of the major issues underlying the Egypt debates, an issue that would grow in importance throughout the antebellum period: the inherent intelligence of the Negro race. Were blacks different in kind from whites, each with their own physical and intellectual characters, as many white naturalists and physicians began arguing in the 1820s and 1830s, or were all peoples fundamentally the same, with variations simply adaptations to local conditions?[85] African Americans and a number of white clergymen, including most famously Frederick Douglass and John Bachman, argued strenuously for the unity of humankind, emphasizing the naturalness of variation in all animal creation.[86] Russwurm had made just this point in an 1828 *Freedom's Journal* article, as had Easton a decade later:[87]

> whatever differences there are in the power of the intellect of nations, they are owing to the difference existing in the casual [sic] laws by which they are influenced. By consulting the history of nations, it may be seen that their genius perfectly accords with their habits of life, and the general maxims of their country; and that these habits and maxims possess a sameness of character with the incidental circumstances in which they originated.[88]

By the 1840s and 1850s, however, others were insisting with increasing frequency and to ever greater public notice on the existence of permanent, meaningful racial differences, especially in intellect. Louis Agassiz, a Harvard professor and renowned naturalist, waded into the fray in 1850, rejecting "a unity of origin" for the human species and concluding that "the races are essentially distinct, and can hardly be influenced even by a prolonged contact with others when the differences are particularly marked."[89] Just a year later, New Orleans physician Samuel Cartwright contributed his own verdict on the Negro's physiological characteristics, arguing that physically and intellectually the race, "tinctured with a shade of the pervading darkness," was inherently inferior to Caucasians:

> It is this defective hematosis, or atmospherization of the blood, conjoined with a deficiency of cerebral matter in the cranium, and an excess of nervous matter distributed to the organs of sensation and assimilation, that is the true cause of that debasement of mind, which has rendered the people of Africa unable to take care of themselves.[90]

When added to Camper's determination of the inferior facial angle of Africans and Morton's contention that blacks had smaller brains than whites, the scientific evidence for the mental inferiority of African Americans by the 1850s seemed overwhelming, at least to many already so disposed. Indeed, Georgia lawyer Thomas Cobb could remark, almost in passing, that "in this opinion of the mental inferiority of the negro, every distinguished naturalist agrees."[91] Not all did, of course—as the African American press, in particular, took pains to point out. Some continued to assert variation's importance and the environment's powerful effects. "My position is *that the notion of inferiority, is not only false but absurd,*" James W. C. Pennington argued in 1841, providing as evidence a list of eminent "colored men"—including the ubiquitous Thomas Fuller—and then asserting on the basis of the biblical account of Creation that "*intellect is identical in all human beings.*"[92] In addition, many seized on the work of the eminent German anatomist Friedrich Tiedemann, who performed comparisons of skull and brain measurements for Negroes, Caucasians, and orangutans and then concluded in 1836 that the brains of Negroes and Europeans did not significantly differ.[93] As an article in the *Colored American* reported in 1840:

> The ablest living anatomist of Germany, Professor Tiedemann, has lately directed his researches with singular felicity to the vindication of the uncivilized man's capacity for improvement. The result of a most exact analysis of causes are thus stated by him. I. The brain of the negro is upon the whole as large as that of the European and other human races; the weight of the brain, its dimension, and capacity of the ravum cranil prove the fact. . . . V. The negro brain does not resemble that of the ourang outang more than the European brain.[94]

Tiedemann's work was attacked, most prominently by the French anthropologist Paul Broca, as was Nott's, Agassiz's, and even Morton's, and the debate over the "Negro" versus Caucasian brain would continue throughout the century.[95] Unresolved though it remained, the controversy did push issues of human mental capacity to center stage, suggesting that questions about human difference and its implications were a matter of scientific, rather than moral or political, inquiry. Agassiz articulated this position clearly in his 1850 essay:

> we entertain not the slightest doubt that human affairs with reference to the colored races would be far more judiciously conducted, if, in our intercourse with them, we were guided by a full consciousness of the real difference existing between us and them, and a desire to foster those dispositions that are eminently marked in them, rather than by treating them on terms of equality. We conceive it to be our duty to study these peculiarities, and to do all that is in our power to develop them to the greatest advantage of all parties.[96]

For a naturalist, of course, advocating the primacy of science was not a radical stance. But more mainstream figures, especially proslavery southerners, Mia Bay and Bruce Dain have convincingly demonstrated, also turned to nature as a powerful support for their position.[97] They were pushed to make this move, Thomas Cobb, for one, suggested, because of the increasingly strident denunciations of slavery by northern abolitionists based on claims about natural rights.[98] "The true defence of negro slavery," the *Richmond Examiner* reported in 1853,

> is to be sought in the sciences of ethnology and natural history. The last defines the negro to be the connecting link between the human and brute creation. The order of nature consists of infinite gradations; there are no abrupt endings and beginnings, separated by an empty interval. . . . From the most powerful family of the white race, we proceed by regular steps to the lowest type of the dark race, which is the negro; and close to him we find the chimpanzee of his native country, the first step in which we call brute creation.[99]

Emphasizing gradation, difference, and level of intelligence, the *Examiner* author captured well the language of race that would gain intensity in the second half of the century, when "race" and "intelligence"—ambiguous and polysemous in the Enlightenment—had hardened in meaning and become tightly linked. Especially as the institution of slavery came under ever greater pressure and then was eliminated, many whites' desire for assurances that race differences were real and permanent accelerated. Writing in 1851, Cartwright asserted what many of them wished fervently to believe, that "there is a radical, internal, or physical difference between the two races, so great in kind, as to make what is wholesome and beneficial for the white man, as liberty, republican or free institutions, etc., not only unsuitable to the negro race, but actually poisonous to its happiness."[100]

Inborn, physiological distinctions, particularly in intellect, became key to this reassurance. Ostensibly immune to humanitarian interventions, intelligence differences evoked a natural hierarchy, thus promising that Negroes would remain forever inferior. John H. Van Evrie crystallized these arguments in *Negroes and Negro "Slavery": The First an Inferior Race; The Latter Its Normal Condition*, first published as a pamphlet in 1853.[101] Douglass and other African American intellectuals recognized the challenge the emerging racial science presented. They sought, in response, to both undermine its claims and argue that any racial differences that might exist were irrelevant to basic human rights.[102] In addition, the reality of miscegenation and unstable racial identities such as "white Negroes" in the antebellum period alone called claims like Van Evrie's into question.[103] Nonetheless, even African American writers did not deny that one could speak meaningfully of a race's mental level; they sought only to demonstrate that Negro intellectual capacities were not inferior.[104]

By midcentury, therefore, the language of intelligence had moved beyond the narrow confines of naturalists' discussions of the relative abilities of animal species and, at least in America, become part of a much broader conversation about assessing human groups.[105] That this was as much a story of culture as of nature, however, is clear when considering developments in France. Although French anthropologists took the science of skull measurement (craniometry) to high technical levels, the place of the language of intelligence in the culture was very different. Relatively unconcerned with demarcating groups within the citizenry, few anthropologists and intellectuals viewed the language of intelligence as having an important internal social function. While it might justify the *mission civilisatrice* in Africa, and perhaps the poor social conditions of the French laboring classes, even in these instances the issue was less the French elite's racial superiority than its cultural eminence. As political as the circle of anthropologists around Broca was, their vision of French republicanism relied little on their analyses of human intelligence.

French Anthropology and Craniometrics

When the Société d'Anthropologie de Paris first met on May 19, 1859, in a small room in the Ecole Pratique des Hautes Etudes, a police spy was in its midst. This was no clandestine affair, however. The members were fully aware that a representative of Napoleon III's government would attend their meetings. Indeed, in order to convene at all, the society had had to enter into an extraordinary agreement with the ministry of public education. As Terry Clark describes it:

> The Imperial government viewed the proposal [to establish an anthropology society] as a likely cover for revolutionary propaganda, or at the very least, an unchristian enterprise; it was strongly discouraged. After considerable petitioning, the Ministry of Education nevertheless authorized the Société to hold meetings, subject to three conditions: that no more than twenty persons attend, that there be absolutely no discussion of politics or religion, and that an Imperial police officer attend every session to assure compliance with regulations.[106]

For almost five years a government agent duly monitored the society's meetings, looking for subversion in discussions about the prehistoric use of hatchets, relative fecundity of primitive races, capacity of crania and the weight of brains as related to intelligence, relative dolichocephaly or brachyocephaly of the ancient French, and so on. Doubtless completely bored, the spy could report back to his superiors that neither religion nor politics was ever mentioned, and that there was little to merit imperial interest in such ceaseless prattling about such esoterica. Eventually police supervision ended, and the government moved on to surveillance of groups it deemed more genuinely threatening.

In one sense, of course, the government was right. Talk of cranial indices and brain convolutions was unlikely to foment revolution. In another sense, however, it missed the point entirely. The society's discussions could, indeed, have been seen as subversive, and in just the ways the Empire feared. They might not have spoken of religion and republicanism directly, but within the idiom of science their discussions of brains, bones, and the perfectibility of races privileged positivism, materialism, and disdain for tradition-based authority. In a political culture where the emperor's concordat with the church was critical to his power and where republicanism and positivism were linked both intellectually and institutionally, the Société d'Anthropologie constituted a kind of oppositional space. Politically, the society's members proved to be stalwarts of the Third Republic when it emerged out of the Second Empire's rubble: Broca became a senator, and one of his chief disciples, Léonce Manouvrier, supported a host of left-oriented Republican causes.[107] Ideologically, the society's walls marked a boundary inside which the program was clear: human nature was to be investigated on avowedly positive (thoroughly empirical) and material principles, and objectified such that talk of vital forces, immaterial agents, and even the soul would be rigorously excluded.[108]

Within this context, the concept of intelligence proved of significant value to French anthropologists. Associated not with *esprit* (soul), but with a naturalistic mental quality present throughout the animal kingdom, and open to empirical investigation, intelligence was used not so much to establish differences between human types—the ever more elaborate technology of measurement produced an almost endless array of distinctions—as to give those differences meaning. Extending early nineteenth-century work on craniometry and race, Broca and other French anthropologists embarked on an extensive program of measuring the skull and brain and relating those measures to degree of intelligence. The goals of this research were twofold: on the one hand to pursue anthropology as the "natural history of man," identifying physical characteristics that differentiated the types of the human species; on the other, to show that the "natural" organization of races and their subdivisions constituted an intelligence-based hierarchy.[109] Although the craniometric program would ultimately fail to establish lasting connections between skull characteristics and human intelligence, what would persist was the conception of intelligence as a quantifiable, hierarchical characteristic that could be assessed via measurement technologies and that constituted a primary means of characterizing groups and perhaps even individuals.

French anthropology's guiding force during the second half of the nineteenth century was Paul Broca. Trained in anatomy and pathology, Broca turned to anthropological problems in the late 1850s, with a series of reports on human and animal hybridity.[110] Rebuffed when he attempted to present these works before the Société de Biologie because of the potential divisiveness of research interpretable as polygenist and possibly materialist, in 1859 Broca joined with other positivist members of the society, mostly physicians, to form a new organization dedicated to the scientific study of the human races, the Société d'Anthropologie de Paris.[111] For much of the next fifty years the society, its publications—the *Bulletins* and *Mémoires,* and the *Revue d'anthropologie* (begun by Broca in 1872)—and its associated institutions, the Laboratoire d'Anthropologie (established by Broca in 1867), and the Ecole d'Anthropologie (started in 1875), dominated French anthropology.[112] Strongly polygenist and positivist in orientation, the Society emphasized, as Joy Harvey has noted, "measurement, observation, factual evidence, and where possible, experimentation," mostly toward the goal of determining the characteristics of *Homo sapiens'* various subgroups and arranging those groups in their "natural" order.[113]

Craniometry, not surprisingly, was a central activity for a society wanting to establish a positivist anthropology anchored in scientific methods and dedicated to studying race. In many respects, these French anthropologists pursued simply a more elaborate version of the craniometric studies undertaken by Blumenbach, Cuvier, and Morton. Like their American counterparts, most French craniometricians were avowedly polygenist and committed to the superiority of white male Europeans.[114] Typically their studies compared measurements of some racial group or subgroup to the standard for modern Europeans, most often emphasizing a characteristic readily linked to relative intelligence.[115] In his 1886 research on prehistoric crania, for example, Broca's student Paul Topinard performed 101 measurements each on forty-four skulls. He then contrasted the mean volumes of the sample's males and females with accepted figures for contemporary Europeans, concluding that the males were distinctly inferior to their modern analogues and the females superior; in addition, based on wide variations in the cranial index (ratio of skull width to length), he determined that the group was a mélange of races.[116] While the variety and precision of Topinard's quantifications far surpassed anything Morton had ever achieved, the approach was similar: measure crania carefully so as to situate a group in some taxonomic pattern. Broca had explicitly endorsed such an approach some twenty years earlier, maintaining that "craniology not only furnishes characteristics of the first order for the differentiation and classification of the subdivisions of the human species, but also furnishes precise data about the intellectual value of these subgroups."[117]

To generate the "precise data" necessary to distinguish groups and determine their "intellectual value," the renegades from the Société de Biologie emphasized the investigation of human physical properties through the technologies of instrumentation and statistics. Broca's creation in 1867 of an anthropological laboratory, first associated with the Ecole de Médecine and later with the Ecole Pratique des Hautes Etudes, exemplifies the style of French anthropology that craniometry embodied and promoted. While Broca tolerated field studies, especially of non-Europeans, he preferred laboratory-based measurement of skulls employing the instruments he and his fellow anthropologists devised. Proper method, Broca argued, was critical to establishing anthropology on a scientific basis: "The method of individual observations," Broca explained, "collected through simple, uniform procedures, sheltered from the imagination, and repeated *on a large number of individuals chosen by chance*, is thus the true basis for anthropological research."[118] To achieve such pristine data, Broca declared, anthropology must use instruments:

> The goal of these instruments is to substitute for evaluations that are in some sense artistic, that depend on the acuteness of the observer and the exactness of his gaze—and sometimes even on his preconceived ideas—uniform and mechanical procedures, that permit the results of each observation to be expressed in numbers, that establish rigorous comparisons, that reduce as much as possible the chances of error, and finally and especially, that group the observations in a series, submit them to computation in order to obtain mean measurements, and thus escape the deceptive influence of individual variations.[119]

This fear of subjective observation and possibly anomalous variations, Lorraine Daston and Peter Galison have shown, became pervasive in any number of sciences during the late nineteenth century, convincing many practitioners that mechanical objectivity, as Daston and Galison have dubbed the new ideal, was the only scientifically viable method.[120] In anthropology, articles describing new, more mechanical instruments filled the journals, measurements were carried out to two-decimal-place or greater precision, and mean measurements became the preferred manner for presenting data.[121] Broca's position at the center of almost every major French anthropological institution and his control of the world's largest skull collection—more than seven thousand by his death in 1880—gave him enormous influence over deciding what standards to adopt and how to impose them.[122] His success, however, was limited. Although he and Topinard published numerous guidelines for craniometric investigators—ranging from Broca's "Instructions générales pour les recherches et observations anthropologiques" (1865) to Topinard's anthropological bible, *Eléments d'anthropologie générale* (1885)—and worked to make different anthropologists' data compatible, even such fundamental points as the positions on the skull from which basic cranial measures should be taken remained unresolved.[123] Crania and brains could simply be measured too variously, and the vast increase in the separate measurements performed meant that achieving consistency and comparability among studies required imposing the use not only of standardized instruments but standardized measurement procedures. What exactly marked the endpoints of the anterio-posterior curve? Where precisely did the brain end and the nervous system begin?[124]

As these measured features multiplied, the possibility that characteristics would vary consistently across measurements proved ever more elusive. On the basis of Topinard's 101 measurements, for example, it might have been possible to develop a statistical portrait of Marne neolithic crania stable enough to distinguish them from another group, say the Basque skulls that Broca investigated in 1862–63, although large intragroup variation could make that difficult.[125] Ranking these groups, however, would have presented greater problems: if cranial volume were used, one group would be superior; if anterio-posterior diameter, the other. This led some anthropologists to adopt an approach similar to Morton's when facing the complexities of ethnology: simply describe the group under investigation, as if it were some new specimen of flora or fauna, and make no attempt to fit it into any overarching hierarchical scheme. Thus, especially in the century's final two decades, French anthropological journals were replete with articles that provided precise quantitative descriptions of some collection of skulls, but avoided comparing the data with those of other groups.[126] Other anthropologists, most notably Manouvrier, as Jennifer Hecht has demonstrated, questioned the usefulness of such ceaseless quantifying altogether, suggesting instead that French anthropology focus on the qualitative aspects of human cultures.[127]

Such a reaction to the thicket of numbers was not, however, typical. Most anthropologists remained convinced that it was possible not just to differentiate groups, but to find meaningful measures reflecting the natural organization of the human races.[128] When the Society split at the end of the 1880s over the issue of applying anthropological science to social problems—radical republicans, led by Gabriel de Mortillet, broke with more moderate ones loyal to Broca's precepts—neither group rejected the importance of physical measurements.[129] Relying on the explanation of racial and group differences embodied in both the chain of being and the new transformist biology, French anthropologists of all camps generally assumed that the single most influential characteristic distinguishing human groups was intelligence, itself understood as whatever general intellectual power it was that made white male Europeans obviously more civilized and advanced than Ethiopians or Hottentots, not to mention other animal species.[130]

Does Measuring Crania Equal Measuring Minds?

Although committed to establishing a connection between some physical feature of the brain or skull and the degree of a group's intelligence, French anthropologists were at least initially little concerned with *proving* where a group stood in the intelligence hierarchy. They already "knew" which groups were superior (Europeans, males, scholars) and which inferior (Africans, females, workers.) Thus Gustave Le Bon relied on women's presumed intellectual inferiority to substantiate his findings about the relation of cranial volume to intelligence, and even the monogenist Armand de Quatrefages declared that "there are the superior races and the inferior races," adding, "for a long time I have considered the adult Negro as a being whose intelligence has remained, by a sort of arrest in development, at the point that we observe among adolescents of the white race."[131] Rather, the goal of craniometry was to prove that particular physical features could be connected to intelligence, and in ways matching these "known" hierarchies.

French anthropologists singled out four brain or skull characteristics for extensive investigation: cranial volume, brain mass, brain convolutions, and cephalic index.[132] The attractions of cranial volume and brain mass derived from the widespread beliefs that the brain was the mind's organ and that organ size correlated with functional strength. Brain convolutions attracted attention because simpler organisms seemed to have fewer convolutions than more advanced organisms; interest grew after Broca contended that the number of convolutions related directly to brain volume.[133] The cephalic index was taken as a more indirect indication of brain size. Proposed by the Swede Anders Retzius in 1842 and refined by Broca during the 1860s, the index was defined as the ratio between the width of the head and its length, essentially a measure of the relative size of the anterior lobes. Retzius used this index to divide crania into two fundamental types: brachycephalic, or round-headed, and dolichocephalic, or long-headed.[134] Retzius's contention that dolichocephalic blonde-haired Aryans, such as the Swedes, were the most intelligent and progressive Europeans and were so because of their anterior lobes' extensive development, engendered enormous interest in the index, and much controversy.[135]

Throughout the second half of the century, French anthropologists tried and, ultimately, failed to find consistent connections between each of these measures and the degree of intelligence.[136] The cephalic index, though it was used throughout the century, foundered most quickly. Dolichocephalic superiority, it turned out, worked well for the Swedish and British, but brachycephalic Mediterranean peoples, such as the French, found the measure problematic, as did most craniometricians upon discovering that African bushmen and Fijians were longer-headed than Europeans.[137] Broca did try, as Stephen Jay Gould noted, to rescue the index by arguing that an increase in the posterior lobes (thought responsible "merely" for basic sensory operations) could account for dolichocephalic "primitives." Nonetheless, anomalies continued to accumulate, forcing most anthropologists to conclude that the cephalic index could not reliably place human groups in the "right" order.[138]

This failure to fit presumed intellectual hierarchies also plagued the other measures, particularly the relation of cranial volume to intelligence.[139] Since at least Morton's work, cranial volume had stood as the preeminent characteristic for ranking the races by intelligence. Even in those early studies, however, certain discrepancies appeared vis-à-vis presumptions about intelligence when comparing the mean cranial volumes of Native Americans and Africans. By the late 1870s, anomalies in the data were even more noticeable. At a meeting of the Société d'Anthropologie in 1879, for example, Broca remarked that "the mongolian races are, all things being equal, *less intelligent* than ours. However the *great size* [*grosseur*] of their cranium does not reflect [*ne rend pas compte de*] their intellectual situation."[140] In response to such misgivings, Le Bon attempted to marshal all available evidence for a close connection between skull volume and mental power. Reacting particularly to two criticisms—that volume differences disappeared when body size was accounted for, and that some members of "inferior" races had larger crania than members of "superior" races—Le Bon introduced a new method for representing intragroup variations that discounted both objections and suggested that not mean cranial size but deviation from the mean (variability) was the key to racial superiority: "*what really constitutes the superiority of one race over another,*" Le Bon declared, "*is that the superior race contains many more voluminous crania than the inferior race.*"[141] Le Bon's arguments, however, met strong resistance.[142] Many society members criticized his study for only including groups that fit his hypothesis about cranial variability and intelligence and for failing to account for anomalies at the level of individuals.[143] After surveying decades of such research, in fact, anthropologist Adolphe Bloch had come in 1885 to despair of the entire cranial volume endeavor:

> There is no absolute relation between intelligence and the volume of the cranium, because some very intelligent individuals can have a small skull, while very ordinary individuals can have a very large skull. That is known. From another side, in certain races, said to be of little intelligence, one can find a skull or cranial capacity of a relatively considerable size.[144]

Bloch did not stop at cranial volume; his investigation into the data on brain weight and brain convolutions produced similar conclusions. Discrepancies at the level of individuals proved most damaging. Although Le

Bon had successfully explained why, at the group level, some "inferior" peoples might contain individuals with superior intellects, his analysis of variation was useless when groups were disaggregated, for it was widely conceded that intelligent people could have relatively small brains and, as Bloch pointed out, average people large brains. Indeed, Pierre Gratiolet had noted such objections as early as 1859.[145] By the 1870s, extensive lists of the brain weights and cranial capacities of eminent individuals confirmed the problem. While most had brains larger than the mean, others, deemed equally superior, had markedly little gray matter.[146] Indeed, the rather modest brain weights of some prominent individuals had caused contention within the Society since its second year. Relying on German anatomist Rudolf Wagner's data, Gratiolet argued that the disparity in brain weights in eminent men—ranging from Cuvier's 1,829 grams to Hausmann the mineralogist's 1,226 grams—indicated that, within a normal brain-weight range (for men between 1,200 and 1,900 grams), mass and intelligence did not correlate. The case of the famous German mathematician Karl Friedrich Gauss, whose brain mass was 1,492 grams—337 less than Cuvier's—took on particular importance. Gratiolet would not concede that Gauss was the less intelligent; if anything, he argued it was the reverse: "I will not examine the futile question of what rank in the hierarchy of intelligence poets, historians, philosophers, artists, naturalists, and mathematicians occupy; but I can at least affirm that, in the order of the sciences and of things and of thought, a great geometer is inferior to no one."[147]

Broca's response to Gratiolet was tortured. Broca spent little time questioning Gratiolet's actual data. While he did wonder whether two figures near the bottom of Wagner's list—Hermann, professor of philology, and Hausmann, professor of mineralogy—really deserved to be included, in the main Broca accepted the raw numbers. Instead he put forward a number of ways in which the measurements had to be adjusted—to account for relative age and body size—and insisted that he had never claimed the relation between brain size and intelligence was absolute. Broca's main counterargument, though, was statistical: it was a matter of means, not individuals, as numerous studies of brain size and intelligence at the level of races, he claimed, had proven.[148] Nonetheless, given the assumption that brain size was the *causal* factor determining degree of intelligence, no amount of statistical manipulation would make the problem vanish. And so the entire program of relating intelligence to some physical characteristic began to appear suspect.[149] In the 1860s, Broca had tried to insulate craniometry from this criticism by stating repeatedly that variables other than brain size contributed to overall mental power.[150] By 1886, Topinard had qualified still further, listing five organic factors affecting intelligence, including one—the possibility that two apparently identical brains could possess different properties—that should have undermined the entire enterprise.[151] Manouvrier concluded as much in 1899, declaring the whole project to be ill conceived, by which time most French anthropologists would have been unlikely to dispute Georges Pouchet, when in 1871 he observed:[152]

> One had first dreamed of relating intelligence to the mass of the brain. . . . It became necessary to renounce this opinion, as it could not be maintained: one finds other examples equally illustrative and convincing that contradict it. The number and complicated pattern of the convolutions on the surface of the organ were invoked in their turn without any more success. . . . it will without doubt be the same for the internal structure.[153]

Uttered while craniometric measurement was still in its prime, Pouchet's comments nonetheless could have stood as a fitting epitaph for the entire program.[154]

Anthropological Intelligence and French Culture

Critical though Pouchet was of the various attempts to link a specific physical attribute to degree of intelligence, his analysis never questioned the concept of intelligence itself. And in this he was not alone. In common with his fellow anthropologists, Pouchet's interest lay in the possibility of taxonomizing human groups, not exploring the intellect's nature. For them, craniometric research had three functions: first, to strengthen the authenticity of the presumed intellectual hierarchies; second, to assign unknown, usually prehistoric peoples, their places in the civilization/development continuum; and third, to bolster the biological, hereditary underpinnings of the intellectual spectrum. As long as human mental power could be conceptualized as a singular entity rather than a collection of discrete faculties, anthropologists felt little incentive to pursue the issue further. Thus, for all the discussions at the Société d'Anthropologie about how to measure intelligence appropriately, few attempted to define what they meant by the term; most simply took its meaning as unproblematic.[155] While addressing the relation between cranial volume and intelligence in 1879, Le Bon was, in fact, almost unique in actually trying to clarify the concept, and his very hesitancies indicate the oddity of this endeavor:

> In order to be complete, the preceding study would now need to address the question of understanding what superior intelligence really consists of, but this problem is so complex that I can hardly touch upon it here. . . . If it is absolutely

necessary to state in a few words a formula for measuring intelligence, I would say that it can be appreciated by the degree of aptitude for associating—I did not say accumulating—the greatest number of ideas, and perceiving as clearly and rapidly as possible their analogies and differences.[156]

Le Bon's definition suggests the problem confronting anthropologists. What possible physical correlate could allow direct measurement of "the aptitude for associating" ideas, especially when working with collections of mostly anonymous skulls? While some of Broca's best-known research worked to establish the localized functions of certain parts of the brain, and while most agreed that the anterior lobe was the seat of the highest brain functions, physiologists and anatomists had long since abandoned belief in the existence of direct connections between particular cranial or brain characteristics and individual intellectual functions, as posited in phrenology.[157] Lacking these, skulls or brains alone could say little about the characteristics of the mental operations housed inside; whatever intelligence might be, illumination did not lie in the ever more precise measurement of cranial features to which Broca et al. were committed.[158]

Given this orientation, intelligence took on a particular set of characteristics for French anthropology. First, when French anthropologists spoke of intelligence, they almost invariably referred to it as a unitary, global intellectual characteristic, whether of individuals or whole peoples. Consumed with measuring skulls and brains in order to position the races and subgroups on a single scale of intellectual development, anthropologists had little opportunity or incentive to observe, à la Morton when doing ethnography, intelligence as a complex of activities and abilities. Second, intelligence took on a decidedly physicalist cast. It became something material—visible and even palpable if one examined the right feature of the skull or brain—and thus, like any good biological trait, both heritable and measurable. As such, intelligence could explain the continued superiority of some individuals or groups and the persistent inferiority of others, and could, as well, be linked to an entire system of laboratory work, instruments, and statistics.

Finally, because anthropologists spent little time investigating intelligence per se, they essentially had no other means but common opinion to determine how much of it any particular individual or group possessed. This posed little problem when craniometricians used what they took to be clear-cut examples at the extremes of the intelligence spectrum, such as idiots and eminent scholars, or Negroes and Caucasians. But closer to the middle, consensus disappeared. One anthropologist might argue that Celts were more intelligent than Scandinavians, another the opposing view, and no mechanism existed to adjudicate such disputes. The problem was even more acute at the level of individuals. "Let us first remark," Broca observed in 1860,

> how difficult it is, in most cases, to appreciate the relative degree of intelligence of two individuals. In this comparison, one finds most often some inequalities in the reverse sense, and, as these inequalities themselves cannot be measured, judgment remains doubtful. . . . But, all doubts disappear when the intellectual inequality is very great. There are people so superior, or so inferior, that the original endowment could not be the object of any contestation.[159]

As Broca suggested, the preferred solution was, whenever possible, to focus on extreme cases and to leave the confusing middle both undefined and unexplored. Intelligence, consequently, came to be associated primarily with its most vivid manifestations, and the anthropological literature became filled with polarities—men/women, whites/blacks, intellectuals/laborers, humans/animals—that reflected presumably "unequivocal" demarcations according to intelligence.[160]

Constructed in this way as a marker of large-scale differences, intelligence played only a limited role in broader French social discourse. In part, this reflected the widely shared conception that France was racially homogeneous, and, so, little in need of a language of natural difference to explain internal distinctions. Although a number of French anthropologists investigated whether the French were composed of one race or two, with some hypothesizing that the nation might be a mixture of brachyocephalics and dolichocephalics and that these differences corresponded to class status, such theories failed to capture the cultural imagination.[161] It is true, as William Cohen has observed, that "French middle-class commentators viewed the poor, underprivileged classes in the country and in the cities as forming a different race"; nonetheless few social practices were tied to marking physical distinctions in the population, except perhaps around Jews during the Dreyfus Affair and the ever present differentiations according to gender and age.[162] Rather, regional and class distinctions, and urban/peasant splits, cleaved the French social topography, none founded significantly on physiological difference. Since at least the Revolution, in fact, the French state had worked assiduously for national unity, and during the Third Republic, under the name "solidarism," imagining France in collectivist terms became the quasi-official state philosophy.[163]

Moreover, the political ideology of most Société d'Anthropologie members and the dominant understanding of the new transformist biology together minimized the role for anthropological intelligence in broader cultural discourse. Founded in opposition to the dominant ideology of the Second Empire, the Société d'Anthropologie had quickly become a center for fiercely anticlerical and prorepublican sympathizers. In the guise of discussing human beings and societies as material and natural objects,

French anthropologists sought as well to develop scientific foundations for reconstructing society on republican principles.[164] Central to their outlook was commitment to a kind of elitist egalitarianism, in which experts would reform society on behalf of the common good.[165] As Philip Nord succinctly put it, "the new science of man was, at its origins, a science of the left."[166] The neo-Lamarckian orientation of many in the Broca group simply strengthened this ameliorist strain in its political thought. Although Broca himself, an unorthodox Darwinian at best, believed natural selection played some role in evolution, many anthropologists and most transformist French intellectuals were solidly neo-Lamarckian, less concerned with natural selection than with the inheritance of acquired characteristics, sure that evolution meant gradual progress, and committed to the possibility of human improvement, both biological and cultural.[167] As a result, the notion of intelligence as a fixed mental endowment, however important in anthropology, seemed less immediately applicable to French society.

Nonetheless, for all the Broca group's commitment to republican democracy, in at least two senses their work was also deeply bound up in the Empire's and then the Third Republic's colonial projects.[168] First, materially, many of the skulls and observations that found their way to the society were gathered by imperial administrators or travelers. Without the contact with other cultures and ability to transfer artifacts to the West provided by a century of colonialism, the society's program of categorizing and classifying *Homo sapiens'* many subgroups would have been almost unimaginable. Second, ideologically, the society's commitment to producing a hierarchical taxonomy of the human species, with Europeans inevitably at the top, at the very least accorded nicely with the French approach to colonial administration.[169] France's *mission civilisatrice* was predicated on a paternalism that assumed Western and particularly French superiority and the inferiority of "childlike" native peoples, who might at best be raised to a limited understanding of and appreciation for European culture and governance. Although colonialism was rarely directly mentioned in society meetings, the unquestioned "scientific" representation of especially African peoples as closer in intelligence to apes or children than accomplished Europeans paralleled closely the administrative state's model of indigenous cultures.[170] Colonialism did not give rise to French physical anthropology, but the demands of empire unquestionably linked the two.

Conclusion

The legacy of almost a century of American and French anthropological investigation was twofold. As a research program, craniometry had clearly failed by century's end. No skull or brain characteristics could be tied absolutely to intelligence, and arraying the human subgroups on a single, simple scale seemed an increasingly remote prospect. With Broca's death in 1880 and the ascendance of Manouvrier and others dedicated to a more relativist methodology, French anthropology turned decisively to other pursuits, especially to social and cultural analyses of the world's peoples.[171] Although some American practitioners pursued physical anthropology well into the new century, there too the influence of cultural anthropology proved largely irresistible, spurred on by the enormous influence of Franz Boas and his followers.

Nonetheless, the legacy of American and French physical anthropology for the study of intelligence was profound. Not only did craniometry strengthen the notion that intelligence could be a unitary descriptor of native intellectual endowment that varied by degrees, but it also indicated the possibility of precise, quantitative measurement of those degrees. In addition, anthropological practice suggested that the specific content of intelligence mattered much less than its practical value as a means of demarcating and categorizing.

For Americans, moreover, the language of intelligence became part of a general public discussion about race. Anthropological notions of intelligence particularly enticed those wishing to prove that racial differences in mental ability were real and permanent, and thus that white superiority in a multiracial nation would be assured. Most nineteenth-century anthropologists were also committed to the natural hierarchy of the races, and thus rarely resisted the insertion of anthropological notions of intelligence into nonspecialist debates. In France, conversely, there was little direct resonance between issues vexing the public and anthropologists' claims. "*Intelligence*" remained a more specialized term, valuable in battles between positivists/republicans and spiritualists/monarchists, but not routinely used to establish scales of intellectual difference within the French nation. The system of education already performed that task admirably.

Near the end of the century, however, Alfred Binet and other "scientific" psychologists in France, Britain, and the United States would conclude from the story of craniometry not that interest in quantifying intelligence should be abandoned, but that its measurement should focus on the mind rather than the body. They would carry with them, though, the vision of intelligence as a heritable, quantifiable, global characteristic, applicable to individuals as well as groups, most visible at its extremes, and grasped best through measurement and statistical analysis. In this guise, intelligence would be given concrete embodiment in its own technology, the mental test, where it could be used to produce fine-grained markings of human difference.

Part II

INDIVIDUALIZING INTELLIGENCE THROUGH THE SCIENCE OF DIFFERENCE

Four

Between the Art of the Clinic and the Precision of the Laboratory

INDIVIDUAL INTELLIGENCE AND THE SCIENCE OF DIFFERENCE IN THIRD REPUBLIC FRANCE

IN 1870 HIPPOLYTE TAINE opened his most ambitious intellectual endeavor and the work that represented almost twenty years of labor, *De l'intelligence,* with a definition. "If I am not mistaken," he began, "one understands today by intelligence, what one understood previously by understanding and intellect; that is to say, the faculty of cognizing."[1] For a philosophical treatise, such a move might not seem unusual; philosophers—even the rather wordy nineteenth-century variety—are notorious for their delight in precision, especially in defining key terms. Nevertheless, if the strategy itself is not surprising, then Taine's particular application of it certainly should be. Taine claimed a novelty for his definition that French dictionaries fail to support. From at least the late seventeenth century, one of the primary meanings of "intelligence" was "intellectual faculty or capacity for understanding," a meaning that remained preeminent through Littré's *Dictionnaire de la langue française* of 1874 and well into the twentieth century.[2]

Why, then, did Taine feel the need to define an already well-defined term and suggest as novel a long-standing definition? Partly it was a matter of orientation. While "*intelligence*" was both available and used frequently in nineteenth-century French, outside of anthropology it carried connotations of immaterial principles and metaphysical properties that Taine sought to exclude.[3] More significantly, intelligence was truly a term in some flux at the time. Within "scientific" psychology, education, and medicine, not to mention broader intellectual and bourgeois culture, interest was growing in new ways of defining and describing human differences.[4] Intelligence was one word pressed into service to accomplish this goal.

Cousinean eclectic psychology's sense of intelligence as a universal human attribute equated with the expression of reason in normal, adult, white, civilized males was not abandoned. But added to it was intelligence as a scientifically analyzable, biologically based entity that different groups and perhaps even individuals could manifest to various degrees and in a variety of ways. While the impetus for this change came from several sources, French philosophers of mind and anthropologists were in the forefront of remaking the notion of intelligence and uniting it with their own developing interest in human physiology and difference. French anthropologists, as we saw in chapter 3, moved first, seeing in intelligence a way of describing differences between human groups. The new, self-styled "scientific" philosopher/psychologists followed suit, breaking with midcentury notions of the mind's homogeneity. Turning to clinical observation, they fashioned a psychology of pathological difference that combined associationism with indigenous interests in mental pathology, positivism, transformism, and development. Most were also intrigued by laboratory experimentation and used its tools to develop a science that emphasized instrument-based, often quantitative explorations of mental abilities and the categorization of individuals according to fundamental psychological features. Within both approaches, French psychologists found in intelligence a site for exploring individual differences and making claims to social relevance.

During the four decades when this transformation in French psychology occurred—bracketed symbolically by Taine's *De l'intelligence* and the final revision of the Binet-Simon intelligence scale in 1911—intelligence nonetheless remained a term of limited provenance and shifting definition. In philosophy/psychology, following Taine, intelligence continued to denote primarily one of the three fundamental faculties of mind; in anthropology, a group's overall mental power; and in general discourse, typically either the operations of reason or a community's leading intellectuals.[5] Alfred Binet's creation of a new technology of categorization and differentiation, the intelligence scale, to address certain practical social problems—particularly detection of kinds of mental deficit—did more firmly anchor one set of the concept's connotations, in both meaning and social role. But the specificity provided by the Binet scale remained only one way in which "*intelligence*" was understood and deployed in early twentieth-century France. Binet himself continued to be unsure about the precise nature of the intellect, especially in regard to whether intelligence was unitary or multivalent, and the rest of the French psychological community could come to no firm consensus about either intelligence or its tests. Moreover, the test itself proved of only limited relevance to the state's needs, and intelligence as a concept of only limited interest to the culture at large.

From Cultural Turmoil to a Positivist Moment

On the eve of *De l'intelligence*'s appearance, there was trouble in the Empire. After ten relatively quiet years, popular support for Napoleon III and the Second Empire had begun to wane. Angered by declining wages and bleak economic prospects, workers were among the first to express their dissatisfactions, but many members of the middle class joined in, displeased by the Second Empire's conservativism, growing social inequities, and Napoleon's failures in his confrontations with Bismarck and Prussia. A recession in 1867–68, though mild, accelerated these tensions, and by 1869 France was swept by a wave of strikes and increasingly strident calls for reform from both the laboring and middle classes. At the same time, a movement on the right advocated military action to restore glory to the Bonapartist regime.[6]

Exacerbating this sense of crisis were several developments well outside quotidian politics. In the intellectual/cultural arena, serious dissatisfactions with French society's orientation and character had spread widely. For intellectuals, the dissension swirled largely around the issue of orthodoxy, with Ecole Normale graduates leading the battle to open academia to new ideas and methods, particularly those derived from the sciences and writers ignored within "official" thought. The close connections that Napoleon had forged with the Catholic Church inclined authorities to denounce and combat strongly any approach that seemed tainted with materialism, determinism, or atheism, which to the orthodox these all did.[7] By the early 1860s the Société d'Anthropologie was thus only one group among many pushed into revolt, pursuing their intellectual programs on the margins of academe. Most opposed the Napoleonic regime as well, openly espousing a combination of anticlericalism, republicanism, and positivism that would dominate the Third Republic. These various troubles came to a head during the watershed period 1870–71, when war, revolution, and reaction created havoc, and opportunity, throughout French political and social life. France's humiliating defeat in the Franco-Prussian War, Napoleon III's fall, the establishment of the third French republic, and the rise and annihilation of the Paris Commune transformed French political and cultural life. Many came to doubt the very soundness of society and to believe that change was both necessary and desirable.[8]

Among the resulting approaches to politics, society, and epistemology that gained enormous vitality was positivism.[9] Reflecting an attitude toward knowledge that had been developing since the early nineteenth century, positivism as a philosophical system was formalized in Auguste Comte's *Cours de philosophie positive,* published in sections from 1830 to 1842.[10] The progress of both knowledge and society, Comte theorized, passed through three stages: from the theological, through the metaphysical, to the positive. He modeled the third and most advanced, the positive, on empirical science, understood as the investigation of observable phenomena to generate predictions of practical utility.[11] A positive theory or practice, according to Comte, had four characteristics: it was predictive, empirically verifiable, ultimately practical, and untainted by metaphysics.[12]

Knowledge so gained, Comte contended, would be as certain as human knowledge could become, and would command assent because of its rigorous exclusion of any phenomenon or principle not open to objective verification. As such, Comte argued, positivism could also form the basis of a harmonious social order, where disputes would be settled peacefully by an elite of well-trained, mentally superior individuals adjudicating according to positive, and hence indisputable, principles.[13]

After gaining prominent early admirers, including John Stuart Mill in England and Joseph Fourier in France, Comtean positivism lost favor in the 1840s and 1850s, when Comte's personal excesses and attempt to transform positivism into a religion seemed to discredit the entire program.[14] In the late 1850s and 1860s, however, shorn of its founder, positivism began to reemerge in scientific circles to describe an epistemological approach emphasizing empirical fact, experimental investigation, and distrust of unverifiable—"metaphysical"—presuppositions. Critical to positivism's rehabilitation were the ceaseless efforts of Comte's most important disciple, Emile Littré, and the work of France's most eminent midcentury scientist, the physiologist Claude Bernard. Indeed, Bernard's widely read *Introduction à l'étude de la médecine expérimentale* (1865) alone did much to revitalize the philosophy.[15]

Over the course of the 1860s, positivism also began to spread beyond the experimental sciences. Students at the Ecole Normale and other elite institutions were in the vanguard, adopting positivism as an outlook on knowledge and society.[16] This new generation equated positivism with the scientific method, and many considered it to be the sole legitimate means of acquiring useful knowledge, whether about epistemic or social problems. From Emile Durkheim and Alfred Espinas in sociology to Hippolyte Taine and Théodule Ribot in psychology to Paul Broca and Gabriel de Mortillet in anthropology to Léon Gambetta and Jules Ferry in politics, positivist-oriented intellectuals demanded that modern life's pressing intellectual and social questions be answered by recourse to science and the truths it could guarantee.[17] In so doing, they formed a movement united less around the answers it proposed than the means for attaining them and the faith that "positive" answers did exist.[18]

The turmoil of the 1860s, culminating in the events of 1870–71—especially defeat by the nation regarded as the world's most advanced scientifically, Prussia—intensified reactions to positivism among both supporters and opponents, and hastened its diffusion in intellectual circles. French writers routinely compared France to Germany, most often to underscore the inferiority of some aspect of French culture and to call for its revitalization according to the more scientific principles associated with Germany.[19] For French mental philosophy, the effect was to open up space for new approaches to problems of mind, especially those that could embody positivist methods.[20] No work exploited this new space more successfully and exemplified better the influence of these cultural forces than Taine's *De l'intelligence*.[21]

The Curious History of *De l'intelligence*

After opening *De l'intelligence* with a definition, Taine proceeded to explain his project:

> we are concerned here with our cognitions (*connaissances*), and not with anything else. The words *faculty, capacity, power,* which have played such an important role in psychology . . . do not designate a mysterious and profound essence, which persists and remains hidden under the flow of transient facts. This is why I have treated only cognitions, and if I have dealt with faculties, it has only been to show that in themselves, and as distinct entities, they do not exist.
>
> Such a precaution as this is very necessary. By means of it, psychology becomes a science of facts.[22]

From the outset, therefore, Taine delineated his position on one of the major issues confronting French mental philosophy: he firmly rejected Cousineanism and unequivocally argued that for psychology to be a true science it must adopt a positivistic approach to the mind, one in which all entities beyond the level of discrete phenomena would be eliminated as metaphysical.[23] In what followed, Taine presented a detailed account, based on English associationism, of how the workings of the intellect, even at its most rarefied, could be reduced to the material interplay of elemental sensations combined according to the associative laws.[24] Disparaging as rationalistic eclecticism's reconstructions of the operations of abstract mind, Taine emphasized instead observation and clinical investigation—the hallmarks of positivism—as the methods appropriate for mental philosophy. "The novelty of my book [*De l'intelligence*]," Taine wrote Jules Soury in 1873, "consists in its being entirely composed of a number of small but significant facts and cases, individual observations, and descriptions of psychological functions, atrophied or hypertrophied."[25]

At the time of *De l'intelligence*'s publication, Taine was forty-two.[26] Born to a middle-class family in Vouziers (Ardennes), Taine had moved to Paris with his mother in 1840 after his father's death, and attended the Lycée Bourbon. Spectacularly successful at lycée, Taine continued on the route to academic prominence, entering the Ecole Normale in 1848, seemingly assured of a glittering career.[27] Taine's growing and public contempt for Cousinean philosophy, however, doomed him to failure on the culminating examination, the *agrégation,* in 1851, and for the next thirteen years he remained on the academic world's margins, too controversial for the Second Empire's ultraorthodox official culture.[28] During this period Taine immersed himself in science—especially physiology and mental pathology—attending lectures at the Faculty of Medicine, Sorbonne, and Salpêtrière asylum. Convinced of the scientific method's broad applicability, in 1863–64 Taine published a series of literary analyses from a scientific perspective, in which he emphasized the influences of heredity, environment, and historical development (*race, milieu,* and *moment*).[29] At this time Taine was also preparing to write *De l'intelligence,* projected as the first in a series of three works on the human mind (to be followed by explorations of the will and the emotions).

Although *De l'intelligence* was extremely successful for a philosophical work—there were seventeen editions between 1870 and 1933, including an English translation in 1871–72 and a major revision in 1878—Taine never returned to psychology per se.[30] The Franco-Prussian War and the Commune diverted Taine's attention to politics, and for the rest of his career he attempted to apply his psychological ideas and scientific knowledge to explaining the evolution of French society, specifically what he took to be its pathological outbreaks of chaos and anarchy, for him epitomized by the French Revolution and Commune. Conservative and anti-egalitarian, though ultimately republican, Taine celebrated order, civilization, and the dominance of society by the upper and educated classes. "By playing upon the sense of despair, national humiliation, and cultural decline," Susanna Barrows has noted, "Taine, 'the pathologist of French society,' offered an explanation of how and why France had lost it primacy among nations."[31]

Taine made much of his scientific background and empiricism. Nevertheless, *De l'intelligence*'s approximately one thousand pages strike today's reader rather differently: as yet another systematic philosopher's account of mind in which even concrete examples—drawn principally from mental pathology—were used mainly to illuminate the logical structure of normal adult human reason.[32] In essence, Taine—like his forerunners Alexander Bain, John Stuart Mill, Herbert Spencer, John Locke, and the Abbé de Condillac—discussed the mind abstractly, as a mechanism for acquiring and combining sensations. "There is nothing real in the self," he declared, "except the file of its phenomena (*événements*); these phenomena, diverse in aspect, are the same in nature and reduce to sensation; sensation itself . . . can be reduced to a group of molecular movements."[33]

In Taine's view, the mind's operations commenced with fundamental building blocks, the elemental sensations—initial perceptions caused by stimulation of sensory nerves during some encounter with the external world. These sensations were then combined with each other or previous sensations according to the associationist laws of contiguity and resemblance.[34] Associations were retained physiologically by being located in discrete neural connections, and psychologically by being assigned names. Two factors regulated the initial bond's strength: the intensity of the original pairing and the frequency of its repetition. The product of this association of elementary sensations was the sensation proper. For example, encountering a person called Paul would stimulate various elementary sensations, which would be linked by contiguous association into a set of sensations, named "Paul," and then stored in memory, available for later recall.[35]

Further sets of sensations could also be linked, either to each other, by contiguity, or with sensations stored in memory, by resemblance. These larger-scale associations were themselves given names, though ones more general than those denoting discrete sensations. Thus, the observation of another set of sensations, "Marie," might lead via comparison with those remembered as "Paul" to the more general name "person." This process of generalization through associative connections could continue ad infinitum, producing a hierarchy of ever more abstract ideas. Displaying his allegiance to positivist principles, Taine granted these generalizations no particular ontological significance, characterizing them simply as collections of sensations sharing a common element. As such, he explained, general ideas required continual refinement through further experience and multiple comparisons to attain some stability. Moreover, he argued that in the act of recalling or retaining ideas, there was a veritable struggle for survival; only those that best corresponded to the environment would prevail.[36]

Taine's conception of the intellect, highly reliant on the physiological reductionism and mental chemistry of English associationism and French sensationism, positioned the intelligence as a simple *machine intellectuelle,* a mechanical apparatus that translated sensory inputs into hierarchically arranged ideas.[37] But where previous mental philosophers had relied principally on purely logical or theoretical arguments, Taine's innovation was to introduce, wherever possible, empirical evidence: illustrations of and substantiations for his ideas drawn from physiology and mental pathology. In addition to giving *De l'intelligence* a highly empirical cast, this connection of intelligence with physiology and pathology emphasized the material nature of the mind. Where Cousinean eclecticism represented the intellect as a collection of powers, agencies, faculties, and the like, operating according to innate and ultimately unspecifiable rules, Taine posited an intelligence that was the simple sum of individual perceptions, connected or disjoined according to a few universal laws. Thus, although *De l'intelligence* retained the basic equation of intelligence with the faculty of knowing, it altered the definition's meaning by arguing for a mechanical and materialist understanding of what it meant to know.

Ultimately, philosophers aside, the technical details of Taine's depiction of intelligence mattered less than his general orientation: intelligence as a concept imbued with the empirical, the mechanical, and especially the scientific. Margery Sabin has analyzed the reverberations of this change for the French literary world. "'L'intelligence,'" she notes, "comes [in France] to represent the very spirit of the scientific mind, methodically gathering all kinds of information to bear upon its investigations of the objective world."[38] The attraction, and threat, of intelligence so conceived lay in its complete naturalization of the mind's operations. As a machine intellectuelle, Taine's intelligence had no place for metaphysical agents or spiritual truths. Only empirical sensations—combined, reordered, and in a constant struggle for ascendancy—had any reality. Human nature was thus reducible to physical laws, and individual psychology, at least theoretically, was as fully deterministic as planetary motion. Such a portrait left scant room not only for human agency, but also for a transcendently founded morality and, to the orthodox most horrifying of all, God. As the *Edinburgh Review* observed, Taine "disclaimed all adherence to Comtism . . . but his language was not less skeptical; it was a distant echo of the philosophy of the eighteenth century, which destroyed all beliefs and planted nothing in their place; it was an avowal of the supremacy of matter over mind."[39]

In this guise, *De l'intelligence* contributed to the full-scale combat raging for the soul of French intellectual culture. Although monarchists and Bonapartists like those who had opposed Taine throughout his career remained ensconced in power during the 1870s, the fall of Napoleon III emboldened positivist-oriented politicians, including Gambetta and Ferry, and intellectuals, such as Taine and Paul Bert, to work openly to transform French political and intellectual life.[40] Liberal republicanism and anticlericalism eventually prevailed, but only after many years; meanwhile, the new scientism's opponents tenaciously defended their institutions and ideologies against such challengers. Taine and his fellow positivistic mental philosophers posed a particularly dangerous threat to the conservators of the old orthodoxy, concerned as they were with the specter of the French Revolution. For the mind was the seat of the soul; to lose it to the forces of materialism and determinism, identified with Enlightenment and revolution, seemed to many tantamount to legitimating the repudiation of religion and morality, and thus to undermining the social order in favor of upheaval and anarchy.[41]

As a result, the more traditional intellectual elite took exceptional steps to keep such scholars from gaining positions of authority. Until 1878 they maintained sufficient control of the Académie Française to block Taine's election, and hounded his chief disciple, Théodule Ribot—whose own associationist and positivist work, *La psychologie anglaise contemporaine*, also appeared in 1870—continually for his "suspect" ideas. Indeed, on submitting his positivistic doctoral thesis, *L'Hérédité psychologique* in

1873, Ribot endured a vigorous campaign to deny him a teaching post.[42] "I believe I wrote you," Ribot noted in a letter to Alfred Espinas on March 15,

> that my defense [*soutenance*] is being made an affair of state. Lorquet told me that they have spoken of passing me almost behind closed doors, that is to say without preliminary announcements. They fear positivist manifestations! (which is fantastic) and (which is more serious) the bawling of the newspapers in one or the other sense. Caro calls my thesis "a provocation in 600 pages." All of which, as you can guess, is far from making me acceptable to the Minister [of Public Instruction].[43]

Even in 1885, a proposal that Ribot be allowed to teach a course on the new psychology at the Sorbonne provoked outbursts from prominent philosophers, fearful that such material would have adverse effects on their students' souls, and his appointment to France's most prestigious chair of psychology, the new position in experimental psychology at the Collège de France, was bitterly contested as late as 1888.[44]

Philosophy of mind, these worries make clear, was no idle matter in early Third Republic France.[45] Seeming to carry the banner of science and progress against the dead weight of church and dogma, the scientific, secular, empiricist, and vaguely materialistic *De l'intelligence* and *La psychologie anglaise contemporaine* were transformed by many dissatisfied intellectuals into clarion calls for reforming French academe and society on positivist principles.[46] "The thought of this powerful mind," Anatole France recalled on Taine's death in 1893, "inspired in us, about 1870, an ardent enthusiasm, a sort of religion. . . . What he offered us was the method of observation, the notion of fact and of idea, philosophy, history; at last here was science."[47]

Especially under the influence of Ribot, an entire school developed that appropriated the Cousinean term "psychology" for their own "scientific" endeavors and used Taine's *De l'intelligence* as a starting point for investigating problems of the mind.[48] Accepting Taine's basic tenets of associationism and positivism, the new psychologists added yet another connotation to "*intelligence*" by combining it with the notion of difference, an orientation that would dominate their work through the century's close.[49]

"Scientific" Psychology and the Method of Difference

Ribot's numerous studies, as well as those produced by his fellow "scientific" psychologists—associated mostly with the clinic of Jean-Martin Charcot—provided the intellectual basis for the new French psychology, one oriented particularly toward questions of difference and pathology.[50]

Ribot explained his dissatisfaction with "ordinary" (Cousinean) psychology near the start of *La psychologie anglaise contemporaine:*

> Proceeding further, we could show that ordinary psychology, in restricting itself to man, has not even embraced all of man, that it is not at all concerned with the inferior races (blacks, yellows), that it is content to affirm that the human faculties are identical in nature and vary only in degree . . . that in man it has emphasized the faculties fully constituted and is only rarely concerned with their mode of development. . . . [P]sychology, instead of being the science of psychic phenomena, has taken for its subject simply adult, white, and civilized man.[51]

By 1870 Ribot was convinced that the differences separating child from adult, nonwhite from white, and aborigine from civilized were at least as important as their similarities. Ribot's influences were many, including Herbert Spencer's evolutionary associationism and Charcot's pathological neurology, not to mention positivism and cultural anxieties about the crowd, the "other," and the nation's vitality.[52] Out of these Ribot and his followers fashioned a new methodology for philosophy/psychology, the "method of difference," in which individual variation became a central concern.[53] Ribot championed this methodology vigorously to the emerging community of scientific psychologists, thereby transforming Taine's approach to intelligence into an entire program for reinterpreting human nature.[54]

At its simplest, the method of difference meant seeing the natural world as replete with variation, and according ontological status to anomalous as well as typical phenomena. Rather than assume that differences were isolated and independent, however, the method of difference sought to relate them to each other or some established norm. It thus required that psychologists learn essentially a new way of seeing, that they perceive the psychical world as a realm where variations existed and mattered. Mental philosophers had long acknowledged that no two people were exactly the same, although the eclectics had considered such differences of little import. In so doing, however, Ribot and other critics believed that the eclectics had sacrificed true science for philosophical generalities. "If one believes," Ribot proclaimed, "that the psychologist must set aside all of these accidental variations in order to arrive at the final and absolute state, then one is transforming a concrete study into an abstract one, one . . . resembles a zoologist who takes as the basis for his research the ideal type of animality."[55] Following a path similar to French anthropology's, Ribot argued that the truly scientific psychologist must focus on the real, not the ideal, that is, on "accidental variations" as evaluated through the eyes of the careful naturalist.

French scientific psychologists invoked two models, neither exactly complementary nor contradictory, to explore mental variations. One, derived from evolutionary transformism, envisioned variations as discrete manifestations in a continuum of difference, like gradations in size from smallest to biggest. The other, drawn from pathology, interpreted variations as nature's experiments, windows into the workings of the normal through alterations caused by disease or heredity, as evidenced, for example, by an organ increased or reduced in size. Both were in accord, however, in assuming that variations had epistemologically significant stories to tell.

The first sense of difference, emphasizing developmental or hierarchical arrangements of variations, was most influentially articulated by the English evolutionary associationist Herbert Spencer. An avowed evolutionist years before Darwin's *Origin of Species* appeared, Spencer was perhaps the most popular systematist of the nineteenth century. In 1855 he published *The Principles of Psychology,* in which he constructed a theory of mind by wedding associationist principles to his notion of universal organic development.[56] Spencer's theory of development was, at base, simple: because effects slightly exceed their causes, there is an inexorable and progressive tendency throughout the organic world for the simple to become more complex. This process, regulated by the organism's need to correspond closely to its environment, produced ever greater heterogeneity as organisms evolved ever more elaborate and distinct mechanisms for harmonizing with the world around them.[57]

One of nature's central features for Spencer was thus its extraordinary variety. The increasing heterogeneity did not, however, mean increasing chaos, because individual variations were always intrinsically connected both to the environment and each other by the process of evolution itself. Spencer envisioned evolution as developmental, progressing in continuously linked stages, with each more complex stage subsuming its predecessor. His model was the "great chain of being," reinterpreted as a sequence of progressive transformations from the simplest one-celled organisms to civilized human beings; at each stage, while variety and complexity increased, the hierarchical order was maintained.[58] "From the lowest to the highest forms of life," Spencer declared, "the increasing adjustment of inner to outer relations is one indivisible progression."[59]

Turning to psychology proper, Spencer discarded the notion of mind as a static and essentially historyless organ. Rather, mind for Spencer had a past: it had evolved from the most basic reflex actions to the most complex abstract reasoning in a developmental sequence embodied in the chain of being.[60] Rejecting the Lockean notion of a tabula rasa, though not Locke's emphasis on experience, Spencer argued that an organism's and especially a species's history—the sum of its experiences—became physically instantiated in connections among the fibers of the brain, and thus could be transmitted to successive generations in an evolutionary process that progressively improved the species.[61] Spencerian psychology necessarily subsumed the entire sentient world: every type of creature, because it must manifest mind at some developmental stage, could be used to lay bare some facet of the mind's operations.[62]

Intelligence played a critical role in Spencer's understanding of mind as a product of evolution. While he accepted the traditional definition of intelligence as the faculty of reason, and even the tripartite division of mind into intelligence, emotions, and will, Spencer sought to recast the intellectual faculty in evolutionary/associationist terms. For Spencer, intelligence consisted "in the establishment of correspondences between relations in the organism and relations in the environment." Like a good associationist, Spencer dispensed entirely with the Common Sense notion of specific mental faculties, favoring instead accumulated experiences manipulated via the associationist laws. Moreover, in Spencer's progressive evolutionism, accumulated knowledge included not just what the individual itself had learned, but every experience that, by increasing the "conformity of the inner to the outer order," had improved the species' fitness for survival, become part of physical memory, and so entered the genetic stock.[63]

Dynamic and progressive, intelligence was also linear. Evolution, for Spencer, meant progress up a particular path, defined as the correspondence between organism and environment. Because all organisms occupied places on a single scale, Spencer argued, all could be compared; because some had evolved much more complex relations with their environments, they must be considered more intelligent. When explaining the hierarchy of the human races, for example, Spencer observed:

> The minds of the inferior human races cannot respond to relations of even moderate complexity; much less to those highly-complex relations with which advanced science deals. . . . We must therefore conclude that the complex manifestations, intellectual and moral, which distinguish the large-brained European from the small-brained savage, have been step by step made possible by successive complications of faculty.[64]

Intelligence, in other words, was not something absolute, but a characteristic existing in gradations from inferior to superior manifestations, whether at the level of organisms or of ideas themselves, where he posited a continuity between basic instinct and most elevated reason.[65]

The significance of Spencer's characterization of intelligence was twofold. First, his rejection of faculties made speaking of intelligence as a unitary entity more credible. Having abandoned his earlier interest in phrenology, Spencer completely swept aside notions of intermediary mental powers acting on individual or racial experience. Intelligence became the sole legitimate way of discussing the mind's ability to order, analyze, and act upon knowledge. It also became something substantial, the mental equivalent of an organism's physical organization. And second, by positing intellectual differences as a primary characteristic defining his scale of progressive evolution, Spencer raised intelligence to a position of central importance. Whether discussing connections between one-celled organisms and humans, or among the various human races, Spencer used intelligence as the feature binding these groups together on a single, graded scale. The scale of nature, for him no less than for physical anthropologists, was thus something real, comprehensive, and fundamental, and intelligence itself became crucial to the articulation of a scientific, progressive psychology.

Spencer's psychology struck a chord in France, especially among Ribot and his compatriots.[66] Writing to Alfred Espinas in 1867 after first reading Spencer's *Principles of Psychology,* Ribot declared "it is one of the most original and most interesting works that I know. It is psychology studied in the *positive* manner."[67] By 1870 Spencer had come to dominate Ribot's associationist psychology: Ribot gave Spencer pride of place in *La psychologie anglaise contemporaine* and with his friend Espinas translated the second edition of *The Principles of Psychology* in 1874–75.[68] Indeed, Spencer remarked in the preface to his 1872 edition that, thanks to Taine and Ribot, *The Principles* had been more warmly received in France than in England.

The positivist aspect of Spencer's work—its emphasis on science, the explanatory power of evolution, and the primacy of phenomenal fact—contributed greatly to its reception by Ribot and his circle. Spencer's popularity was also encouraged by the particular history of evolutionary thought in France. In a sense, Spencer rode into France on the wave of Darwinism, but Darwinism so transformed by the French context, especially by French neo-Lamarckianism, that it looked more like Spencer than Darwin.[69] The "culprit" in this story was the indigenous strain of French transformism, traceable back at least to Diderot, and most clearly articulated by Lamarck at the turn of the nineteenth century. Emphasizing progressive sequential development and inheritance of acquired characteristics, Lamarckian transformism was, to be sure, a minority position among French naturalists, in part because of Cuvier's unwavering hostility.[70] Nonetheless, echoes of Lamarckianism persisted, becoming much louder after midcentury. When Darwin's *Origin of Species* appeared, it was read in France not as a story about evolution via natural selection, but as a continuation of the Lamarckian tale of progressive transformation.[71] Clémence Royer, in the preface to her 1862 translation of Darwin, implicitly tied the *Origin* to notions of progress, Lamarckianism, and spontaneous generation, connections made more explicit in succeeding editions.[72] Throughout the latter part of the century, in fact, French writers almost automatically linked Darwin's name with Lamarck's.[73] This transformationist context, although it retarded the spread of the natural-selection version of Darwinism, accelerated the acceptance of Spencer, whose psychology was ultimately much more Lamarckian than Darwinian.

Spencer's *Principles of Psychology* provided French psychologists with one starting point for a new, "scientific" psychology. Positivistic, evolutionary, and developmental, it not only acknowledged the importance of differences but provided as well a way of analyzing them: as modifications of a faculty or psychological ability that could be arranged in a sequence.[74] French psychologists thereby could relate instincts to thought, protozoa to humans, and aborigines to civilized peoples, all by seeking, or assuming, the existence of gradations between them.[75] As Ribot observed, "experimental psychology aspires to discover, describe, and classify the diverse modes of sensation and thought, by following their slow and continuous evolution, from the infusiora up to man, white and civilized."[76]

The Pathological Style, Difference, and French Psychology

Transformist ideas were one source for the method of difference; a second was the approach dominant in French biomedical investigations, the traditions of physiological and mental pathology, whose roots stretched back to Cuvier and Pinel and whose best-known contemporary representatives were the physiologist Bernard and the neurologist Charcot.[77] The pathological method investigated nature's anomalies to understand the "normal" phenomena from which they were taken to be deviations. "Diseases," Bernard asserted in 1872, "are only at base vital perturbations furnished by nature in lieu of being provoked by the hand of the physiologist."[78] Head injuries, for example, were regarded as possible gateways into the mind's invisible areas, allowing the pathologist to infer brain functions by observing abilities lost when a given region was affected.[79] In the realm of mental phenomena, it was Charcot, dubbed by contemporaries the "Napoleon of neuroses," who above all embodied the style of pathological investigation that Taine and Ribot employed in their 1870 works and that greatly shaped the new French psychology.[80]

A Parisian carriage builder's son with a flair for observation and theater, Charcot continued in the footsteps of Pinel and Esquirol in transforming the ancient hospital and asylum, the Salpêtrière—a kind of "Mont Saint-Michel of melancholics" and old or poor women, in Mark Micale's vivid description—into "a 'temple of science' and an internationally renowned educational center."[81] Charcot assumed a position at the Salpêtrière in 1862 and quickly established a reputation for brilliant lectures and innovative research on epilepsy, polio, and other neurological disorders. Included in his coterie by the 1880s were neurologists and physicians from around the world—Sigmund Freud, Gilles de la Tourette, and Joseph Babinski being only the best known—as well as most of the individuals who constituted the first generation of French scientific psychologists, Ribot, Binet, Pierre Janet, and Charles Féré, who together established in 1885 the short-lived Société de Psychologie Physiologique. What they saw in Charcot, in addition to his oratorical brilliance, was a commitment to individual clinical observation, especially of patients who presented relatively pure manifestations of whatever mental disease he wished to investigate.[82] Empirical in practice where Taine was only in theory, Charcot's technique was to isolate pathological mental disturbances and then search, especially via postmortem microscopic analyses, for organic, and hence possibly hereditary, physiological causes.[83] In his experimental-clinical setup, Charcot clearly differentiated subject and experimenter, with the subject undergoing manipulative interventions from the experimenter—typified by Charcot's hypnosis studies—toward the goal of investigating the mind's abnormal aspects.[84]

In sum, Charcot's pathological method sought out differences, especially extreme differences, and then related them to the normal, assuming a continuity between exception and rule. It was a method based on particularity, on the minute investigation of individual cases in all their uniqueness, and on detailed case histories derived from intensive expert-subject interactions. For French scientific psychologists, Charcot's and Bernard's approaches helped convince them to eschew the notion that only the "normal" mind was of interest and to investigate a spectrum of mental pathologies.[85] Taine's *De l'intelligence* was an influential early example of introducing pathological phenomena into mental philosophy. Although Taine focused on general characteristics of mind rather than particular diseased or abnormal minds, as did Charcot, Taine paralleled Charcot in arguing that the linkage between normal and pathological, and especially the use of the one to understand the other, was critical to establishing a positive mental science.

In his later psychological works, Ribot maintained Taine's emphasis on employing the pathological to illuminate the normal, as did most of the first generation of French scientific psychologists. However, reflecting perhaps their more intimate knowledge of Charcot's methods—attained from personal experience at the Salpêtrière—they generally concerned themselves less with defining psychology as the science of abstract, universalized mind and more with investigating individual minds in all their variations. Deviations from the normal were interpreted quite broadly, as including any mind that was not adult, white, civilized, and healthy. Thus, theoretically, the entire panoply of mental variation was open to investigation through the method of difference.

Difference, continuous variation, hierarchical gradations, pathology, clinical observation, and experimentation—these were the principal elements from which Taine, Ribot, and their colleagues fashioned the "new" psychology. For Ribot, this approach would lead to his classic 1880s monographs on the maladies of the memory (1881), will (1883), and personality (1885), each based on innumerable physiological and pathological facts meant to illuminate the mental phenomenon in question.[86] For Ribot's student Pierre Janet, it led in 1889 to studies of psychological automatism (cataleptic and somnambulistic phenomena), especially via hypnosis; for others, to studies of disturbances of the emotions or attention.[87] But at first, very few psychologists followed Taine's lead directly and studied that difference presumed to separate animals from humans, aborigines from civilized peoples, and children from adults: intelligence. Perhaps the principal exception, and that in only a short lead article in Ribot's new journal, *Revue philosophique,* was produced by Taine himself.

De l'intelligence was not the place where Taine most explicitly or completely developed his ideas about differences in human intelligence. Fundamentally, Taine's 1870 tome was an account of the mechanical workings of abstract mind, of human intelligence regardless of particularities. Taine's most sustained analysis of difference and intelligence came as the conclusion to his 1876 article "Note on the Acquisition of Language Among Children and in the Human Species." The analysis, based loosely on an investigation into language acquisition by a girl "neither precocious nor slow," emphasized development, both within the individual and throughout nature. Taine began by describing the process of language acquisition, envisioning a series of stages that was a close analogue of the sequence described in *De l'intelligence* by which the adult mind created increasingly abstract ideas. Taine then proposed a whole series of other analogies that linked children with aborigines, simple nervous systems, small cerebral areas, and lower animals, connecting them via developmental, graded sequences to their presumed opposites: adults, civilized peoples, complex nervous systems, large cerebral areas, and human beings.[88] (See table 4.1.)

The power of these analogies lay in Taine's developmental model. By starting with the child-adult connection and then noting parallels between a child's ideational abilities and those of a simple organism or aborigine, Taine suggested that all were developmentally linked and existed in continuously differentiated quantities. "What distinguishes man from animals," he noted,

> is that, starting like the animals with interjections and imitations, he [man] now arrives at principles (*racines*) at which the animals do not arrive. But there is here only a difference of degree, analogous to that which separates a well-endowed

TABLE 4.1
Taine's Intelligence Spectra

Few associations	Many associations
Small cerebral area	Large cerebral area
Simple organisms	Humans
Children	Adults
Aborigines	Ancient Greeks
Simple nerves	Complex nerves
No language	Abstract language
Idiots	Geniuses

> race, like the Greeks of Homer and the Aryans of Veda, from a poorly endowed one, like the Australians [aborigines] and the Papua, analogous to that which separates a genius from an idiot.[89]

Taine thus suggested a program of difference that attempted to link variations not only at the species level—by seeing them as slices in a progressive, developmental continuum with intelligence as the determinative variable—but also at the level of individuals within a group.

The decision to place the genius and the idiot on a single scale marked a significant extension of the method of difference. By employing the analogy of child development, in which an individual organism progressed up a well-defined intellectual hierarchy, Taine could see intellectual ontogeny recapitulate phylogeny. He thus could imagine graded differences as existing within groups in addition to between them. And what is more, Taine suggested at the very end of his piece that pathology could be used to substantiate this scale:

> Thus the monkey is on the same scale as the human being, but many rungs below. . . . If one searches for the psychological condition of this superiority, one will find it in a much greater aptitude for general ideas. If one searches for the physiological condition, one will find it in the much greater development and much finer structure of the brain. The proof of this is that, if this double condition is lacking, the individual can acquire neither language nor the distinctive talents of which we have been speaking; he stops below the rung of humans; this is the case for cretins, [and] idiots.[90]

Taine was still hesitant about the status of steps in the gradation; his description of the cretin and the idiot connoted as much beings truly other and separate as ones differing only by degree from the rest of humankind. But it was a beginning, and one that later French scientific psychologists would develop more fully.

The combination of difference and connectedness was one of the hallmarks of the new scientific psychology, providing a framework for reinterpreting the nature of the individual according to particularities, including intelligence. During the half century following *De l'intelligence*'s publication, a generation of psychologists would begin to elaborate these possibilities, mainly by following one of two approaches. Some would conduct research based on the pathological style of understanding difference, investigating mental processes, including intelligence, via intensive clinical case studies. These explorations tended to emphasize the complexity of the mind's various functions and processes. Others, however, indebted more to the transformist approach to linking variations, focused on intelligence per se, mostly as a group-level attribute that they sought to apply to individuals. The laboratory proved particularly congenial for these investigations, as did techniques of measurement and quantification that promised mechanical objectivity. The resulting mixture of approaches, however, meant that the French psychological community could come to no firm consensus about what intelligence truly was.

A Laboratory Psychology for France

Infatuated with the possibility of remaking psychology into a "positive science," virtually all late nineteenth-century French psychologists were profoundly affected by the vision of science as a laboratory-based, experimental enterprise, an approach they associated in psychology particularly with the Leipzig psychologist Wilhelm Wundt.[91] Like their colleagues throughout the West, many French investigators were struggling to make space for psychology within (or outside) its parent discipline, philosophy, and turned to Wundt as a model of how to create a physiologically oriented laboratory science of the mind.[92] Ribot initiated the development of French laboratory psychology when he declined an offer in 1889 from the Ecole Pratique des Hautes Etudes (EPHE) to lead its newly created psychological laboratory at the Sorbonne, and suggested instead someone more experimentally oriented. The choice of Henri Beaunis, a physiological psychologist, and especially Beaunis's decision to accept Alfred Binet as an assistant, marked the real beginnings of French laboratory psychology. Soon thereafter, psychological laboratories were established at the Salpêtrière by Pierre Janet in 1890, the University of Rennes by Benjamin Bourdon in 1896, the Asile Villejuif by Edouard Toulouse in 1898 (attached to EPHE in 1900), and the University of Montpellier by Marcel

Foucault in 1906.[93] While a number of these institutions were short-lived (Janet's and Foucault's) or dedicated to pathological psychology (Janet's) rather than German-style physiological pursuits, the others became active centers of experimental research in the German mode. Investigations into the basic intellectual processes, sensation, memory, and a range of other mental phenomena flourished, facilitated by the growing collection of instruments—including Hipp chronoscopes, d'Arsonval chronometers, myographs, audiometers, and dynamometers—that soon cluttered the small new laboratories.[94]

Much of the experimentation was decidedly Wundtian.[95] As Kurt Danziger has characterized this approach, it emphasized laboratory exploration of normal cognition, especially the perception of simple sensory stimulations, by experimentation with individual perceptors, in order to investigate universal, intersubjective psychological phenomena.[96] Experimenters relied on introspective reports and used a variety of instruments designed to provide precise time measurements for reaction-time studies or to record a range of physiological responses. A hallmark of Wundtian experimental procedure, Danziger has argued, was the indistinct boundary between experimenter and subject. Pairs of psychologists typically worked together, one describing introspective reactions and the other manipulating the experimental apparatus and recording descriptions. Which could be called the experimenter was arbitrary; indeed, the essence of the social relation was the two participants' relative equality.

When French psychologists turned to experimentation near the end of the century, many of their studies followed the Wundtian model closely. Thus, Bourdon, who studied with Wundt in 1886–87, extensively investigated the psychology of perception, especially of vision, using German instruments and techniques. Toulouse collaborated with Henri Piéron and Nicolas Vaschide on experiments analyzing elementary sensations, in addition to producing *Technique de psychologie expérimentale* (1904, 1911), a manual for transforming psychology into an instrument-oriented experimental science.[97] And in the 1890s Binet's studies included numerous experiments in which he and associates recounted their introspective observations in response to stimuli, while collaborators monitored instruments and observed their partner's reactions.[98]

The experimental work for which the French became best known, however, adapted rather than simply adopted the Wundtian style. As Henri Bergson suggested in 1901, French experimentalism combined two distinct approaches, the experimental and the clinical:

> Psychology has always aimed at being experimental "and" comparative, but for many centuries it had not had at its disposition any precise method of experimentation and measuring, any means of distinguishing what in thought

> [*vie consciente*] is really elementary and what is really compound, and finally any firm procedure of analysis. This method came to it from two different sides, from the *clinic* and from the *laboratory*. . . .
>
> These two psychologies, the one founded on pathological observation and the other on direct experimentation, are now united: united with them as well is the ancient psychology of introspection. Today there is only one psychology, a true positive science, which has, like the other positive sciences, its own methods of investigation and also its own instruments and laboratories.[99]

French experimental psychology, especially as defined by Binet, Toulouse, and their associates, maintained Wundt's emphasis on a laboratory setting, precision instruments, and induced phenomena.[100] Their goal, however, as Bergson suggested, was not only to illuminate the mind's universal characteristics, but to explore as well manifestations of specific mental attributes in particular subjects.[101] It was clinical as well as experimental. Individual variations drew their attention, with the extent of the variation vis-à-vis some norm increasingly attracting greater interest than the underlying process itself. Because of this focus, many French studies were based on comparing a few or even many participants largely unknown to the researchers, and were oriented more toward higher mental functions where differences were believed to be most acute.[102] Thus social relations in the experimental setup shifted—experimenter and subject were now clearly differentiated—and psychologists increasingly relied on instruments, quantification, and other symbols of mechanical objectivity, expending, as Jacqueline Carroy and Régine Plas have noted, "infinite efforts in the attempt to prevent errors, bias and illusions."[103]

The work of Binet and Victor Henri in "psychologie individuelle," a field that in many respects they pioneered in France during the mid-1890s, exemplified this French style of experimental psychology. In papers appearing between 1894 and 1898, Binet and Henri both together and separately pressed for a new psychology that would "substitute for vague notions of man in general, of the archetypal man, precise observations of individuals considered in all the complexity and variety of their aptitudes."[104] Binet and Henri carried out this program principally in studies of school children, presenting groups with a series of "mental tests"—including studies of memory, imagination, attention, and moral sentiments—and then recording and quantifying individual responses.[105] Binet justified the focus on the higher mental processes theoretically by claiming that children and adults varied little in tests of elementary mental functions, and practically by suggesting that more advanced abilities were of greater interest to educators, physicians, and the state.[106] In addition to obtaining quantitative measurements wherever possible, the investigators also analyzed the data, not so much to better understand the nature of the mental processes as to rank the groups according to their proficiency.[107] The purpose of such rankings, Binet and Henri explained, was to produce socially useful knowledge:

> our principal goal will be to indicate the problems with which individual psychology must be occupied, to illuminate the practical importance that it presents for the pedagogue, doctor, anthropologist and even the judge, and finally to indicate by what means one could attempt to resolve these problems once posed.[108]

The place of intelligence in the project of French laboratory-based psychology, however, was curiously ambiguous. On the one hand, as the sine qua non of the higher intellectual processes, intelligence was of intense concern to experimental psychologists. On the other hand, given notions still prevalent that intelligence referred to a collection of independent mental powers, studies tended to focus on specific capabilities rather than the unitary concept. When Binet and Henri listed the ten mental processes of prime concern to the new psychology, for example, they omitted intelligence entirely, and Binet himself characterized intelligence in 1890 as "reasoning, judgment, memory, [and] the power of abstraction."[109] Binet and Henri did remark in an early study that their experience indicated the virtual impossibility of experimenting on isolated faculties—convincing them that mental powers must be investigated as an ensemble—but they made no immediate attempt to follow up on this.[110] Only near the century's end, in fact, did Binet and others begin large-scale experimental investigations into intelligence understood in the anthropological sense, as a thing in and of itself, and that came first in the context of exploring the craniometric techniques associated with that separate discipline.

French Psychology and the Anthropometry of the Mind

With his then-colleague Nicolas Vaschide, Binet turned to craniometric investigations of intelligence in 1898, when they published a review article in *L'Année psychologique* on the relations between intelligence and head size and form.[111] Both Binet's growing commitment to individual psychology and the lack of clear results from the mental tests Binet and Henri had developed in 1895 may have stimulated Binet and Vaschide to search for physical correlates of individual intellectual capacity. They may also have hoped that establishing a connection between intellectual ability and physical characteristics could yield important social benefits. As Vaschide would explain in 1904:

> if it were possible to recognize from infancy, by means of special physical signs, those of superior intelligence, one could push their education much further, prepare them specially for high culture, to the end that, on becoming adults, they would be an intellectual elite capable of advancing society in all branches of its activity.[112]

Surveying the anthropological studies produced over the preceding forty years by Gall, Parchappe, Broca, and Topinard, among others, Binet and

Vaschide concluded that head volume and intelligence were clearly if not absolutely correlated at the group level, but that craniometric methods had not been adequately applied to living persons or specific individuals.[113] On and off for the next twelve years, Binet, Vaschide, and other researchers would strive to do just that in the context of experimental psychology. Binet devoted a series of articles in 1900–1901 to cranial measurement and intelligence, and in 1904 Vaschide collaborated with Madelaine Pelletier on *Recherches expérimentales sur les signes physiques de l'intelligence,* summarizing their extensive work in this area.[114]

As we saw in chapter 3, by the late nineteenth century physical anthropology was the human laboratory science par excellence in France. When French experimental psychologists began to investigate, as Binet put it, "the question of knowing under what limits and conditions the dimensions of the head could provide information about the intellectual capacity of a particular individual," they quickly adopted the anthropologists' approach, especially as it accorded nicely with experimental psychology's prevalent methods.[115] Thus in his craniometric studies, Binet appeared as the quintessential experimentalist, using calipers and other instruments as suggested by Broca and Topinard to make exquisitely precise measurements of the heads of highly intelligent children (as classified by their teachers) and very deficient ones (those in asylums). Binet was convinced that he would discover a strong correlation between certain cranial measurements and intelligence. Instead, like the anthropologists, Binet found little to encourage his preconception: the unintelligent, he discovered, varied in head size much more than the intelligent, and the difference in means was, for most measures, almost insignificant.[116] By splitting his groups still further and concentrating solely on the absolute extremes, Binet was finally able to discern some appreciable difference—on the order of 3mm—in favor of the mean head size of the highly intelligent. But it is clear that this seemed to him at best weak support for craniometric investigation of intelligence even at the level of groups, and that the method was wholly inadequate for individuals.[117]

More significant was Binet's methodology. Losing sight of his initial interest in the individual, Binet followed anthropological practice: he divided his sample population into groups according to intelligence and then calculated and compared means for those at the spectrum's extremes.[118] This raised almost immediately, however, one of the central problems plaguing all research into intelligence, encountered by craniometricians upon first measuring skulls: How did one independently determine who was smart? Anthropologists, as we have seen, relied largely on broadly disseminated ethnocentrism: whites were obviously the most intelligent race, blacks the least, with other races ranged between. For psychologists working with small subgroups of children, however, the issue was more vexed. Binet turned to assessments by teachers or others who knew the individuals, but it was with great reluctance. He was never satisfied with this procedure, which he felt was fraught with subjectivity, depending too critically on the opinion of those neither scientifically trained nor versed in the ways of intelligence.

The notion of intelligence itself, in fact, presented the greatest difficulties. Except within anthropology, individuals or groups were rarely ranked according to some global sense of mental ability; rather, assessments were typically rendered on the basis of specific talents or performances.[119] Binet's work would contribute to changing this state of affairs, but only later; in turn-of-the-century France, intelligence was *not* a familiar criterion by which to categorize. Indeed, Binet admitted that this was one of his studies' chief difficulties: "I requested that only the elite and backward children from 11 to 13 years old be chosen for me. Unfortunately, these expressions are so vague that many teachers did not succeed in understanding my thought."[120] Vaschide and Pelletier encountered a similar problem in their own attempts to link intelligence with physical characteristics, and in 1904 envisioned a possible solution:

> It would be necessary to have a scale of tests of comprehension in which the progressive difficulty would be graduated with a precision sufficient, so that one could be certain that a test classed under the no. 2, for example, would be of a level of comprehension more difficult than that which is assumed for no. 1 and less than for no. 3. Such a scale does not yet exist.[121]

They would prove remarkably prescient. Within a year, in conjunction with Théodore Simon, Binet would attempt to resolve the issue of determining the level of an individual's intelligence along strikingly similar lines. Before turning to that story, however, we must examine the other way in which French psychology investigated individual intelligence: case studies.

Case Studies and Human Intellect

As the French experimental style developed, a number of researchers bridled at what they perceived to be the arbitrary limitations, artificialities, and superficialities of the instrument-based laboratory approach. Preferring the clinical-pathological style of careful, often long-term observation of individual subjects, they celebrated the case-study method and the richness and complexity of psychological phenomena. "We find already," one such researcher observed, "that the laboratory is a place too narrow, too artificial, and that it would be valuable to draw nearer to reality and make more studies from nature, taking man as he lives . . . not as a subject."[122]

What may be surprising, however, is that many of those most active in constructing and promoting laboratory psychology were simultaneously its most vociferous critics. It was Alfred Binet who uttered these disparaging remarks about the laboratory, and he was not alone among experimentalists in expressing misgivings about the very project they proselytized. Edouard Toulouse was equally at home conducting aggregate experimental investigations in his laboratory at Villejuif and celebrating the intensive study of a single individual.[123] Indeed, one of the peculiar features of French psychology was that most prominent practitioners felt deeply divided about their own procedures.[124] Until his death, Binet vacillated wildly on the value of objective methods, statistical studies, and mechanically produced information, lauding their scientific objectivity at one moment, condemning their shallowness at the next. In general, many among the first generation of French scientific psychologists—equally well trained in clinical observation and experimental manipulation, and acutely aware of the different investigative programs each implied—found it difficult to abandon either, and instead often sought to merge the two.

This hybrid approach was most visible in studies of the intellect. In one sense, case studies of higher mental processes markedly resembled laboratory investigations involving numerous experimental subjects. Thus when Edouard Toulouse examined the writer Emile Zola in 1896, he employed many of the same instruments and tests used routinely (albeit to a different purpose) in psychological laboratories, including his own. Where experimentalists typically kept the instrument constant and varied the subjects, however, as in Binet and Henri's examination of three hundred school children, Toulouse kept his subject, Emile Zola, constant and varied the instruments. The result was a much more complex portrait of Zola's intellect than would have been possible had Toulouse relied on a single investigative modality. Toulouse himself likened his approach to that of the quintessential clinical practice, the physician's:

> In order to do truly useful work, it is necessary to employ as much as possible the same methods of examination used in the hospital, those of the doctor who is not content with posing a list of questions, but who interrogates very closely, examines, palpates, scrutinizes and verifies most often with his usual instrumentation.[125]

While the medically trained Toulouse had firsthand knowledge of such methods, most of his fellow psychologists learned the clinical style of investigation by working with Charcot at the Salpêtrière. Binet, for example, emerged from his experience of close research on individual subjects with a deep appreciation for the power of pathology to illuminate the normal, and for the value of intensive observation and experimentation. This clinical emphasis on obtaining an intimate feeling for the organism, as Raymond Fancher has pointed out, was most completely manifested in Binet's case studies of the 1890s and early 1900s.[126] In investigations of chess players, calculating prodigies, and several authors, Binet sought through extended observation and careful questioning to understand the intricacies of the mental processes underlying their talents.[127] Doubting the value of quantitative results alone, Binet stressed the need for exacting observation and detailed description:

> Mere numbers cannot bring out . . . the intimate essence of the experiment. This conviction comes naturally when one watches a subject at work. . . . The experimenter judges what may be going on in [the subject's] mind, and certainly feels difficulty in expressing all the oscillations of a thought in a simple, brutal number, which can have only a deceptive precision. How, in fact, *could* it sum up what would need several pages of description![128]

Binet's very choice of subjects—individuals manifesting exaggerated instances of whatever quality he wished to investigate—reflected one fundamental tenet of the clinical-pathological method. As Binet himself explained, "it seems to me that people of talent and of genius serve better than average examples for making us understand the laws of character, because they present more extreme traits."[129]

One of the most striking examples of the power, and limits, of Binet's use of case studies of the exceptionally talented was the two-year investigation that Binet and his colleagues conducted of a young Piedmontese man, Jacques Inaudi, known for his prodigious feats of mental calculation.[130] In addition to interrogating him about his calculating methods and watching him perform in front of an audience—Inaudi marketed his mathematical talents on the French entertainment circuit—they also took him repeatedly to their laboratory. There they timed Inaudi as he executed various calculations; tested his memory for numbers; measured his reaction times for various sensations; investigated his calculating techniques and way of storing data; monitored his ability to mentally rearrange sequences of numbers; compared him to other calculating prodigies, cashiers from the department store Bon Marché, and Sorbonne students; and frequently asked him to describe the interior mental processes that the various assigned tasks provoked. With this welter of information, Binet both provided a general categorization of Inaudi's style of calculation—that Inaudi was an "auditif," one who operates on numbers via sounds rather than visual presentation—and also registered the complexities and particularities of a specific individual.

The study of the "great calculator" yielded no simple, all-encompassing depiction. Over his fifteen encounters with Binet and his assistants, Inaudi took on a richness and complexity that problematized placement in simple categories, even that of "auditory type." Not only was much revealed about Inaudi's past, family, and present pursuits, but even when experimenters focused exclusively on his memory, not one feature was registered, but many.[131] Binet himself would not have denied this; indeed, he frequently castigated other psychologists who recurred too quickly to reductive, quantified assessments rather than attending to the rich detail afforded by the close study of an individual subject.[132]

In addition, Inaudi arose out of the multiple modalities in which he was observed. Examined not through a single lens, but through a variety of instruments, tests, and observation procedures, Inaudi's particularity and individuality became experimentally real. Everyone knew that Inaudi was a calculating marvel, but the specifics of that talent and Inaudi's other mental features required the tests and instruments in Binet's laboratory to be fully revealed. In the process, Inaudi became many things: the individual who scored off the scale on the test for number memory but performed less exceptionally on number span; the individual whose vocal cords moved almost imperceptibly as he calculated; and the individual whose reaction times for sound were little different than for sight or touch. It is not that they were different Inaudis, but rather that they did not result in an Inaudi easily summed up by one measure. Even when Binet compared him to other calculating prodigies, Inaudi proved to be superior in some respects, inferior in others, and simply different in many more. The very multivalency of the approach to investigating him revealed Inaudi's specificity and heterogeneity. Examined closely, Inaudi's intellect proved anything but simple to characterize.

Binet arrived at similar conclusions in his most sustained use of the case-study approach, his incredibly painstaking and thorough research on his daughters Alice and Madeleine. Begun in 1890, when the children were respectively three and five, it culminated in his 1903 work *L'Etude expérimentale de l'intelligence,* which many contemporaries deemed his masterpiece.[133] A mixture of clinical and experimental studies, *L'Etude expérimentale* distilled years of observation of Alice and Madeleine to generate conclusions about the specific types of intellect each daughter manifested. In contrast to Binet's craniometric studies of the same period—full of instruments, quantities, and rapid measurements of a large number of individuals—*L'Etude expérimentale* condemned "la méthode de la statistique" as generating mediocre results, and praised, especially for studying the superior functions, the slow, intimate observation of a few well-chosen subjects. "If I have been able to arrive at any light by the attentive study of two subjects," Binet observed, "it is because I have watched them live and have scrutinized them over several years."[134]

With this close scrutiny and intimate familiarity came a particular relation to the object of investigation: intelligence.[135] While studying his daughters' minds, Binet was most concerned with intelligence's nature, and how different people could manifest intelligence of different types. Although Binet was still willing to classify his subjects' intellects, and indeed categorized Madeleine as "l'observateur" and Alice as "l'imaginatif," he made no attempt to rank these categories or imply that one was superior; each style of intelligence, according to Binet, had its own strengths and weaknesses. Binet's other case studies were similarly oriented: his goal in each was to understand the wide variety of ways in which even the highly proficient achieved their skill. Examined through the multiplicity of instruments, tests, and observation procedures that characterized the mixture of clinical and experimental styles, intelligence became a complex, multivalent phenomenon. When conducting investigations with large numbers of subjects, Binet looked through intelligence to see his subjects; when performing case studies, Binet looked through his subjects to see intelligence. And what he found from this perspective was not one thing, but many.

The Binet-Simon Scale: Between Clinical Tool and Laboratory Instrument

By 1904, when Binet began the work that would culminate in the first measuring scale of intelligence, he had already been investigating the higher mental powers from a variety of perspectives for several years. His studies of individual differences, of correlations between craniometric measurements and intelligence, and of individual intellectual character had left him with a wealth of empirical data about and insights into the higher mental functions. Binet's creation of the intelligence scale in 1905 was in many ways an uneasy consolidation of these approaches.[136]

Binet's stated impetus for developing the scale was practical, his appointment in 1904 to the Léon Bourgeois ministerial commission established to address the needs of the *anormaux*.[137] "In October, 1904," as Binet recounted it,

> the Minister of Public Instruction named a commission which was charged with the study of measures to be taken for insuring the benefits of instruction to defective children. . . . They decided that no child suspected of retardation should be eliminated from the ordinary school and admitted into a special class, without first being subjected to a pedagogical and medical examination from which it could be certified that because of the state of his intelligence, he was unable to profit, in an average measure, from the instruction given in the ordinary schools.
>
> But how the examination of each child should be made, what methods should be followed, what observations taken, what questions asked, how the child should be compared with normal children, the commission felt under no obligation to decide. . . . It has seemed to us extremely useful to furnish a guide for future Commissions' examination.[138]

There is little evidence that Binet's concerns were broadly shared by the other commission members, although the commission's final report did suggest "drawing up a scientific guide" to help authorities decide questions of possible mental debility.[139] Undeterred, Binet advocated fundamentally redefining the problem by relocating it from the medical and educational communities to the psychological laboratory. Accordingly, in collaboration with Théodore Simon—with whom he had started working in 1899 when Simon was an intern helping "backward and idiot" children at the asylum in Perray-Vaucluse—Binet turned to the world of precision measurement. He was determined, he explained, to create an instrument that could identify unambiguously those who were of subnormal intellect. "It is a hackneyed remark," Binet declared near the beginning of his first essay on the measuring scale,

> that the definitions, thus far proposed, for the different states of subnormal intelligence, lack precision. These inferior states are indefinite in number, being composed of a series of degrees which mount from the lowest depths of idiocy, to a condition easily confounded with normal intelligence.[140]

Binet and Simon hoped to remedy this lack of precision by creating a series of tasks that, they argued, would differentiate clearly among the four major classifications of intelligence then common: idiocy, imbecility, *débilité* (in English feeblemindedness or, after 1910, moronity), and normalcy.[141] Binet developed a series of thirty tests, ranging from number 1, the ability to follow a moving object, to number 30, the ability to explain the differences between various pairs of abstract terms, such as "esteem" and "affection" or "weariness" and "sadness," and arranged them from simplest to most difficult. Calibrated through application to "normal" children, the series contained sets of tests that, Binet contended, could be passed by individuals of one level of intelligence, but not of a level below. For example, in distinguishing between normal five- and seven-year-olds Binet highlighted the comparison-of-known-objects test, number 16, in which the subject must identify a difference between a fly and a butterfly, a piece of wood and of glass, and paper and cardboard. All normal seven-year-olds, Binet insisted, could answer these questions correctly, while normal five-year-olds could not. A similar approach was used for degrees of subnormal intelligence: imbeciles, for example, were described as unable to pass tests involving comparisons and repetitions of numbers, and thus tests 8 through 12 in the scale separated them from the more intellectually advanced *débiles* (morons).[142]

At its simplest, therefore, the Binet-Simon was constructed as a series of barriers that would stratify a population into its "natural" intellectual levels along the scale of mental development.[143] While Binet originally conceived of the test as limited to subnormals, the test's "genius," as it were, was its equation of subnormal intelligence with arrested normal intelligence and thus its creation of a scale that in principle could apply to normal children as well as the feebleminded. The consequences of this equation were profound, especially in later versions of the scale. But before considering these further developments, it is important to examine the character of the scale that Binet and Simon actually created in 1905.

One of the words that rang loudest in Binet's description of the scale's purpose was "diagnosis." In part, this was strategic, a way of reaching his intended audience, the physicians responsible for determining the intellectual levels of those recommended for asylum admission. But in large measure the term sprang from another source, Binet's background in the clinical approach to psychology learned at the Salpêtrière and practiced in his case studies.[144] Binet's 1905 scale was not at all like the modern multiple-choice intelligence test. First, the scale was focused on the subnormal, and thus on the pathological character of human intelligence, the clearest indicator of the test's clinical orientation. "What the physician seeks with the greatest care," Binet asserted, "is the differential diagnosis of idiot, imbecile and moron [*débile*]."[145] Not only was the scale never intended to generate some number characterizing an individual's mental power, but it could not possibly have been mass administered in a multiple-choice format; indeed, a central rationale for the scale was to facilitate "holding the subject in continued contact with the experimenter." Otherwise, Binet argued, "the test loses its clinical character and becomes too scholastic."[146] Binet thus envisioned the scale as a kind of stethoscope of the mind: it would be used to examine a possibly feebleminded individual in order to isolate and amplify—by means of the questions of which the scale was composed—relevant phenomena, which had then to be interpreted by the investigator to construct a detailed portrait of the subject's intellect. The subject would not so much *take* the Binet-Simon scale as *be observed* by an expert using it to fashion a diagnosis.

Perhaps because of its orientation toward diagnostics, the 1905 measuring scale lacked a highly elaborated method for scoring replies or a system for in any sense *calculating* the subject's "true" intelligence. Although he suggested a notation system to track responses, Binet emphasized writing detailed descriptions, similar to clinical observations, of all of a subject's reactions to a test, and he further counseled that certain wrong answers—those that were absurd—could reveal far more than mundane wrong answers or even many correct ones. Above all, Binet stressed the need for judgment: the scale, he argued, must not be treated like an automatic recording instrument, but required the active involvement of an experienced experimenter.[147]

Adhering to a fundamental tenet of French clinical psychology, that the pathological differed only in degree from the normal, Binet imagined the measuring scale as a means of determining a precise relation between abnormal states of intelligence and normal intellectual development. "Here, in our studies on children," Binet commented in 1905, "it is not only a comparison that is necessary, it is a physiological, anatomical and anthropological table of standards [*barême*] to which one must return every time with each new subject to determine in what measure this subject is inferior to the normal."[148] The crucial word here is "barême," which signified a standard that captured some aspect of what it meant to be "normal" and that could be used to analyze deviations from this normal state.[149] Having constructed the scale to reflect the performance of the average Parisian child, Binet could treat the scale as a pathologist treated a healthy organ: the model against which the degree of inferiority of an abnormal organ could be judged.[150] Binet's "barême," however, suggested a second feature as well: the possibility of measuring the degree of deviation from these objective standards. And Binet's scale itself did not just compare diseased with healthy, it also measured the degree of intellectual development.

The scale's instrumental dimension was apparent in several of its most salient features. First, Binet justified the scale's development as a way of standardizing physicians' assessments of the intellectual levels of the subjects they were diagnosing. Arguing that methods for assessing intelligence varied excessively from doctor to doctor, Binet declared that "the simple fact, that specialists do not agree in the use of the technical terms of their science, throws suspicion upon their diagnoses, and prevents all work of comparison." Binet's solution was to develop a "*precise basis for differential diagnosis,*" quantitative in nature, that would not be liable to "subjective processes"; in other words, a measuring *scale.*[151] Second, the measuring scale was constituted through the invariable performance of a series of specified actions. No room for personal discretion or variations was acceptable, as Binet made clear when praising an earlier experiment in assessing intellectual deficit: "[Blin's is] a first attempt to apply a *scientific* method to the diagnosis of mental debility. The method consists of a pre-arranged list of questions which are given to all in such a way that, if repeated by different persons on the same individual, constantly identical results will be obtained."[152] Two features critical to a scientific instrument were evident in this description: first, repeated measurements must give essentially the same results; and, second, different operators should not significantly change the instrument's measurements. Finally, for all of Binet's emphasis on providing insight into the examinee's mind, the 1905 scale's instrumental nature was manifested in its breadth of application and purpose. Ultimately, Binet designed the scale for use on a large population of anonymous individuals not to illuminate their intellects' subtleties but to measure and classify their mental power against a linear scale. What was constant about the 1905 scale was its questions, not its subjects.

In subsequent revisions, Binet and Simon changed both little and much about their scale. Little, in the sense that the scale remained a mixture of the clinical and the laboratory-experimental; much, in the sense that the balance between these styles shifted markedly, as the scale's instrumental nature was increasingly accentuated at the expense of its clinical dimensions. Without access to his long-vanished personal papers, we cannot determine definitively what motivated Binet to revise the scale. Undoubtedly part of the impetus came from perceived imperfections in the test once it was used more extensively. Also, physicians' decided lack of interest in the 1905 test and Binet's own shift in attention toward education must have helped convince him to reorient the measuring instrument away from the asylum and toward the school, where the availability of psychological expertise would have been minimal and demand for clear-cut classifications the greatest. Whatever the reasons, Binet's approach in 1908 and even more so in 1911 was to reduce the scale's clinical facets, significantly increase its instrumental dimensions, and thereby make the data more "objective" and mechanical. The 1908 revision revealed this tendency clearly. Gone was the informal hierarchy of the 1905 scale; gone, indeed, were most of the individual tests. Replacing them was a new collection of fifty-six tests, arranged in sets according to mental age, and calibrated so that most normal children of the relevant age could pass them while at most half of the age just below could do so.[153] In addition, Binet added to the 1908 scale an explicit procedure for determining a numerical intellectual level, the "*niveau mental,*" and more rigid criteria for a correct response. These changes substantially increased the emphasis on precision—as finer discriminations, more careful calibrations, and even quantitative evaluations came to predominate—and expanded the scale's target population to include normal children as well as subnormals.[154]

The most telling evidence, however, for the scale's increasingly instrumental nature was Binet's own equivocal reaction to his changes. On the one hand, Binet the committed empiricist seemed to delight in refinements that made the scale more precise and less subjective; on the other, Binet the clinical observer seemed dismayed with some of the "improvements" he had wrought. For example, when describing one of the tests for eleven-year-olds—the sixty-words-in-three-minutes test—Binet remarked, "To employ a series of words, to give abstract words, are good signs of intelligence and of culture. But here we consider *only* the *number* of words."[155] Strikingly, although Binet recognized the value of deeper analyses of his subjects' responses, as he had done when using the test to study his daughters, in 1908 he rejected performing such a study and contented himself with simple quantification. Forced to choose, he chose efficiency and precision over more in-depth interpretation.[156]

The 1911 revision simply confirmed this tendency. Binet further standardized the scale by establishing five tests for each age level and increased emphasis on the output's quantitative dimension by further subdividing the mental-level score to tenths of a year and extending the scale to normal adult intelligence.[157] Binet continued to insist upon the need for judgment and description—especially in recording subject responses—over the simple quantification of answers, but his own use of data from the scale belied his words, as he focused almost exclusively on the mental-level statistics.[158] Thus, by 1911 the Binet-Simon measuring scale had essentially been transformed. In contrast to the 1905 scale, which permitted both the arbitrariness of personal observation and its potential richness, the 1911 scale provided decimal-place accuracy in its measurements and the possibility of testing children and adults at all levels of intelligence, but only by limiting the results to little more than a single number on a unidimensional scale. Retaining only traces of its original character, it had inched much closer to becoming the automatic recording instrument that Binet both admired and feared.

Binet-Simon and the Meaning of Intelligence

The change in the nature of the Binet-Simon scale brought with it a shift in the notion of intelligence underlying its operation. In 1905 the scale combined, perhaps uneasily, two visions of intelligence. On the one hand, intelligence was multivalent, a complex phenomenon whose different features individuals could manifest in a multitude of ways. Thus, Binet and Simon placed a premium on the interaction between investigator and subject, and on focusing and intensifying the investigator's appreciation for the particularity of a subject's intellect.[159] Counterpoised to this notion, however, was an idea of intelligence as a single entity possessed to different degrees by all human beings. A means of classifying rather than an object of investigation, intelligence in this guise was embodied in the 1905 scale's establishment of a single standard against which minds could be measured and a metric categorizing intellects according to amount of deviation from this norm.

With successive revisions, this second sense of intelligence increasingly predominated, and by 1911 intelligence had largely been flattened out in the Binet-Simon scale.[160] In place of the very different intelligences of his daughters or even of intelligence as the composite name for a diversity of faculties, Binet had substituted the notion of intelligence as a singular entity of varying power.[161] Applying this anthropological understanding of intelligence to individuals, the scale equated normal intelligence with the ability to make judgments, comparisons, and decisions in line with broadly accepted norms, and statistically with the sample population's mean performance. Each of these characteristics was apparent in the 1911 scale's output: a single number, the mental level, derived from performing tasks—such as making up a sentence containing three given words—with the same success as the average for the Parisian children on whom the instrument was standardized.

Four features of the scale merit particular attention. First, the equation of normal intelligence with average intelligence was especially significant. Playing on the dual meanings of "normal" as "average" and "healthy," the scale united the pathological method with the statistical, although only at the cost of tying the concept of normal intelligence firmly to the specifics of the populations from which it was derived.[162] Binet himself conceded as much, noting that his sample population was "of average social standing" and then cautioning "that these indications are very important; because the intellectual level of the children is modified according to the wealth of the population."[163] The problem would become more pressing once the scale was decoupled from its originating context: What would it mean to apply an instrument constructed for one group to another? Did different populations imply different normal intelligences? Could meaningful comparisons even be conducted across culturally distinct groups, however defined? Binet had no unequivocal answers for these issues, and he was not alone; psychologists would return to these problems repeatedly throughout the history of intelligence and its tests.[164]

Second, intelligence as defined by the Binet-Simon scale was both developmental and universal. Predicated on intelligence cumulatively increasing with age, the scale's construction reflected a developmental continuum: Binet and Simon included only those questions that produced age stratifications when tested on "normal" children. At the same time, the universal quality of intelligence—the sense that intellects differed by degree, not kind—inhered principally in the way the instrument was employed. Contravening Binet's explicit requirement that test populations be homogeneous, examiners in fact made no adjustments for specific groups. Whether the population was Parisian schoolchildren or institutionalized adults, every examinee was judged against the same standard.[165] In addition, the resultant classification was part of a set, either of mental levels or descriptive categories, meant to cover the entire intellectual spectrum. Before the Binet-Simon scale a few specific kinds of individuals—idiots, imbeciles, and geniuses—had been typed according to intellect, with the vast majority left unclassified. After creation of the scale, however, classification became universal: every child and adult was theoretically subject to its categorizing imperative and assignable to some classification.[166]

Intelligence, in a sense, thus helped regularize the population, as it had when anthropologists applied it to groups, by allowing all individuals to be assessed according to a "normal" developmental sequence and placed in precise relation to one another.[167]

Third, Binet-Simon intelligence was explicitly fashioned to fulfill a variety of social roles.[168] The 1905 scale began as a new means of diagnosing idiocy, imbecility, and feeblemindedness.[169] Binet and Simon's initial goal was to replace what they took to be the arbitrary classificatory methods of doctors and educators with a procedure that was more objective, precise, and above all scientific. In subsequent years, their vision of the possible applications of metric intelligence grew. Binet in particular became increasingly convinced of the importance of a psychology "oriented toward practical and social questions," advising readers of *L'Année psychologique* in 1908, for example, that such material would henceforth have a preponderant place in his journal.[170] And Simon, who worked for most of his career in asylums, was always concerned with producing an intelligence relevant to French institutional life. Reflecting these interests, by the scale's second revision Binet and Simon's conception of the social value of assessing intelligence had extended well beyond classification of the feebleminded; they envisioned it as providing a way of objectively allocating resources, including human beings, to meet a variety of state, social, and individual needs:

> the practical applications of this study [of intelligence] are evident in recruitment for classes of the abnormal, in the formation of classes for the supernormal, in the determination of the degree of responsibility of certain feebleminded [*débiles*], etc., without even taking account of the great interest that a parent or a schoolmaster could find in knowing if a child is intelligent or not, if his scholastic performance [*succès*] is related to his idleness or intellectual incapacity, and towards what kind of career it is fitting to direct him.[171]

As the range of these potential applications suggests, by 1911 the metric scale had largely been reconceptualized into a mechanism for assisting institutions in managing individuals. Having lost much of its feel for the idiosyncratic and personal, the Binet-Simon scale had become instead a measuring tool, advertised as able to assign or explain social roles, to mediate impartially between the state's rationalizing imperatives and the citizenry's diversified talents, and to shift the basis for social decisions from subjective choice to scientific determination.[172] While the actual uses to which the metric scale was initially put in France were limited, the scale's interweaving of intelligence and social policy would ultimately cast a long shadow.

Finally, it is important to clarify what the Binet-Simon scale did *not* say about intelligence. Nothing about the scale or its presumed uses suggested intrinsically that intelligence was biological in origin, hereditary, or remained at a fixed level. A number of Binet's contemporaries, especially among American psychologists, interpreted the scale's output in just this fashion. Binet himself, however, insisted that such readings were ill advised and unwarranted.[173] "Some recent philosophers," Binet remarked in 1909, "seem to have given their moral approval to these deplorable verdicts that affirm that the intelligence of an individual is a fixed quantity, a quantity that cannot be augmented. We must protest and react against this brutal pessimism; we will try to demonstrate that it is founded on nothing."[174]

The connection others fashioned between intelligence and the notion of a biologically based, hereditarily transmissible mental faculty of predetermined power should not, however, have been surprising. Binet and Simon's methodology and their overriding concern with intelligence as a classificatory tool smacked strongly of Binet's previous work in craniometry, in which intelligence was accorded just such characteristics. Indeed, the Binet-Simon scale itself might best be seen through its successive revisions as representing the tendency for Binet's concern with the pathological and the individual to be subsumed into a framework structured by the instrumental and statistical outlook of craniometry. While Binet might complain that no evidence supported a "brutally pessimistic" reading of intelligence, his own fusing of the two approaches to psychological investigation made this reading not only likely but, perhaps, almost unavoidable.

Like a Prophet in His Own Land: The Ambiguous Status of Alfred Binet

Who was this figure who pushed so hard to transform French notions of the mind? And how did he become one of France's most famous psychologists and yet remain rather marginal to academic psychology? Born in Nice on July 11, 1857, the only child of a well-to-do family—his father was a physician, his mother an artist—Alfred Binet moved with his mother to Paris in 1875 to attend the elite lycée Louis-le-Grand.[175] Success there, however, did not induce Binet to follow the path of most would-be entrants into the French academic/intellectual world. Avoiding that incubator for French academe, the Ecole Normale, he instead pursued first law and then medicine. Neither proved particularly congenial, and by 1880 Binet had abandoned both for the reading room of the Bibliothèque Nationale and the pleasures of the new psychology. After some early, largely unsuccessful dabblings in psychophysics, Binet was introduced in 1883 to Charles Féré and Charcot's clinic at the Salpêtrière. There Binet commenced his work in psychology in earnest, collaborating with Charcot and Féré on investigations into hysteria and hypnotism and quickly becoming embroiled in the bitter dispute that Charcot was having with Hippolyte Bernheim and his followers over connections between hypnotism and hysteria. Forced eventually to concede publicly that many of his reported observations were artifactual, the results solely of unintentional suggestion, Binet emerged chastened and now well schooled in the importance of experimental rigor.[176] Binet soon sought out a new institutional base, finding it later that same year through a chance encounter with Henri Beaunis, director of the first psychological laboratory in France, as Beaunis's unpaid assistant in the Laboratory of Physiological Psychology established at the Sorbonne in 1889.

The Sorbonne laboratory remained Binet's professional home until his death in 1911. Appointed associate director in 1892, Binet became director upon Beaunis's retirement in 1894, the year that Binet and Beaunis founded France's first journal dedicated exclusively to psychology, *L'Année psychologique.*[177] For the next five or six years, Binet and a series of associates—including Victor Henri, Nicolas Vaschide, Charles Henry, Jean Philippe, and J. Courtier—carried out experiments on a wide range of mental and physiological phenomena.[178] By the early 1900s, however, having been rejected for professorships at both the Collège de France and the Sorbonne, Binet's interests turned in other directions, especially toward the investigation of children and the mentally deficient.[179] In 1899 Binet began his long collaboration with Simon; the same year, he also commenced his extensive involvement in pedagogical issues by joining a society dedicated to education reform, La Société libre pour l'étude psychologique de l'enfant. Already looking for a new outlet for his enormous energy, Binet almost immediately became the society's dominant figure, editing its *Bulletin* from 1900 on, assuming the presidency in 1902, and in 1905 with one of the society's members, V. Vaney, founding a laboratory of experimental pedagogy at Vaney's primary school on the rue Grange-aux-belles.[180] By this time, except for Thursday-afternoon receiving hours, from 2:00 to 4:00 P.M., Binet had essentially abandoned his Sorbonne laboratory.

Demanding at best and difficult at worst, shy, reserved, and unapproachable, Binet cut an unusual figure in the Parisian intellectual world. He came from a class—the *haute bourgeoisie*—atypical for Third Republic academics, failed to attend the Ecole Normale or do the agrégation, and evinced little interest in the camaraderie of the salon or conference. All of these distanced Binet from the social connections crucial for advancement in the upper reaches of the Belle Epoque French academic world.[181] Rejections by the Collège de France and the Sorbonne for professorships indicated Binet's outsider status, as did his turn away from the laboratory to pedagogy, a field with few rivals and enormous opportunities for establishing a national reputation. Nonetheless, Binet's prolific publication record and directorship of an important psychology laboratory and journal ensured that he was well known and respected within psychological circles, both in France and abroad. To his study of marginal populations—children and the mentally deficient—Binet brought extensive familiarity with the dominant French, German, and British methods of psychology, broad interests in and ferocious attachment to the empirical/experimental approach, and the freedom to explore new avenues born of his own isolation from academe's professional structure.[182] While this position served him nicely when investigating intelligence, it would prove more problematic for disseminating his ideas broadly.

Thus when Binet died somewhat suddenly in 1911 at the age of fifty-four, he was not in the best position, institutionally, to have his legacy persist. Indeed, his measuring instrument virtually perished with him in France. Leadership in developing new intelligence tests and revised versions of the Binet-Simon scale passed quickly to the United States, and Binet's Sorbonne laboratory was entrusted not to his disciple Simon, but to a younger rival, the thoroughly experimentalist Henri Piéron, who had little interest in Binet's program of large-scale intelligence assessments.[183] In this Piéron was not alone. While most French psychologists acknowledged that Binet's measuring instrument had some practical value, Binet's legacy was at first visible largely in other facets of his work; few found the scale of more than limited relevance to their own research programs. The outbreak of war in 1914 compounded these problems, as psychological research was largely put on hold, not to begin again in earnest until the 1920s.

Certainly, the Binet-Simon scale's failure to attract more extensive interest before the war derived partly from Binet's marginal status.[184] Never part of Ribot's inner circle and lacking the appropriate credentials, Binet attained neither a permanent academic position nor powerful patrons. He also remained largely unconcerned with disseminating his ideas through collaborative work with other senior researchers. His physical separation from his discipline, expressed in his abandonment of the Sorbonne psychology laboratory, and his choice of a medically trained asylum psychologist as his assistant, emphasized his isolation. Compounding these factors, Binet had created no institutional structure to continue research on the intelligence scale. While Simon and some educational collaborators (Vaney, Morlé, Belot) persisted in using the instrument, none was in a strong position to develop intelligence testing further.

The equivocal response to the Binet-Simon version of intelligence also had roots in the seismic shift taking place in French high culture generally and French psychology specifically around the turn of the century. Celebration of the positivistic and scientistic that had characterized the early Third Republic was now strongly criticized by those proclaiming the "bankruptcy of science" and promoting reliance instead on the personal and arational.[185] Paul Bourget's most important novel and the literary sensation of 1889, *Le Disciple,* was an early example of this alternative approach to knowledge, society, and mind.[186] Initially a positivist and follower of Taine, Bourget manifested his "conversion" to Decadence in *Le Disciple,* particularly through his savage portraits of the scientistic and deterministic Robert Greslou and the positivist philosopher Adrien Sixte, modeled on Taine himself. In 1893 Maurice Blondel's *L'Action* took up the pragmatic philosophy of William James and celebrated action and faith over contemplation and reason, while Ferdinand Brunetière devoted a series of essays, including *La Science et la religion* (1895), to the limits of science and the profound importance of religion in addressing social problems.[187]

The most important representative of this cultural transformation within psychology was undoubtedly the intuitionist philosopher Henri Bergson, a star product of the French educational system.[188] Bergson was initially a committed Spencerian; however, he soon rejected the positivistic aspects of evolution and adopted instead—as explained in his most famous work, *L'Evolution créatrice* (1907)—an approach to the nature of mind that celebrated the intuitional and spiritual, particularly the *élan vital* (life force).[189] By the early twentieth century, Bergson had attracted an enormous following, although his direct impact on academic psychology was more muted. Probably Bergson's most significant influence on French psychology resulted from his support in 1901–1902 of the appointment of Pierre Janet over Binet to the Chair of Experimental and Comparative Psychology at the Collège de France. Bergson's advocacy was critical to Janet's success, and Janet's assumption of such an important position to the perpetuation of the pathological style of psychology in France.[190] In general, however, Bergson stood too isolated from institutional French psychology either to compel its reorientation or to create his own school. Rather, Bergson's popularity and the correlation between his ideas and broader intellectual trends helped to alter the context for psychological studies. Experimentalism began to seem less compelling, and interest renewed in introspective techniques, pathological methods, and case-study-style investigations.

Evidence for this sea change in French academic psychology can be found as early as 1911, with the publication of Nikolai Kostyleff's *La crise de la psychologie expérimentale.* A tutor (*maître de conférences*) employed by the Ecole Pratique des Hautes Etudes, Kostyleff attacked much of the work in French experimental psychology for the previous thirty years as fragmentary and overly concerned with individual capacities at the expense of mental phenomena themselves. Praising Binet's *L'Etude expéri-*

mentale as exemplifying the proper style for psychological research, Kostyleff declared that "the true goal of experimental psychology . . . [lies in understanding] the nature of psychic phenomena, of their localization and of the ties that bind them to the organism."[191] After the interruption of French psychological research by the war, the shift in direction to which Kostyleff pointed became clearly visible.[192] The Binet-Simon scale, especially in its 1911 version emphasizing quantitative classification, accorded poorly with this changed environment.

Probably most influential, however, in shaping the French response to the intelligence scale before the war were the institutional structures and the underlying ideologies that Binet and his followers confronted. First were the problems with the medical community. Binet's reliance on scholastic-sounding questions and an observation period that could easily extend to an hour made the 1905 scale an instrument for which the ideal operator was the psychologist, not the physician. Doctors had enormous social prestige in Third Republic France, and jealously guarded their many prerogatives.[193] Conceding diagnostic authority to a group of upstarts from the Faculty of Letters was not likely, and Binet was not institutionally positioned to prosecute such a battle successfully.

The same might also be said of Binet's relations with the educational hierarchy. For educators, Binet's outsider status was compounded by his importation of unfamiliar methodologies to judge educational practices. As Theodore Zeldin has noted, Binet's insistence on subjecting traditional pedagogical methods to experimental analysis and verification alienated numerous instructors. The antagonism was heightened in 1907, after Binet engaged in a bitter dispute with Jules Payot, editor of the teacher's journal *Le volume,* over whether pedagogical reform should be based on teachers' experience or experimentally derived fact.[194] Although Binet had supporters in education, the shift in decision-making from educators to psychologists that the Binet-Simon method entailed was unlikely to be broadly accepted.[195]

Moreover, at the ability spectrum's upper end, there were specific reasons why the Binet-Simon scale attracted little interest from French educators. By the late nineteenth century the educational system had already developed an elaborate system for winnowing the "best" from the school-age population and directing them to special institutions. The system of competitive examinations and elite schools selected and made available to the French state young people whose very success defined them as among the nation's most talented.[196] Although the percentage of primary-age children in the educational system increased throughout the century—finally reaching close to 100 percent after the law of March 28, 1882, mandated primary instruction for all—this expansion had little effect on the system's upper tiers and their effectiveness in producing the next generation of technocratic elite.[197] The Binet-Simon, associated as it was at least initially with detecting mental deficits, appeared to offer little of relevance to prosecuting this function more efficiently or effectively. In addition, as Theodore Porter has argued, the French administrative state's underlying ideology was opposed to the reduction of expert qualitative judgments to simple quantitative measures.[198] Thus it is not at all surprising that the intelligence scale attracted little interest from those involved in secondary or more advanced levels of instruction.

The Normal, the Abnormal, and the Binet-Simon

Openings for Binet and his tests did seem to lie at the intellectual spectrum's lower end, in identifying and dealing with individuals deemed unsuccessful in coping with education's rigors, the *anormaux* or *arriérés*.[199] Comprising individuals manifesting any of numerous physiological or psychological conditions—including idiocy, cretinism, imbecility, epilepsy, hysteria, blindness, deafness, intellectual weakness, laziness, extended absence from school, and instability—anormaux was a catchall classification for students who caused, or might cause, problems for the normal system of primary instruction and discipline.[200] Many educational commentators believed the 1882 law mandating universal primary instruction brought the "problem" of the anormaux to a head, because it guaranteed the anormaux an appropriate education and forced into the classroom children who might previously have been kept at home or allowed to perform simple manual labor.[201]

As a result, by the end of the century, articles wondering what to do with the anormaux in the classroom appeared increasingly frequently in the French educational press. The issue was partly couched in humanitarian terms, as how best to provide those struggling in school with the education they deserved. In large measure, however, it was framed as one of protecting the "*normaux*" from being held back or corrupted by this discordant element in the classroom. Thus, an article in the *Journal des instituteurs* from 1899 warned that "the presence of the *anormaux* is a cause of danger for their normal comrades," and the 1904 *Rapport de la Commission Spéciale sur la Création d'Ecoles pour les enfants anormaux et les indisciplines* advised that "the presence of one or two of these children [anormaux] suffices to contaminate an entire class."[202]

Unlike the United States or Britain, French culture was never consumed with panic over "the menace of the feebleminded." The French state was always more concerned with promoting births than preventing them, and the neo-Lamarckian understanding of evolution and heredity dominant in France predisposed those concerned with "improving the race," as William Schneider has demonstrated, to look more to "puericulture" (better heredity through improved prenatal care) than to the negative eugenics of sterilization programs for the "inferior."[203] Nonetheless, long-standing native worries about degeneration, depopulation, and the rise of various social pathologies from alcoholism to criminality to pauperism were quickly associated with the anormaux.[204] Some argued that such social ills, especially alcoholism, caused abnormality; others that the anormaux, if not properly trained and disciplined, would perpetuate these problems; and many adopted both positions.[205] When combined with the strong tendency during the Third Republic, as Robert Nye has argued, to medicalize social problems by treating them as ills awaiting diagnosis and cure, and with the commitment of many republicans to the philosophy of social solidarity (articulated forcefully by Léon Bourgeois) and equality as uniform treatment before the law, the government's eventual turn to the issue of the anormaux was probably inevitable.[206] Joseph Chaumié, the Minister of Public Instruction, responded to the confluence of these forces when he established the Bourgeois Commission (chaired by Léon Bourgeois) on October 4, 1904, and charged it to study how "the prescriptions of the law of 28 March 1882 on obligatory primary education could be applied to abnormal children of the two sexes (the blind, deaf-mutes, the backward [arriérés], etc.)."[207]

Consisting of twenty-one members (mostly educators or education officials, as well as Binet), the commission met repeatedly in late 1904 and 1905, and issued its report early in 1906.[208] Largely based on a subcommittee report that Binet drafted as reporting secretary, the Bourgeois Commission recommended that the anormaux be educated through *classes spéciales* annexed to ordinary primary schools and, in certain situations, through separate institutions to be known as *Ecoles de perfectionnement.* The commission emphasized that there should be "a material and complete separation between children of the normal group and those of the abnormal group," although the curriculum should parallel that of the normal primary school as much as consonant with the goal of developing "the social utility of the *anormaux.*" To determine which children should be removed from ordinary instruction, the report suggested that "an [examination] commission composed of an inspector of primary schools, a doctor, and a director of a special school" be established and that they do "a medical and pedagogical examination of any child, when requested by the family, teachers, inspectors of the primary schools, or physicians and directors of asylum schools." All children assigned to special education were to be examined by a doctor every six months, and pedagogical and personal progress were to be recorded at least every three months. Finally, as noted earlier, the commission also recommended that the minister have "a competent person draw up a scientific guide designed to facilitate subsequently the work of the examination commissions which must decide on the mental debility of a child."[209]

Thus, for all his efforts and enthusiasm, the role the report laid out for Binet in particular and psychologists in general was limited. The power to determine who was anormaux was ceded not to psychologists, but to a commission composed of physicians and school administrators, the dominant voices around Bourgeois. Binet was left solely with the carrot of developing a guide to aid such commissions. Binet ran with that opportunity, to be sure, and produced the first version of the Binet-Simon scale even before the report went public. By the time the legislature had enacted the law of April 15, 1909, on the education of the anormaux, however, even this slender reed was gone. Article 12, which specified how children would be selected for special education, simply duplicated the Bourgeois Commission's recommendation that the commission determining which children would and would not receive special instruction be composed of a physician, school inspector, and director of or teacher at an Ecole de perfectionnement. It made no mention of special methods for assessing students or a role for psychologists, highlighting instead the medical examination's central place.[210] Indeed, the only contribution envisioned for psychology was laid out in supplementary legislation, which mandated that the examination to become certified to teach arriérés children include a written test partly assessing knowledge of the "psychology and pedagogy of *arriérés*."[211]

Binet's failure to convince the Bourgeois Commission that psychology could vitally contribute to identifying the arriérés reflected not only Binet's personal standing and that of his fledgling discipline, but also a central feature of how French intellectual and institutional culture understood the normal and abnormal. In practice, anyway, the two were treated as fundamentally different states, each with its own experts and each analyzed and understood distinctly. Numerous pathologies or conditions were thought to be possible causes of abnormality, but as far as the education system was concerned, the etiology was insignificant. Blindness, idiocy, insubordination—all were reason to classify a child as anormaux and to demand his or her removal from the classroom, so that the education of the normaux would not be disrupted.[212] To be normal, therefore, was to be able to follow the daily scholarly regimen established by the Ministry of Public Instruction and the ever-burgeoning collection of educational laws, which mandated, among numerous other requirements, that classroom seating be continually adjusted to reflect relative performance on the latest examination.[213] Normal children were thus fundamentally the province of instructors and school officials, not experts on pathology.

The anormaux, however, were another matter. Since the beginning of the century, and right through the Third Republic, the profoundly impaired were unquestionably the concern of the medical establishment. The physician's duty, it was understood, was to detect pathologies of all sorts, to cure what they could, and to ensure that, whatever else happened, the problem remain contained.[214] Thus, severely abnormal individuals—the "incurables"—not kept at home were routinely sent to asylums, where doctors would oversee treatment and maintain custodial vigilance. By the turn of the century, the "incurables" requiring institutionalization included idiots, cretins, imbeciles, epileptics, moral imbeciles, and hysterics. Following Charcot, most physicians assumed that these very different pathologies could be traced to a similar cause: some physical lesion in the brain. Basic etiology aside, however, what was most apparent to doctors about the abnormal was its heterogeneity. Abnormal referred to a collection of very different pathologies, grouped together only because none of them was characteristic of the normal.

When the effects of compulsory education and the social worries about degeneration converged at the end of the century, in the form of concern about an invisible group of anormaux in the classroom, the arriérés, it brought together the differing perspectives of doctors and educators. Binet had hopes that his new instrument might exploit the tensions between these groups, by providing a method that combined the medical community's diagnostic orientation and interest in the pathological with educators' pedagogical approach. Instead, the law of April 15, 1909 maintained the conceptual separation of normal and abnormal, now applied to the arriérés themselves. Children able to be in primary school, though not to follow the standard curriculum, were to remain in school, in *classes de perfectionnement,* where they would be taught the same material as their "normal" comrades, though more slowly and in a way adapted to their needs. Those unable to be educated thus would be assigned to écoles de perfectionnement, separate institutions where they could develop the limited skills of which they were deemed capable in order not to become burdens on society. Constant surveillance through frequent medical and pedagogical examinations would ensure that the classifications were correct and would allow reassignments, typically when a student could no longer benefit from more scholastically oriented instruction.[215] The law's insistence on physical separation between ordinary classes and classes de perfectionnement and of both from écoles de perfectionnement symbolizes the conceptual distinction being drawn between normal and abnormal. While the project of educating the anormaux was founded on the belief that a child's intelligence would continue to develop and could be nurtured, both educators and doctors emphasized the very different needs and futures for these various groups.

Where physicians and educators saw differences in kind, however, Binet saw differences in degree. All children could be tested with the Binet-Simon scale, and the normal child's performance was not only a standard (barême) to judge who was deviant, as in the scholastic definition of the arriérés, but also by how much. An enfant anormaux was one whose pathology was conceived of in developmental terms, as not having reached the same level as his or her peers. Following Binet's lead, for example, Vaney declared in a 1908 article in Binet's pedagogical organ, the *Bulletin de la Société libre pour l'étude psychologique de l'enfant,* that arriérés were those at least three years behind their classmates scholastically and at least two years in terms of Binet's definition of intellectual development.[216] Binet insisted on this criterion numerous times himself, arguing that a clear standard was necessary to prevent classifications from differing wildly between examiners.[217]

Binet's push to standardize the anormaux's categorization, however, accorded poorly with the needs of those most engaged with understanding the abnormal in all its varieties. Physicians, accustomed to seeing the pathological as heterogeneous, were unlikely to find Binet's attempt to subsume all abnormality under the single standard of arrested intellectual development persuasive. The scale was simply not designed to differentiate among kinds of, to them, discrete conditions such as imbecility, cretinism, and epilepsy. At the same time, many instructors could see little practical point to an instrument that, as Binet himself admitted, produced classifications largely in keeping with those arrived at via scholastic and medical means, and so continued to rely on their own judgments derived from classroom familiarity with the whole child.[218] Despite Binet's insistence on the value of scientific assessments, many teachers and parents simply remained convinced that personal experience and determinations founded on months of observation were superior, a position consonant with the high valuation placed on expert judgment by the nation's technocratic elite.[219]

Both practically and theoretically, therefore, the Binet-Simon scale and the version of intelligence it embodied proved ill suited to the institutional and intellectual realities of French culture. As Monique Vial has argued, the republican ideology of most educators encouraged, from the start, reluctance to segregate children into special classes.[220] In addition, the legislators fashioning the 1909 law were concerned primarily with the extremely backward, the undisciplined, and those who were behind out of ignorance, not with those of weak intelligence.[221] The Binet-Simon scale's ability to disaggregate the normal and to make the slightly abnormal visible thus offered few attractions. To educators, such work was either already being done every day (in the case of the normal schoolchild) or should not be done at all, where invidious distinctions might undercut republican equality. And to physicians, the instrument's function of classifying the abnormal both threatened their professional prerogatives and was founded on a belief in the unity between normal and abnormal deeply at odds with mainstream practice and thought.

Conclusion

In his 1911 article on "*intelligence*" in Buisson's pedagogical dictionary, Roger Cousinet observed:

> An intelligent person is a person who understands. And there exists in this domain every difference in degree from the imbecile who does not understand anything . . . up to the very intelligent person who is capable of understanding ways of life and forms of thought much different from those to which he is accustomed.[222]

Intelligence as a characteristic existing in gradations encompassing every human mind had thus made its presence felt in France by the early twentieth century. Certainly, the Binet-Simon scale specifically, and French psychology more generally, did not accomplish this alone. The long history of French infatuation with neo-Lamarckian understandings of the scale of nature, which presumed a hierarchy of intellect, predisposed many intellectuals to find gradations within humanity plausible, as did the comparative method emphasizing gradations between the pathological and the normal. In addition, the French educational system, concerned above all with the formation of a talented, homogeneously trained elite, embodied the principle that all minds were comparable and that some were distinctly superior. The origins of this superiority, however, were of little import to pedagogues, who were concerned primarily with its practical manifestations.

Such was not completely the case, however, for French psychologists. Starting with the notion of group-level global intelligence articulated by the craniometricians, Binet and his coworkers applied that concept to individuals by means of their new techniques for revealing difference. Intelligence became both an inherent feature of individual minds and a way of categorizing those minds. It thus united the long-standing and independent classifications of idiot, imbecile, and genius with the ill-defined default state of everyone else, now disaggregated into a spectrum of normalcy, so that an unbroken continuum of intelligence could be posited. The data from examinations using the Binet-Simon scale showed the power of this approach: with sufficiently large groups of examinees, subjects did fall into a nicely stratified array that matched independent judgments of overall intelligence. When French psychologists observed individuals more closely in case studies, however, global intelligence began to dissolve, separating into a collection of strengths, weaknesses, and characteristics that even individuals at the same measured level of intelligence could manifest in an extraordinary variety of ways. Intelligence became more complex, and for many French psychologists a subject of investigation interesting in its own right. Most of their American colleagues, however, would never take this lesson fully to heart.

Five

American Psychology and the Seductions of IQ

In 1901 the first edition of Princeton psychologist James Mark Baldwin's *Dictionary of Philosophy and Psychology* appeared. A massive two-volume affair, it contained elaborate explanations of most of contemporary psychology's central terms. About the word "intelligence," however, the dictionary was surprisingly reticent. No separate entry existed; rather the term was listed simply as a variant of "intellect," defined as "the faculty or capacity of knowing; intellection or, better, COGNITION denotes the process."[1] Thirteen years later, however, the Binet-Simon Intelligence Scale alone generated a 254-item bibliography, with another 457 items added four years after that.[2] Something extraordinary had happened to "intelligence"; in a flash it had become a term of central importance within American psychology and, to a certain degree, American culture. Out of a variety of different ways of conceptualizing intelligence in play at the beginning of the century, one dominant theme had emerged. Intelligence was understood as a differential, quantifiable, unilinear entity that determined an individual's or group's overall mental power. And just as importantly, this new understanding of intelligence also defined the professional identities of what became one of psychology's largest practitioner subgroups—a shift that set the stage for intelligence's emergence during World War I and thereafter as a concept able to knit together notions of merit and mind in a variety of settings.[3]

The attraction of this metric version of intelligence for so many American psychologists lay in its potential as a new psychological object, one able to accomplish the very practical task of differentiating human beings on seemingly scientific grounds. Over the nineteenth century, the meaning of "intelligence" had narrowed, as certain senses ("news," "knowledge," "divine being") dropped out of common parlance, and intelligence as "ability" rose. By the early twentieth century this more restricted denotation was accorded a new explanatory power. Drawing on notions about difference developed in psychology and anthropology, many scientists argued that intelligence was a measurable, statistically distributed, hereditary trait, biological in origin, that characterized the nature and value not only of groups, but also of individuals. Criminality, poverty, and vice, they contended, derived less from personal moral failure than congenital mental inadequacy; moreover, an individual's intelligence could be used to explain that person's actions and place within the social order. The plausibility of such physicalist explanations for behaviors was strengthened by the growing acceptance of Darwinian and Spencerian evolutionary theories, by the belief that physical stigmata indicated mental disorders and criminal tendencies, and by scientific justifications for racial and sexual discriminations.

The context within which the meaning of intelligence was remade, however, was in no way defined solely by the needs of the new scientific psychology for its own investigative objects. Confronted with the late nineteenth century's immense social transformations—including industrialization and its consequent urbanization, record immigration from central and eastern Europe, emancipation of formerly enslaved African Americans, and the continuing expansion of the electorate—many members of the middle class in particular believed that elements of mass society were developing around them, and they sought new methods for understanding and assessing this changing social world. Convinced that the ongoing industrial/technological revolution had proven the value and power of scientific approaches to problem solving, the managerial, bureaucratic, and professional classes (including but by no means limited to self-identified Progressives) turned to science, sure that it could provide as well the means of comprehending society and its inhabitants.[4] The period's dramatic growth in the human sciences was one manifestation of this faith; the vigor with which many advocated objective, technocratic solutions to social problems was another.

These moves toward scientific or naturalistic explanations for the behavior of those deemed "other," however, also raised a host of concerns about society's future—the flip side of the era's obsession with progress. Worries about degeneration pervaded Europe and America, underscoring the idea that nations possess better and worse biological stocks, with the weaker feared ascendant.[5] At the same time, the language of medical pathology was increasingly employed to understand social problems, privileging thereby analyses rendered in naturalistic terms. The culmination of this wedding of anxiety and biology lay in turn-of-the-century eugenics. Coined by Francis Galton in the 1880s to describe the need for active intervention into a population's breeding patterns, eugenics implied that biology determined quality and that a civilization's success depended on enhancing reproduction of the "best" elements and retarding that of the least desirable. Widely accepted in Britain and the United States, eugenics had a significant impact throughout Europe and the Americas, meeting a perceived need for objective markers able to define superiority or inferiority, and thus to justify existing social hierarchies.[6]

For American psychology, one consequence of these shifting grounds for understanding human nature and society was to encourage a more biological orientation toward mental phenomena. Led by William James and his student G. Stanley Hall, academic psychology turned toward laboratory experimental science for its definition, if not always its practice, emphasizing the physiological investigation of mind. The implications for understanding the intellect were not, however, immediately clear. While many assumptions supporting the long-standing picture of mental abilities, based on Common Sense philosophy, were questioned, no widely accepted alternative immediately replaced them. Most psychologists continued to study the mind as a collection of elementary processes, now investigated via laboratory techniques and explained in terms of associationism rather than independent faculties. Some, to be sure, did employ the notion of an individual's overall mental power, but before 1910 few empirical methods were available to investigate such an entity, and none that met with much success.

In 1910, however, Henry H. Goddard published an article recounting his experiences with the 1908 Binet-Simon Intelligence Scale, offering American psychology a new way of generating data about higher-order mental functions. Goddard—psychologist at the New Jersey Training School for Feebleminded Girls and Boys in Vineland—had a seemingly ideal population for seeing intellectual ability as a simple, linear capacity.[7] While "normal" adults were thought to manifest complex sets of abilities and deficits, the feebleminded were defined solely by the level of mental ability they did, or rather did not, exhibit. In a sense, the same was also true of children, who were readily arrayed along a developmental continuum, with each stage considered superior to its predecessor. Moreover, both the feebleminded and children were housed in institutions—asylums and schools—that gridded their subjects according to intellectual attainment, and were sites where more precise-seeming classificatory methods could find a receptive audience. To the degree that the Binet-Simon scale and its variants could capture such long-standing (if often contradictory) medical, legal, and educational categorizations and thereby make differential intelligence unambiguously visible to experts and lay observers alike, it could provide American psychologists with a concrete instantiation of metric intelligence and a means of extending that construct to new arenas.

Given early twentieth-century American psychology's strongly empirical orientation and its interest in practical applications of expertise, the seductions of metric intelligence ultimately proved irresistible.[8] Although some psychologists, such as, notably, Edward L. Thorndike, were skeptical about this version of intelligence and pushed alternatives, most saw in metric intelligence a language and technology of comparison that made plausible their claims of authority over individual assessments and determinations of merit, including occupational placements and educational tracking. This notion of intelligence was also readily used to explain deviance, degeneration, and similar "pathological" manifestations of the "other," allowing psychologists to speak directly to such contemporary public anxieties as the "menace of the feebleminded," unrestricted immigration, and the propagation of the unfit.[9] Psychologists' success in linking instruments such as the Binet-Simon to populations who could make global intelligence visible and to various social concerns—including detection of biologically threatening social groups and the efficient organization of the school and workplace—helped to propel this version of intelligence to the profession's forefront.[10] The emergence of the Stanford-Binet, developed by Lewis M. Terman in 1916, as *the* benchmark instrument for measuring intelligence simply consolidated metric intelligence's position within the profession. That test's success, along with the coining of the term "IQ," crystallized ideas about the meaning of intelligence and its relation to merit, and established the *assessment* of intelligence, rather than its *investigation,* as a major pursuit of American psychology and a growing interest of American culture.[11]

Setting the Stage: E. L. Youmans and the Culture Demanded by Modern Life

In 1867 Edward L. Youmans—editor of *Popular Science Monthly,* noted Anglophile, and enthusiastic disciple of Herbert Spencer—wrote an obituary for the Common Sense philosophy. Characterizing it as an approach in which "living reality, as a subject of study, disappeared from view," Youmans explained that "there remained only *mind as an abstraction,* to be considered as literally out of all true relations as if the material universe had never existed." In Youmans's view, understanding real, complex individual minds should be the fundamental task of a properly oriented "scientific" philosophy. Charting a path similar to that of Hippolyte Taine and Théodule Ribot in their turn to associationism, Youmans too adopted the English philosophy and emphasized humans' material reality. "Man, as a problem of study," he observed, "is simply an organism of varied powers and activities; and the true office of scientific inquiry is to determine the mechanism, modes, and laws of its action." Throughout his essay, Youmans sought to represent the mind as so intimately tied to physiology that traditional mind/body dualism—which he thought essential to the Scottish position—would lose its plausibility, and human nature, fully biologized, would seem directed by the same natural laws governing the rest of the physical world.[12]

Neither a psychologist nor a scientist of any sort, Youmans contributed little directly to the new approaches that would eventually supersede Common Sense. But, as editor of *Popular Science Monthly*—late nineteenth-century America's most important and widely read nonspecialist science journal—he was uniquely positioned to advocate the infusion of science into American culture. And so Youmans filled his journal with articles trumpeting the latest scientific advances, reprints of essays from British scientific journals, and, especially, pieces touting Spencerian biological evolution. Given the profound social and cultural changes wrought by the Civil War, the era's dazzling material and technological progress, and the emergence of highly rationalized forms of bureaucratic corporate capitalism, Youmans's scientism proved an appealing idiom for making sense of late nineteenth-century America.[13] Indeed, with the rapid acceptance of Darwinian or Spencerian theories of biological evolution, under the rubric of social Darwinism both defenders and critics of Gilded Age society scrambled to appropriate biological science to justify or vilify a range of business practices and social policies.[14] At the same time, the period's intense labor strife, which many middle-class Americans associated with the millions of new immigrants from eastern and southern Europe, ignited a nativist backlash that also turned to biology to explain the threat the immigrants constituted, connecting the newcomers in the process with similar worries about the newly freed and enfranchised African American population.

Youmans himself took aim at the traditional liberal-arts curriculum, which, lacking a strong scientific foundation, he regarded as irremediably defective.[15] Youmans was not the first to advocate changes in collegiate education. From the early nineteenth century, the classical curriculum—with its emphases on ancient languages and mathematics—was routinely challenged, particularly by those calling for greater concentration on science and modern languages.[16] The simultaneous creation in 1847 of the Lawrence Scientific School at Harvard and what would become the Sheffield Scientific School at Yale marked not only a concession to these pressures, but one of the first stages in the remaking of postsecondary education.[17] The process accelerated after the Civil War. The Morrill Act of 1862 established land-grant colleges to promote agriculture and the mechanic arts in each state; Harvard president Charles W. Eliot implemented the first system of elective courses in 1874; and Johns Hopkins University was founded on the "German model" in 1876. Soon, many schools modified, or abandoned, the one-size-fits-all model of the classical curriculum. The number and range of courses expanded, especially in the sciences, and universities began requiring advanced training of instructors and granting more graduate degrees. Consequently, by century's end American colleges and universities provided far more opportunities for advanced teaching and training, including in new fields such as psychology, than at any time previously.

Underpinning these structural changes in higher education was the appearance of a secular, scientific source of authority to challenge the dominance of evangelical Protestantism. The vogue for analysis and action anchored in scientific principles was widespread, reflected throughout *The Culture Demanded By Modern Life* and in many characteristic features of contemporary intellectual culture—including Progressivism, eugenics, social Darwinism, scientific management, the settlement-house movement, the social research survey, and even Social Gospel Christianity.[18] Antebellum America's personalized moral universe—whose guiding principles included faith in hard work, honesty, integrity, temperance, and divine justice—did not, of course, disappear in the post–Civil War period, nor had the values of science and technology been ignored earlier. Rather, their relative positions of authority shifted. Postbellum Americans by and large balanced their Christianity with increased appreciation for the values and concerns associated with empirical science. Thus emerged a Protestant or moralized scientism, whose critical feature was its support for scientific inquiry while still sustaining traditional morality and religion.[19]

The reverberations of this new modus vivendi, especially for questions that touched on the nature of human beings, were pronounced. Charles Darwin's publication of *Origin of Species* in 1859 and then *Descent of Man* in 1871, of course, initiated shock waves that spread throughout the Atlantic world. The theory that humans had evolved from "lower" animals and had done so not according to a divine plan but through the random actions of natural selection acting on minute variations alarmed many in the older generation of America's cultural and scientific leaders, seeming to them both empirically suspect and morally unconscionable. Louis Agassiz, Harvard professor and perhaps America's most prominent natural historian, led the scientific attack on Darwinism on the grounds of its empirical insufficiency, while moral philosophers such as John Bascom of Williams College decried Darwin's theory for its rejection of the evidence that nature manifested the handiwork of an intelligent Creator.[20] Particularly problematic for virtually all the critics was the implication that human "mental and moral powers must be shown to be," in Bascom's words, "identical in kind with those of the brute." Although humans had long been recognized to be part of the natural order, they had also been conceptualized as simultaneously separated from it, if not in body then certainly in mind. To most commentators, Darwinian evolution seemed to threaten this special status, and thus to be open to be "used by atheistic thinkers for atheistic ends."[21]

Nonetheless, through the efforts of a younger generation of scientists such as Asa Gray, also of Harvard, and Edward Drinker Cope, and social commentators including Youmans, John Fiske, and Chauncey Wright, evolutionism in one form or another quickly became an established, and in some circles dominant, way of understanding the natural world's dynamics.[22] Yet, while typically this meant adopting unreservedly the notion that human minds were as much part of the natural order as bodies, the theory's atheistic implications were blunted by underplaying evolution's random character in favor of a more Spencerian, progressive version, which left open the possibility that the evolutionary process might be the expression of a divine will acting through natural processes.[23]

At the same time as evolutionary theory was roiling the intellectual waters, other scientific work also succeeded in accounting for many human physical and mental features in naturalistic terms. Most notable were the statistical approaches to human characteristics associated first with the Belgian astronomer Adolphe Quetelet.[24] Quetelet's concept of *l'homme moyen* and famous demonstration in 1844 that the chest measurements of 5,738 Scottish soldiers followed the law of errors and were distributed in a bell-shaped curve became widely touted as exemplifying *the* method for the determination of quantitative truths about human populations.[25] Quetelet's findings that human physical traits not only varied, but did so in a predictable way, were widely reported and much commented upon, especially in publications such as Youmans's *Popular Science Monthly*.[26] While Quetelet made some moves toward suggesting that mental and moral attributes were distributed in a population in the same way as human physical characteristics, others, such as Emile Durkheim in France and Francis Galton in England, went much further. Durkheim argued that statistical regularities such as the constancy in the annual number of people who committed suicide by various methods in France could form the basis for an entire science of society, one that emphasized the predictability even of moral phenomena at the level of populations, however random individual actions might be.[27]

Even more influential in America was the work of Darwin's cousin Francis Galton, who turned his obsession with quantification into a program of applying Quetelet's distribution law to a range of human physical and mental measurements. In his two major works, *Hereditary Genius* (1869) and *Natural Inheritance* (1889), Galton sought to demonstrate that some people (like himself) were naturally more talented than others, that this talent, like any other biological characteristic, was inherited, and that talent was distributed in a population according to Quetelet's bell-shaped error curve.[28] His goal was to use science to explain a social phenomenon: why some people succeeded and others did not, and to argue that individual success was largely a product of biology rather than of effort or environment. Inspired by Galton's work, William James and the Spencerians Grant Allen and John Fiske engaged in a spirited debate in the 1880s over the origins of individual greatness.[29] Both sides agreed that biology was the key factor in the production of great individuals, but they split over the role of society: James emphasized the power of the extraordinary individual to shape social evolution, while Allen and Fiske hewed more closely to Galton and accounted for such individuals as predictable deviations from the mean characteristics for the nation as a whole.

During the 1880s and 1890s, this growing interest in biological approaches to understanding human populations was variously manifested in America. When debating the nature of women's abilities, it was known as the "woman's question"; it became the "Negro problem" when analyzing the prospects for postemancipation African Americans.[30] Framings emphasizing a group's inherent natural potential were also central to the discourses of "race suicide," employed by nativists alarmed about the new immigration from eastern and southern Europe and Asia, and imperial paternalism, invoked by expansionists to justify maintaining the dependent status of indigenous peoples of newly annexed territories.[31] And cutting across all of these, of course, was social Darwinism. What these formulations shared was the focus on presumed biological nature as the key to a group's place within the social order. Conservative social-Darwinist Charles Sumner, for example, extolled capitalist competition as a way of separating the biological wheat from the chaff—notoriously condemning aid to the poor as a program for maintaining the unfit to the detriment of the population as a whole.[32] Opponents of higher education for women routinely cited the presumed greater fragility of a woman's physiology—particularly the injurious stress that would be placed on her reproductive system—as a reason for limiting the kinds of intellectual training that women received and the roles in society they should fill.[33] Feminists themselves often sought to counter such positions by employing their own biologically framed arguments. Turning to the language of evolution, Charlotte Perkins Gilman, for one, contended that society's progress would actually be greatly enhanced if women were no longer "protected" from the competitive pressures of higher education and the workplace, and instead allowed to evolve along with men by developing their inherent potentials to the fullest.[34]

While not completely absent from antebellum discussions, especially around issues of race, this recourse to biology to ground social claims became a marked feature of late nineteenth-century American culture.[35] Appropriated by commentators of all political stripes, the language of science, particularly biology, was applied to a range of social issues, used to suggest that nature itself could provide indisputable answers to vexing moral questions; thereby, the conviction that human beings and their institutions must be subjected to empirical investigation was underscored.[36] Early on, Youmans's call for scientific study of human nature and his emphasis on the individual—body and mind—as biological organism captured central features of this orientation. A decade later, the challenge of extending the laboratory's purview to that most morally freighted of arenas, the human mind, was taken up by a new generation of self-avowedly scientific philosophers, when William James, G. Stanley Hall, and George Trumbull Ladd, among others, fought to establish a new physiological psychology in America.[37]

Between Philosophy and Physiology: The "New Psychology" in America

"The new psychology," G. Stanley Hall declared in 1885,

> which brings simply a new method and a new standpoint to philosophy, is I believe Christian to its root and centre; and its final mission in the world is not merely to trace petty harmonies and small adjustments between science and religion, but to flood and transfuse the new and vaster conceptions of the universe and of man's place in it . . . with the old Scriptural sense of unity, rationality, and love beneath and above all, with all its wide consequences.[38]

Save for the reference to "the new psychology," there is little in Hall's characterization to distinguish his pursuits from those of the mental philosophers who had been teaching in American colleges for the preceding century.[39] And yet Hall, his teacher William James, and numerous other late nineteenth-century self-professed "scientific psychologists" were convinced that the approach to mind that they advocated represented a radical, and necessary, departure from the American Common Sense tradition.[40] There was, no doubt, an element of self-serving rhetoric in their protestations, one consequence of their ongoing struggle to convince universities to establish faculty positions and ultimately whole departments dedicated to psychology as a discipline.

Nonetheless, James and Hall and the new psychology's many other advocates were not just engaged in self-promotion. Taking up, however unwittingly, Youmans's challenge to establish a new kind of mental science, the new psychology's advocates looked to the physiological laboratory as a model for transforming psychology into an experimental science. New methods and instruments in a new setting, however, also meant defining appropriate objects for analysis, ones amenable to the rigors of laboratory measurement by professional scientists. Thus, in creating the discipline of psychology out of mental philosophy, practitioners had to reconceptualize the focus of their investigations—whether mind or behavior—as well. James helped initiate the process in America, drawing on techniques and approaches developed in Europe; others quickly followed suit in a more thoroughgoing manner, slowly replacing the Common Sense mind with one whose structures and functions were amenable to empirical investigation, or, among behaviorists, with individual actions untroubled by notions of mind at all.

No separate academic discipline of psychology existed when James first became attracted to problems of mind.[41] Like many of his contemporaries intrigued by such questions, James pursued his interests by working at the boundaries between two distinct domains: philosophy and physiology.[42] Philosophy's relevance to the study of human nature was obvious: for centuries philosophers had speculated about the mind and its operations, and mental philosophy was already entrenched in the American college curriculum. However, the problem with traditional mental philosophy for the scientifically inclined, as Youmans revealed, was its indifference to empiricism, especially evident in the scant attention it paid to the physical aspects of cognition and feeling. Physiology, on the other hand, had contributed little to understanding mental processes, but employed experimental and laboratory techniques and seemed able to explain many fundamental features of the brain and nervous system.[43] Thus, for those seeking to set philosophy on a more scientific footing, physiology proved an alluring starting point.

Although he had formal training only in medicine, by the mid-1870s James had acquired a respectable command of physiology, thorough understanding of Darwinian evolution, and passing familiarity with contemporary philosophical debates.[44] He had also absorbed the postbellum infatuation with biological approaches to human nature, and determined to investigate psychological issues from an avowedly scientific perspective. In 1880 this orientation received institutional sanction with his appointment as an assistant professor in Harvard's philosophy department, charged with instruction in physiological psychology.[45] From the outset, in his Lowell Lectures of 1878, James insisted on the close connection between physiology and psychology, advancing an associationist interpretation of the mind's operations rooted in the brain and nervous system's association of sensory impressions with ideas.[46] His position had scarcely altered with publication of his landmark *The Principles of Psychology* (1890) or his 1892 article "A Plea for Psychology as a 'Natural Science,' " where he defined psychology as the "science of the correlations of mental with cerebral events."[47] Yet, unlike some of his more reductionist colleagues, James never rejected out of hand psychological theorizing based on speculative philosophy. Endeavoring always to keep his psychology free of the imputation of materialism or mechanical determinism, James remained open to the possibility that truth could be wrung from a variety of sources.[48]

James's moderate stance lent his work an eclectic feel that in many ways represented American psychology.[49] Virtually all American psychologists agreed on the importance of approaching mind from a biological perspective; almost everyone also believed that at least some laboratory experimental work was essential for training a modern psychologist, if not the sine qua non defining psychological investigation entirely.[50] Beyond that, however, widely varying styles and techniques prevailed. Some, including Edward W. Scripture at Yale and Edward B. Titchener at Cornell, hewed to the Wundtian line. Interested in the mind's fundamental content and structures, they collaborated on laboratory-based experiments with a few well-trained assistants, generated measurements with precision instruments, and often relied on introspective reports as either experimental data or interpretive aids.[51] In their approach, the human mind was an object characterized as having universal features, and psychology's role was to elucidate those elements through intensive investigations of the most basic processes.

Others—including the Chicago functionalist school of John Dewey and James R. Angell, Hall or Baldwin when pursuing genetic psychology, and the learning theorists Edward L. Thorndike and Robert S. Woodworth—were concerned more with mind-in-action, as it was termed, than mind-in-content.[52] Although bound to no single style, American functionalists tended, as Kurt Danziger has argued, to base their studies on large numbers of subjects and to investigate individual variations in psychological processes; the development and distribution of mental functions, he notes, rather than their analysis, was the fundamental focus for such practitioners.[53] They were also particularly concerned with practical applications, especially those relevant to the school, asylum, or workplace. Dewey sought to resolve many of Chicago's most pressing social problems through the application of his functionalist pragmatic psychology; Hall was a guiding force in the Child Study movement; Hall's student Goddard spent his career working with the feebleminded; and Thorndike took a position at Teachers College, Columbia, where he trained a generation of educational psychologists, almost single-handedly inventing the subfield. "All natural sciences aim at practical prediction and control," James observed, "and in none of them is this more the case than in psychology to-day. . . . What every educator, every jail-warden, every doctor, every clergyman, every asylum-superintendent, asks of psychology is practical rules."[54]

As John O'Donnell has persuasively argued, this turn to functionalism—with its strong interest in the practical—became a dominant feature of American psychology by the turn of the century.[55] Eschewing the strict version of brass-instrument experimentation associated with Wundt, Titchener, and the other structuralists, functionalist psychologists embarked on various research programs that adopted the evolutionary development of mind as their starting point. Educational, comparative or animal, and individual psychology all flourished, having as their object the investigation of consciousness as an active agent seeking to adapt an organism to novel situations in its environment. "Instead of focusing on mental contents," O'Donnell observes, "functional psychology would examine mental activity defined as the acquisition, retention, organization, and evaluation of experiences that, in turn, were viewed as guides to adaptive behavior."[56]

Nonetheless, for all of this interest in broadly defined, comprehensive mental functions, American psychologists, as Baldwin's dictionary suggests, at first paid little attention to intelligence. The index to James's *Principles,* for example, had only two short entries for the term, and the word itself appeared only about fifty-five times in the text, referring mostly to "consciousness" or "understanding" (twenty-eight times), or overall mental power existing in degrees, particularly with regard to animals (twelve times).[57] E. W. Scripture's 500-page polemic advocating a strictly experimental approach to psychology, *The New Psychology* (1897), went James one better, failing to mention intelligence even once in its index.[58] Higher mental functions, whether called intelligence or talents or faculties, were notable mainly for their absence from late nineteenth-century American psychology. If pressed, of course, most functionalist psychologists in particular would have acknowledged that human beings were thinking creatures and that understanding the more sublime forms of reasoning was among psychology's urgent tasks. Yet that impetus alone generated few research programs devoted to the mind's higher capabilities.

The first impediment to the psychologizing of "intelligence" was the word intelligence itself, still used during this period most often to refer to consciousness in general or to signify a group or species-level mental attribute extending along the entire developmental scale.[59] Concerned with the mind's operations in normal adults, most American psychologists had little use for a term wedded to such a global characteristic of mind. The one subfield that did use intelligence in any significant way was comparative or animal psychology, perhaps the only area where group-level differences were ontologically significant.[60] As pursued by the Briton C. Lloyd Morgan or Americans such as Thorndike, John B. Watson, and Robert M. Yerkes, comparative psychology relied on the evolutionary principle of continuity and connection among organisms to search, as Darwin had in *The Expression of the Emotions in Man and Animals* (1872), for basic features of mind that could be found throughout the animal kingdom.[61] Intelligence had long been regarded as one of the principal characteristics both uniting and differentiating various animal species, and thus was of particular interest to many animal psychologists, especially those conducting research on learning.

The second reason for the disinterest in the higher mental powers among most psychologists is directly traceable to the rejection of Common Sense philosophy. The new psychology was to be based, its proponents contended, not only on embracing experimental investigation, but also on rejecting the metaphysics of traditional mental philosophy.[62] And the primary aspect of the old philosophy to avoid was mention of faculties, powers, and the like, entities acting without a defined material base. James was vehement on this point: "Psychological analysis," he declared, "does not deny supreme importance of particular localities for expression of faculty but it shows how each faculty is built of images, motor and sensory, combining in peculiar proportions, and it assumes that the elementary functions of the brain are the production of these images."[63] Most of his colleagues concurred, adopting a modified version of British associationism that left little room for mental powers besides perception and memory.[64] To James and most other new psychologists, the mind acted as an integrated whole, expressing specific capabilities only as the particulars of the situation warranted. From this perspective, "faculties" and "powers" not only appeared fictitious, but also retained spiritualist connotations that hinted at an imperviousness to empirical analysis.

Thus most research in psychology, whether conducted by structuralists or functionalists, tended to focus on measuring elementary mental processes, particularly sensory perceptions. Such basic abilities constituted, psychologists believed, the fundamental elements out of which higher processes were constructed. For Wundtians, the focus on basics was a matter of pride; as Titchener advised Francis Galton in 1894: "To get entirely rid of any hint of a 'faculty' theory, we avoid now such terms as 'imagination,' & even 'memory.'"[65] Many functionalists were less sanguine about this state of affairs, but felt limited by the experimental equipment and practices available.[66] "Thus far," Hall lamented in 1885, "[these methods have been] chiefly applied to the study of elements fundamental to consciousness rather than to its more complex processes."[67] However, no one had yet devised any reliable means of exploring the higher processes, and the problem seemed technically daunting. How could an instrument measure the complex interplay of sensations, associations, and manipulations that went into reasoning? And without the ability to do that, how could higher-order functions such as intelligence, even if real, ever be subjected to the experimental protocols necessary to make them meaningful within empirical science?

The new psychology thus stood at something of a crossroads as the century approached its final decade. While the institutionalization of psychology as both an academic discipline and even a field of practical applications was proceeding apace and the methods appropriate to it as a scientific endeavor were slowly being formalized, in important ways its objects of investigation remained unsettled. Structuralists were clear enough that absolutely elementary processes must constitute the core of the discipline; the expanding group of functionalist-oriented psychologists, however, were less certain, convinced that many important aspects of an individual's adaptation to his or her environment and of variation among individuals could not be explained on the basis of elementary sensory perceptions alone. Yet, finding ways to investigate scientifically the higher-order mental functions proved to be no easy task. Hall thought he had found one answer with the questionnaire method, which he used in the 1890s and early 1900s when conducting his research for the child study movement. But the amassing of thousands of anecdotal reports on various aspects of child behavior by untrained observers seemed to most psychologists to be thoroughly unscientific, more an example of Baconianism gone wild than professional research conducted according to rigorous procedure.[68] Other psychologists believed that the development of a new homegrown experimental apparatus, the mental test—designed for use with large numbers of ordinary subjects—might provide just the new kind of scientific instrument necessary to explore the more rarefied aspects of the mind. This approach, too, would be found wanting, although it would help establish the possibility that comprehensive mental functions such as intelligence might be viable psychological objects open to investigation and measurement.

Cattell, Galton, and the Rise and Fall of the "Mental Test"

In 1890 James McKeen Cattell published his first article as a new assistant professor of psychology at the University of Pennsylvania. The article, "Mental Tests and Measurements," was an auspicious beginning.[69] A description of ten primarily anthropometric tests that Cattell had administered to Penn undergraduates, it both coined a new term, "mental test," and sparked interest in measuring phenomena such as reaction times, sensation areas, and letter memory to explore the mental functions and their variations. Cattell's goal was to translate the methods of physiological psychology into procedures that would allow him to investigate how individual minds varied. Quickly adopted by a number of other members of the burgeoning community of academic psychologists, mental or anthropometric testing seemed full of promise as a method for investigating inter-individual differences in the higher mental functions according to the strictures of experimental science. By the decade's end, however, such optimism evaporated, as the program of measuring basic mental processes explored through the performance of "normal" subjects was shown to have little statistical relation to activities demanding more complex cognition, such as success in school.[70] In the meantime, for all of the interest in individual differences, the largely Galtonian approach that Cattell and others adopted during the 1890s—one emphasizing the use of a variety of measurement and statistical techniques to generate and manipulate copious amounts of data about variations in sensory and motor capacities—served mostly to constitute groups as entities about whom psychological knowledge could be generated. With the focus on groups, attributes particularly associated with group-level differences—including race, gender, and degree of intelligence—became prominent features of psychological research, and intelligence understood as overall mental power was rendered into an object amenable to investigation within a scientific psychology.

Like many of his peers, Cattell followed a circuitous path to the study of mental functions.[71] Attracted to philosophy as an undergraduate—though more drawn to the Comtean positivism he read on his own than to the Common Sense he was taught in the classroom—upon graduation in 1880, Cattell made his way to Europe, where he attended the lectures of the Göttingen philosopher Hermann Lotze and, after Lotze's death, those of Wundt. Awarded a fellowship in philosophy at Johns Hopkins in 1882, Cattell dedicated himself to laboratory work, first with H. Newell Martin in physiology, and later with Hall in Hopkins's newly established psychology laboratory. Denied a second year of the fellowship (it went instead to John Dewey), Cattell returned to Leipzig to continue training with Wundt. Although Cattell mastered Wundt's experimental apparatus during his three years in Leipzig, he never accepted Wundt's precept that experimental psychology center on introspective analysis.[72] Rather, true to his early infatuation with Comte, Cattell preferred the unadorned generation of quantitative data produced under controlled conditions.

Cattell left Leipzig for good in 1886, passing most of the next two years in England, where he came under the spell of Francis Galton. Much has been made of Galton's influence on Cattell and, by extension, on the mental testing movement in America. In Michael Sokal's view, Galton "provided [Cattell] with a scientific goal—the measurement of psychological differences between people—that made use of the experimental procedures he had developed at Leipzig."[73] Certainly Cattell was enormously impressed by Galton, Darwin's cousin and a polymath obsessed with quantifying human characteristics and analyzing human difference. By 1886 Galton was already famous for his 1869 work on hereditary genius—in which he argued that brilliance ran in families and was as much a biological trait as any feature of the body—and would soon also become known as a founder of eugenics.[74] During the 1880s Galton was especially concerned with collecting anthropometric measurements on the English population, planning an immense study of human variation. Galton's approach was thoroughly statistical, and concerned more with aggregating data at the level of groups than with the meaning of differences for particular individuals.[75]

Cattell's own work under Wundt had already been moving toward the analysis of variations, helping to explain, no doubt, why Cattell found Galton's approach so attractive. Galton's influence was more direct regarding the anthropometric techniques themselves: as Cattell explained in his 1890 article, he followed "Mr. Galton in combining tests of body . . . with psychophysical and mental determinations." Starting with dynamometer measures of the pressure exerted during a hand squeeze, Cattell's sequence of tests went on to measure, among other phenomena, threshold perceptions (just noticeable differences) in skin sensation, reaction time to sounds, accuracy of judgment of time intervals, and maximum letters recalled on one hearing. In all such tests, Cattell stressed the importance of "experiment and measurement" as the keys to making psychology into a science, emphasizing that "a uniform system be adopted, so that determinations made at different times and places could be compared and combined."[76] Perhaps most important was the lesson Cattell absorbed from Galton about how to define the scope of investigations. Where Cattell had worked with a limited set of subjects in Wundt's laboratory, Galton envisioned hundreds if not thousands of participants, drawn from a range of backgrounds.

When Cattell returned to America in 1889, he brought with him this appreciation for statistics and sense of how to define a subject pool, sensibilities that informed the mental-testing research he would subsequently carry out.[77] After inventing the new subdiscipline in 1890, Cattell's most important contribution was research conducted in collaboration with Livingston Ferrand on one hundred Columbia University undergraduates, published in 1896. A compilation of anthropometric and mental measurements, including height and weight, size of head, keenness of sight, sensitivity to pain, reaction time, and memory, the data were presented as composite means for the entire subject group along with two measures of variation (the average variation of individuals relative to the subgroup in which they were placed, and the average variation of the subgroups relative to the entire group). Although Cattell and Ferrand declared that their data allowed an individual to be ranked vis-à-vis other members of the group, they provided no such rankings themselves. In addition, they made no attempt to compare their group with any other, or to investigate more complex mental functions. Their emphasis was on the practicality and precision of the techniques they used, which led them, they explained, to reject analyses of attention or suggestibility or other higher-order processes, because suitable tests had yet to be devised and validated in the laboratory.[78]

In all but one respect, most studies within the anthropometric mental-testing tradition followed this pattern. In 1891 Joseph Jastrow analyzed the responses of fifty University of Wisconsin undergraduates to the task of writing down the first one hundred words that came into their heads. He too used simple averages, evinced no interest in individual differences, and focused on a basic mental task. He deviated from Cattell, however, by using his data to hypothesize intellectual differences between men and women, concluding that "women repeat one another's words much more than men" and that women and men prefer different types of words (women: the ornamental and concrete; men: the useful and abstract).[79] In 1896 Mary Whiton Calkins challenged his inferences, which suggested women's intellectual inferiority, on the basis of her investigation of Wellesley undergraduates, but both studies shared a common technique: aggregating individual performance into group data, and then using the data to compare one group with another.[80] Perhaps most importantly, this procedure—in contradistinction to mainstream experimental practice—focused analysis on aggregates and suggested that mental tests were particularly appropriate for revealing differences or similarities between groups.[81]

During the next few years, a number of psychologists exploited mental testing's suitability to group-level analyses. Some were content to define their groups rather arbitrarily, as Cattell and Ferrand did when focusing on Columbia University undergraduates. Most, however, differentiated their subject populations according to criteria that seemed more natural, more reflective of "real" distinctions than the accidents of institutional affiliation. Jastrow and Calkins turned to gender, R. Meade Bache and Anna Tolman Smith used race, W. Townsend Porter and J. Allen Gilbert relied on age/grade levels, and a few adopted another characteristic particularly associated with group- or species-level differentiations: intelligence, understood as overall mental power.[82] Still largely absent from other experimental studies, intelligence began to appear in research based on mental testing as a fundamental criterion for defining subject groups. In 1893 Porter was among the first American psychologists to construct a study specifically around global mental ability.[83] Convinced that there was "a physical basis for precocity and dullness," and that "mediocrity of mind associated in the mean with mediocrity of physique," Porter calculated mean weights for schoolchildren of various ages and grade levels.[84] He then showed that for a given chronological age larger students tended to be placed in higher grades than smaller ones, and concluded that larger children must be more intelligent. The study's obvious flaws (including that grade placement might have been influenced by a child's size) do not require much comment; what is significant was his choice of subject population, schoolchildren, and use of intelligence differences to characterize groups.[85] Schools constituted ideal sites for such differentiation: their graded structure was easily read as implying that those in higher grades were more mentally developed than those below.

Gilbert relied on the idea of group intelligence in one of the most sophisticated and thorough mental test analyses, "Researches on the Mental and Physical Development of School-Children" (1894). Employing tests similar to Cattell's, Gilbert divided his subject population by age, sex, and level of intelligence. The age and sex could be readily determined; to classify the students' "general mental ability," he relied on teacher judgments.[86] Gilbert arrived at no spectacular conclusions, but he did demonstrate that classifying according to intelligence could seem no less natural than by age or sex.[87] Indeed, only four years later, when Jastrow surveyed approaches to the testing of mental capacity, he highlighted questions of the development and distribution of the mental powers as important for the field. However, Jastrow placed these issues within the context of understanding the "sensory, motor and intellectual endowments, as they occur in the average individual," referred most frequently to multiple mental capacities rather than a singular entity such as intelligence, cautioned that tests of specific abilities should be preferred to "general ones," and despaired that tests of the higher powers could ever be made practicable.[88] Other psychologists concurred, emphasizing the importance of analytical and rigorously experimental methods that made investigating complex mental features difficult to imagine, much less prosecute.[89]

For all of the initial enthusiasm for anthropometric mental testing, it soon essentially died on the vine. By 1900 psychologists had largely stopped conducting such studies, abandoned the research agenda, and lost interest in the data already produced. Anthropometric testing failed to flourish for several reasons. Sokal has argued convincingly that the field's abandonment by two of its most prominent practitioners—Jastrow and Scripture—helped drain it of vitality.[90] In addition, two devastating studies—one by Stella Sharp in 1899 and one by Clark Wissler in 1901—destroyed such testing's intellectual legitimacy.[91] Wissler's findings were especially damaging. Analyzing Cattell and Ferrand's data with the aid of correlational techniques developed by Galton, Wissler demonstrated that Cattell's test results correlated extremely poorly with each other and with the class grades of the study's participants.[92] Whatever Cattell was testing, it seemed disconnected from the higher-order processes he had hoped to capture. As Sokal notes, "Wissler's analysis struck most psychologists as definitive, and with it, anthropometric mental testing, as a movement, died."[93]

Finally, the nature of the subject population itself may have factored into Cattell-style mental testing's demise. Save for the work on schoolchildren, most of the field's studies focused on the ordinary adult mind, a type defined largely in terms of a variety of heterogeneous characteristics.[94] Although talk of faculties and powers had diminished, there were still few ways to conceptualize such minds in more global, holistic terms. In addition, tests of reaction times, or memory, or word associations, performed under the rubric of mental testing, seemed only distantly related to analyses of the higher mental powers. As Cattell noted in 1896, "If we undertake to study attention or suggestibility we find it difficult to measure a definite thing. We have a complex problem still requiring much research in the laboratory and careful analyses before the results can be interpreted, and, indeed, before suitable tests can be devised."[95] Unlike normal organs, normal people were hard to characterize; there was no agreed-upon standard ability or set of abilities all had to share. Statistics on mean reaction times or numbers of words remembered might be scientifically interesting, but as isolated data they provided little better picture of an individual's overall mentality than a phrenological reading with its twenty-seven to thirty-five different factors assessed. Normal human mental functioning just looked too multidimensional. Thus when Jastrow surveyed the field in 1898, he listed a variety of different areas—including memory, attention, association, and imagination—as aspects of higher-order functioning that mental tests would have to measure.[96]

Non-normal minds, however, were another story. In the extreme states of genius or idiocy, or in the developmental stratifications of childhood, individuals were already being characterized according to their overall mental ability. In addition, by 1900 worries about immigration, miscegenation, and degeneration, not to mention the spread of eugenics, had made just such minds the object of intense political/cultural scrutiny. Thus when mental testing did reemerge in American psychology near the end of the decade, it did so by moving in from the periphery, specifically through the study of children and the feebleminded.[97]

The Binet-Simon Comes to America

On March 13, 1908, the American psychologist Henry H. Goddard of the New Jersey Training School for Feebleminded Girls and Boys arrived in London, his first stop on a tour of major European institutions for psychological research and work with the feebleminded. After two stimulating weeks in England, Goddard proceeded on to Paris, where his visit proved more disappointing. Two days of sightseeing produced little comment in his diary, and he had mixed reactions to a day spent visiting Désiré Magloire Bourneville's school for feebleminded children. On his fourth day in Paris—March 31—Goddard recorded: "Called on Janet. Visited Sorbonne. Binet's Lab is largely a myth. Not much being done—says Janet."[98]

Goddard's diary entry was both true and ironic. Although Pierre Janet was no particular friend of Binet's, his description of the Sorbonne laboratory was literally accurate, if somewhat disingenuous. By 1908 Binet had indeed almost completely abandoned the Laboratory for Physiological Psychology. Had Goddard been directed, however, either to Binet's pedagogical laboratory at the Rue Grange-aux-Belles or to Théodore Simon's offices at the Perray-Vaucluse asylum, he might have discovered a much more congenial situation. There, as we saw in chapter 4, Binet, Simon, and their colleagues were actively pursuing just the sort of diagnostic program oriented toward distinguishing "normal" intelligence from subnormality that was a major concern of Goddard's, and were doing so by means of their new psychological instrument, the Binet-Simon Intelligence Scale.[99] As he explained in 1916, however, Goddard had not heard of Binet and Simon's work before his trip to Europe, and indeed only discovered the 1905 scale near the end of his tour, when he met Ovide Decroly in Brussels. The event would soon make not only Goddard's reputation, but the measuring instrument's as well.[100]

Intrigued by the possibility of having a more rapid and accurate way of diagnosing mental deficit, Goddard commenced experimenting with the 1905 Binet-Simon tests almost immediately upon his return, but found them of limited value. When the 1908 revised scale was published, he hesitated, recalling later that it had "seemed impossible to grade intelligence in that way. It was too easy, too simple."[101] Despite his misgivings, Goddard did administer the revised scale to residents at Vineland, and reported himself astonished by the scale's ability to produce classifications that matched assessments arrived at after long familiarity with the patients:

> while giving the tests we had come more and more to feel that Binet had certainly evolved a very remarkable set of questions, and that they did work out with amazing accuracy, and I believe it is true that no one can use the tests on any fair number of children without becoming convinced that . . . the tests do come amazingly near what we feel to be the truth in regard to the mental status of any child tested.[102]

Goddard was not alone in being impressed by the Binet-Simon's speed and accuracy.[103] By 1916, when Stanford psychologist Lewis Terman completed his landmark Stanford Revision of the Binet-Simon Intelligence Scale (the Stanford-Binet), intelligence had become something of an industry in American psychology. Articles either about the Binet-Simon scale or reporting data generated with it filled professional journals; rival scales competed for clientele and professional dominance; and a great deal of hand-wringing was in evidence about the technology's improper use by the "ill trained."[104] Many American psychologists became infatuated with the test and adopted with few reservations what their French counterparts largely found either uninteresting or problematic. In the process, and almost unwittingly, these psychologists also allowed the instrument's hierarchical, unidimensional vision of intelligence to shape their own conceptions. For them intelligence would become what the tests purported to measure, a natural characteristic with which individuals were variably endowed and that helped to explain why some succeeded where others failed.

Goddard began the process of Americanizing the intelligence test. He was the first to publish results obtained using the Binet-Simon scale, and remained associated with it in the public mind.[105] Nonetheless, others quickly followed. Terman and his student Hubert G. Childs began to investigate the 1908 scale in 1910, conducting extensive experiments on schoolchildren in California that would result in their first revision of the scale, published in 1912.[106] Edmund B. Huey praised the test in 1910 as "the most practical and promising means yet made available for determining the fact and for measuring the amount of mental retardation" and produced a translation, as did Fred Kuhlmann in 1911.[107] Other psychologists also jumped into the fray.[108] Although complaints about details of the French scales abounded and almost every psychologist altered the tests at least slightly to meet their own needs, the Binet version of the scale remained dominant.[109] Many psychologists were uneasy about this fact, aware that a measuring instrument designed for Parisian children might be inaccurate for Americans, with their different set of culturally specific background knowledge, a criticism Leonard P. Ayres leveled explicitly in 1911.[110] In 1916 these worries were mostly alleviated with the publication of Terman's Stanford-Binet, an overhauled version of the Binet instrument standardized on an American population.[111] While some psychologists—most notably Robert M. Yerkes—continued to express misgivings about the Binet approach and to propound alternatives, the majority of the testing community adopted the Stanford-Binet as the standard test for measuring intelligence, a position it maintained until well into the century.[112]

The Binet test had found a home in American psychology, but, in one of the few parallels with its history in France, not in the discipline's mainstream. Binet and Simon had refined the scale while working at Rue Grange-aux-Belles and Perray-Vaucluse, not at the Sorbonne. Similarly, intelligence testing's leaders in America—Goddard, Terman, Yerkes, Huey, Thorndike, and Kuhlmann—worked mainly with children or the feebleminded, populations who, to that point, had had little place inside the laboratory. As

Terman recalled years later: "I was quite aware of the fact that many of the old-line psychologists regarded the whole test movement with scorn. . . . I had the feeling that I hardly counted as a psychologist unless possibly among a few kindred souls."[113] According to practitioners of laboratory psychology, the Binet-Simon scale suffered from poor grounding in psychological theory and inaccessibility to rigorous experimental procedures. As Iowa State psychologist Edwin Starbuck admonished Goddard in 1912:

> the Binet people, including yourself, are using methods that are not sufficiently self-critical, and that furthermore they do not establish reliable norms. . . . The burden of my song is simply that the well-established tests on sense discrimination, simple and compound reaction, motor tests in which a definite standard can be followed, and all such things where there is some definite measuring stick together with controllable mean variations, would seem to be the most hopeful kinds of things to work at.[114]

Goddard's response to such criticisms is revealing. While he did not deny the value of the traditional "brass-instrument style" of psychology that Starbuck advocated, Goddard did insist on its inappropriateness for his population. Most feebleminded individuals, Goddard contended, simply could not perform the laboratory tests developed to examine normal adults.[115]

Different populations required different experimental instruments, and perhaps produced different results as well. In his "great human laboratory of nearly four hundred subjects of both sexes, all ages from five to forty (a few older) and of all degrees of mental defect," as he described the Vineland Training School, Goddard faced one of the few population groups already conceptualized and stratified in terms of intellect.[116] Individuals were consigned to Vineland and similar institutions—whose number had been expanding rapidly over the nineteenth century—because they could persuasively be characterized as idiots, imbeciles, or otherwise "incapable of competing on equal terms with his normal fellows or managing himself or his affairs with ordinary prudence," as the American Association for the Study of Feeble-Mindedness declared in 1910.[117] Well before the development of the Binet test or any other such instrument, diagnoses of idiocy, imbecility, and the like were routinely carried out, even when the dominant mental philosophy was Common Sense, whose faculty approach left little room for such holistic concepts as intelligence.[118] Defined as lacking almost all reason, idiots were deemed readily identifiable, as the panoply of regulations and institutions that had developed around them from the early modern period on demonstrated.[119] Less extreme states of mental deficit, such as imbecility, though harder to characterize precisely, were also distinguished from normal intellect. Thus by the late nineteenth century intelligence as an overall measure of individual mental power had at least one well-established social role: a criterion for deciding, or at least justifying, who was remanded to asylums.

Confronting these everyday practices of categorizing mental defect, Goddard's project was to match Binet assessments with those generated by these traditional means and to show the superiority of the Binet-Simon scale in clarifying formerly confusing diagnoses and identifying the feeble-minded more reliably.[120] In pursuit of this practical goal, Goddard and most other testers were content to minimize their reliance on test items derived from theoretical models of the operations of the mind in favor of the heterogeneous mixture of tasks that Binet had advocated, in essence substituting statistical criteria for ones derived from laboratory science as the way of validating their scales. What seemed critical, in the context of increasing worries among many Americans about the "menace of the feeble-minded," was to respond to the escalating demand for determinations able to identify the mentally pathological.[121] As Alexander Johnson, of the Indiana School for Feeble-Minded Youth, warned in the pages of the *American Journal of Sociology:*

> From these 90,000 neglected or abused feeble-minded persons [those not in institutions] have come, or will come, most of the next generation of idiots, imbeciles, and epileptics, and a vast number of the prostitutes, tramps, petty criminals, and paupers.
>
> Of all the dangerous and defective classes this is the most defective, and the most dangerous to the commonwealth, the most to be pitied in themselves, and the most costly to the taxpayer.[122]

Subject population mattered. The study of normal adults, even when approached via anthropometric mental testing, had yielded comparative data about physical features and elementary mental traits, but rarely aggregate rankings.[123] The feebleminded, however, manifested differential intelligence by definition. Few psychologists or institutions were yet interested in the individual, personal features of the mentally subnormal, and the social pressure to identify them rapidly and accurately was intense. Thus, the principal concern was with diagnosis, with standardizing placement of individuals into pathological/therapeutic categories where their deficit would constitute their primary identity.[124] In the Binet-Simon scale Goddard and others found a technology they believed could produce such classifications objectively and accurately.[125] "I never cease to marvel," Goddard remarked in 1916, "at the wonderful insight Binet had and the amazingly accurate results that he obtained. In other words, I have found from nearly ten years living with the feeble-minded that Binet was correct in his theories of the feeble-minded and of their psychology, to a much greater extent than is given to most mortals."[126] The feebleminded thus provided Goddard and other Binet testers with a well-defined population against whom they could calibrate their instruments, allowing them to translate many practices of brass-instrument experimentation into this new psychometric domain, while also speaking to an important social concern.[127]

Imagine what might have happened, however, if Goddard had taken the Binet scale not to Vineland, but to an experimental psychological laboratory and used it in the Wundtian manner. With a subject population now composed of his fellow psychologists, Goddard would probably still have found differences in performance, but interpreting those results would have been much more problematic. No ready-made and socially sanctioned intellectual stratifications of his colleagues would have been available against which to judge the scale's determinations. Moreover, given the close relations typical among investigators working in the Wundtian mode, it seems likely that Goddard would have proved more interested in the particularities of his colleagues' intellects than in ranking their mental power, as Binet had been when studying his two daughters.[128]

In an indirect way, Goddard's own evidence supports these speculations. One of his files contained the case history of a twelve-year-old boy recommended for psychological examination because of repeated run-ins with the law. The boy was administered the intelligence scale, and in the words of the report, "it was a surprise to find that he was able to pass successfully nearly all of the Binet tests."[129] The examiner did not stop there, as he would have had the boy been proven feebleminded. Rather, the examiner used performance on several specific tests to explain why the child was so often in trouble. Inability to resist suggestion, it turned out, was the diagnosis, revealed by the boy's willingness to heed his examiner's slightest promptings (and thus do well on the test!). In this situation, when dealing with a person classified of normal intelligence, significance was accorded not to the Binet categorization, but to the intellectual idiosyncrasies it revealed. Simple quantitative assessment of mental power, in other words, did not say nearly enough when investigating a subject judged of normal intellect.

The feebleminded thus proved to be, in many respects, the ideal population to render degrees of intelligence visible and the intelligence test itself practical. However, they were not the only group relied on to serve this function. Following Binet, American intelligence testers focused as well on schoolchildren, the other population conventionally stratified into groups connoting differential mental ability, their grade levels.[130] The task then became, as we shall see in Terman's creation of the Stanford-Binet, to use the manifest mental development that children undergo to match test results with grade-level rankings. The Binet scale, built on this model of age-based mental development, provided the blueprint; Terman and his students adapted old tests and developed new ones so that the scale could be standardized against the performance of "unselected" American schoolchildren and precisely match the mean rate of their intellectual development. For assessing schoolchildren, as for the feebleminded, the goal was to produce an instrument capable of providing differentiations of presumed social value and the key was to seize on a preexisting stratification roughly tied to mental power and then to make it more precise and scientific by subsuming it into the output of the intelligence scale. In the process, the test would become as applicable to the normal as the subnormal, and intelligence itself would become a characteristic of ever greater professional and public interest.

Fitting the World to the Test: Intelligence Confronts the Normal

Lewis Terman's creation of the Stanford-Binet in 1916 marked a fundamental divide in the American history of intelligence.[131] After 1916 the equation of intelligence with IQ, understood as innate, quantifiable mental ability, gradually became accepted within parts of the psychological community and the broader culture as well. Before the Binet scale's introduction, however, interest in intelligence levels was limited, restricted principally to anthropology, comparative psychology, and discussions of those at the extremes of mental ability. Confined to these fields and associated primarily with diagnosing mental deficiency, mental testing and its version of intelligence could scarcely have had a significant impact on American culture. As Elizabeth Lunbeck has suggested in her pathbreaking study *The Psychiatric Persuasion,* psychiatry's increasing importance was directly related to psychiatrists' attempts to extend their expertise beyond the asylum by creating "a psychiatry of everyday life."[132] Mental testers followed a similar route. Not content with solely the psychopathological, they turned their instrument to the "normal" too by blurring the boundaries that separated well-defined categories of intellect such as idiocy and genius from the largely undifferentiated "everyone else."[133] Normality became a particular range of scores on an intelligence scale, measured by matching age to performance, and as finely graded as the numerical calculation of IQ permitted. In the process, intelligence's status and domain were transformed. Though no one way of understanding intelligence, or instrument for its assessment, completely dominated before World War I, in broad terms the concept of intelligence became ever more tightly associated with quantified measurements and precise rankings of entire populations.

Terman's early work both reflected and helped produce the new vogue for metric intelligence. Born in 1877 in Indiana, Terman passed his formative years in late-Victorian America: working in the fields, attending classes in a one-room school, and reading.[134] After receiving his B.A. from Central Normal School in Danville, Indiana, and serving as a high school principal, Terman went to graduate school in psychology, first at Indiana University and then at Clark University, where he received his Ph.D. in 1905. His first sustained involvement with intelligence came in his doctoral work, published in 1906 as "Genius and Stupidity: A Study of Some of the Intellectual Processes of Seven 'Bright' and Seven 'Stupid' Boys."[135] Like Binet when developing the intelligence scale, Terman turned away from the simple tests of sensory perception and cognitive functioning common in experimental psychology and anthropometric testing. Instead, he relied on eight tests designed to sample more complex abilities. Terman structured his analysis, as had Binet, by focusing on the intellectual spectrum's extremes—his "bright" and "dull" boys. Nonetheless, the differences between Terman's 1906 study and the Binet-style intelligence test were pronounced.

Rejecting as "superficial" Galtonian procedures of collecting small amounts of data from large subject populations, Terman instead intensively studied a very few individuals. Such a method, he explained,

> aims to characterize the mental differences among them, hoping in the end to throw light on ultimate problems of psychological analysis. Instead of applying tests which yield an unequivocal *yes* or *no,* or *so much* or *so little,* it may even put problems which allow of widely different attempts at solution, of a number of possible kinds of errors, and of different methods of correcting these errors. General observation is appealed to, and in all respects such work may utilize rougher data than would be possible in purely quantitative studies.

Terman explicitly cited Binet as one of his models, but it was the Binet of *L'Etude expérimentale de l'intelligence* with its in-depth investigation of Binet's two daughters. Moreover, in Terman's approach, the dull and bright were not different in degree so much as in kind. Terman drew an analogy between the psychologist and zoologist: "While the zoölogist has no longer any difficulty in stating the essential differences between even such superficially similar animals as the whale and the fish, psychologists cannot agree on the features distinguishing the most widely separated grades of intelligence." Throughout his article, Terman emphasized the intellectual extremes' distinctness, drawing on common understandings of the differences between the idiot and genius. But his formulations also revealed that the barrier between the two was not impermeable. Near the essay's end, for example, Terman speculated about "an individual's intellectual rank among his fellows," suggesting that intellectual power varied continuously in a simple linear manner, and thus that the smart and stupid might not be so different after all.[136]

Like many in the new discipline, Terman's work was transformed with the arrival of the Binet scale. For all these psychologists, the separation between the once distinct kinds of intelligence disappeared entirely. In the Binet test, the difference between an idiot and an imbecile, moron, normal person, or even genius lay in the age at which they successfully answered particular questions. Rather than treat the population as composed of fundamentally different, and incommensurable, kinds of people, the test placed all individuals on the same scale, democratically embracing everyone at the same moment that it also made visible profound differences between them. Intellectual level itself became a continuous variable. Indeed, the very power of Binet-style tests derived, in large measure, from their assumption that mental development was uniform, so that all individuals of mental age five, for instance, were considered to have similar intellectual capability, whether chronologically aged twenty and so imbeciles, aged five and so normal, or aged two and so geniuses.[137] The scale worked by flattening ontological distinctions between types of intellect or mixes of attributes and capabilities, treating all minds as varying solely along a single dimension.[138]

Much of the theoretical foundation for this new way of understanding intelligence was provided by the English psychologist Charles Spearman with his concept of general intelligence, first presented publicly in 1904.[139] Employing a statistical technique he developed called factor analysis, Spearman analyzed correlations between the results of a series of elementary sensory-discrimination tests he conducted on schoolchildren and estimates of the individual subjects' intelligence (derived from school performance and the opinions of others who knew them). "*All branches of intellectual activity have in common one fundamental function (or group of functions),*" Spearman concluded, "*whereas the remaining or specific elements of the activity seem in every case to be wholly different from that in all the others.*"[140] In later work, Spearman named this fundamental function "general intelligence," or *g*, and the specific functions, theoretically limitless in number, *s*. For much of his career, Terman would be seen as one of the principal American proponents of *g* and his tests as the best argument for *g*'s existence.

In order to define intelligence this way, as a singular, unidimensional entity, however, the notion of normal intellect had to be sufficiently simplified or even standardized to fit the linear scale connecting idiot to genius, and infant to adolescent. But, because notions of what constituted normality were more diffuse and concerns about preserving the individuality of normal people more intense, stuffing the normal into the bell curve was no easy matter. Normal simply connoted too many attributes. Consequently, throughout the 1910s many psychologists remained uneasy about what was excised when Goddard, Terman, and others defined intelligence solely by such tests.[141] Writing in 1916, Harry Kitson of the University of Chicago echoed some of the anxieties Terman had voiced in 1906:

> It will be noted that this quantitative measure, even in its finished form tells only how much intelligence is possessed. It is best adapted for the grading of persons in groups. In diagnosing individual cases, however, the qualitative phase is the one that is most insistent. Here the problem is to determine kind of intellectual ability, and in this regard I think we can not lay claim to much achievement.[142]

Kitson was not alone in worrying that intelligence tests masked as much as they revealed, especially for diagnostic purposes. In 1912 Clara Schmitt, a psychologist at Chicago's Juvenile Psychopathic Institute, conceded that the Binet tests "point out one of the most important of the elements of general intelligence," but nonetheless argued that often "real disabilities were not disclosed by the tests."[143] At the 1913 Buffalo conference on the Binet-Simon scale, Charles Berry of the University of Michigan contended that the Binet scale did not test all aspects of intelligence, and J. E. Wallace Wallin of the University of Pittsburgh declared that the tests were "useful only for a preliminary survey." Grace Fernald of the State Normal School in Los Angeles concurred, noting that the test "does not apply equally well to some of the street waifs of our large cities because the experience and standards of achievement of such children are almost entirely different from those of ordinary children."[144] Summing up the hesitancies of many clinically oriented psychologists, Lightner Witmer proclaimed in 1915 that "I do not know of any single test on which I can rely for diagnostic purposes."[145]

One of the most sustained and comprehensive critiques, explicitly engaging first with Spearman and later with Terman, came from Edward L. Thorndike and others who shared his belief that global intelligence was more artifactual than real.[146] Unpersuaded by Spearman's statistical demonstration of general intelligence's existence, Thorndike argued on the basis of investigations into animal intelligence and the psychology of learning as well as his own statistical studies that the human intellect was composed of numerous independent abilities, and that proficiency in one area said little necessarily about skill in any other.[147] In Thorndike's view, intellectual ability was contextual and adaptational, and while there might be certain practical situations in which an overall measure of ability could be useful, in the main determining specific aptitudes would be much more valuable to the individual and to educators.[148] Psychologist Carl Seashore and physician J. Victor Haberman, among others, agreed that the importance of measuring general ability, even if it existed, was vastly overrated, and advocated instead that testing focus on the assessment of specific mental functions.[149] Although in a minority even before the war, this faction's hostility to the notion of metric intelligence makes clear the extent to which the concept's success in American psychology was by no means assured in the first decades of the century.

Nonetheless, like Terman, numerous other psychologists, whatever their reservations about specific features of the tests, were enthusiastic about the new instruments' ability to order and illuminate not just the pathological, but the hitherto undifferentiated normal. As W. H. Pyle announced in 1916, "My notion is, that mental diagnosis is of value in connection with *all* the children of the school, and not merely as a means of selecting out the deficient and supernormal."[150] Edmund B. Huey spoke for many psychologists when he observed in 1912 that "in general the scale of Binet and Simon has interested us all in making more methodical [the] study of the intelligence. . . . It gives promise of being developed to a scale which will render much service in the classification and study of normal pupils."[151] Although even proponents conceded that the tests were far from perfect, and many argued that tests of character were necessary as well, they were nonetheless captivated by the potential they believed the instruments had to further psychology as an objective science by revealing hidden aspects of the mentality of every person analyzed with such devices.[152] Thomas Haines praised them as "vastly better than unaided vision for assaying mentality," and Kuhlmann noted that they "offer the best means in existence of determining grades of mental development."[153] Clara Harrison Town was impressed by their simplicity and the accuracy of their estimates of a child's intelligence, while Martha Adler concluded on the basis of her study of pupils in New York City's P.S. 77 "that the Binet Tests form an excellent basis for the classification of young pupils."[154] In general, testing's advocates highlighted the simplicity of the tests' administration, their production of data that fit the statistical demand for score distribution according to the bell-shaped curve, and especially their correspondence with determinations of intellectual level made on the basis of extensive experience with the subjects.

Terman's publication of the Stanford revision of the Binet scale simply accelerated this transformation in intelligence's purview and meaning. Exactingly standardized on 905 "unselected" schoolchildren, approximately equal in sex and age distribution, Terman's revision was technically superior to the Binet-Simon, at least for white middle-class American-born subjects.[155] Terman increased the number of tests to ninety, extended the scale more fully into the adult range (age fourteen and older), and carefully ensured that each test met two criteria: first, the percentage of those passing had to increase continuously with age; and second, when combined with the other tests for its given age level, the median mental age of those passing had to equal their median chronological age.

This second feature became the Stanford-Binet's most celebrated characteristic. Adopting a suggestion made by William Stern in 1912 that the ratio of mental age to chronological age would show "*what fractional part of the intelligence normal to his age a feeble-minded child attains,*" Terman proposed that an individual's score on the Stanford-Binet be reported in terms of "an index of relative brightness," his or her "Intelligence Quotient" (IQ), defined as the ratio of measured mental age to chronological age times one hundred.[156] This quantity, which the Stanford-Binet would make famous, Terman asserted, "has been found in the large majority of cases to remain fairly constant": "The feeble-minded remain feeble-minded, the dull remain dull, the average remain average, and the superior remain superior. There is nothing in one's equipment, with the exception of character, which rivals IQ in importance."[157]

With IQ, intelligence was fully transformed into a stable, quantified characteristic applicable to all human minds. IQ not only reinforced the idea that intelligence was singular and hierarchical, but also ensured that most individuals (at least from the appropriate population) would be categorized as of approximately average intelligence (50 percent with IQs between 93 and 108), and that the numbers at either extreme would decrease sharply. Thus, whatever idiosyncrasies or even growth an individual intellect might manifest would be effaced when producing a Stanford-Binet intelligence quotient, which created a linear index of relative brightness encompassing idiots, geniuses, and everyone in between.[158]

What was this intelligence that Terman built into the Stanford-Binet? Following Binet's methodology of using a variety of tasks to sample mental ability, the ninety tests in Terman's revised version drew on a multitude of functions. Nonetheless, he relied on a few skills repeatedly, providing a composite picture of what attributes intelligent Americans, as the test defined them, should possess. First, since virtually all of the Stanford-Binet's tasks were explained verbally, such Americans had to understand English well and possess a good vocabulary, be they three or thirteen. They were also required to have excellent memories, whether for words, numbers, or images. At almost every level, memory was explicitly tested, with subjects asked, for example, to repeat a series of unrelated numbers or words, or to view for a specified length of time a series of pictures and then describe or copy them. Tests of other capacities, especially making comparisons, manipulating abstract terms, and discerning general patterns, also recurred throughout the scale. Comparisons appeared even in the tests of the very youngest; year-4 subjects, for example, were asked to decide which of two lines was longer. At later age levels, tasks focusing on manipulating abstract concepts became increasingly important. Starting at the year-12 level, the subject's ability to deal with abstractions was explicitly tested by a range of tasks, including knowledge of abstract terms (for example: pity, revenge, laziness, and idleness), puzzle solving, and inductive reasoning.[159] Terman's rationale for his choices, as he explained in his first revision of the Binet scale, was that he believed "that tests of memory, vocabulary, observation, reasoning and reaction to a complex social or moral situation bring out fundamental characteristics of mental ability."[160]

These capacities suggest the major types of mental functions that the Stanford-Binet was designed to measure. Success on the test, however, also required, as Terman put it in an echo of his Victorian upbringing, "character," a certain personal and moral outlook that even he acknowledged was the one characteristic as important as IQ.[161] Obedience to authority, seriousness, and willingness to work hard were presumed from the outset by the very testing situation. While these were requirements of a student facing any test, Terman at times took them further, instructing examiners to fail subjects on a task, even if they obviously knew the correct answer, if the subjects did not fulfill the examination's behavioral norms. When discussing acceptable responses to the question "Which of these two pictures is the prettiest?" (year 5), for example, Terman explained:

> Sometimes the child laughingly designates the ugly picture as the prettier, yet shows by his amused expression that he is probably conscious of its peculiarity or absurdity. In such cases "pretty" seems to be given the meaning of "funny" or "amusing." Nevertheless we score this response as failure, since it betokens a rather infantile tolerance of ugliness.[162]

Here, the issue clearly was not one of knowing the "right" answer, but of providing the appropriate response. Reasoning correctly and answering many different questions, therefore, was not enough; Stanford-Binet intelligence demanded as well the willingness to communicate those answers as prescribed. In addition, fecundity or creativity, though conceivably important components of intelligence, were accorded little value on the test.[163] The single, precise, and legitimate answer was sought and rewarded, even where the question, such as how to act in a particular situation, might be a matter of judgment. Intelligent people as defined by the Stanford-Binet knew the culturally sanctioned answer, and responded accordingly.

The importance of knowing conventional social wisdom was apparent at almost all age levels. When four-year-olds were asked "What ought you to do when you are hungry?" for example, Terman commented that "Have to cry" was incorrect; the only acceptable answers referred to getting something to eat. A six-year-old asked "What's the thing to do if it is raining when you start to school?" must respond with "take umbrella," "put on rubbers," or some such; to say "Go home" was not to prove oneself intelligent. Similarly, for eight-year-olds, the correct answer to the question of what they should do if they broke something belonging to someone else was not "confess"—much less "lie"—but "apologize" or "make restitution."[164] In these examples, and even more so in the fables-interpretation task of year 12, the point was to provide the response dictated by middle-class propriety, the one requiring a child to see and act upon his or her duty.

The influence of convention was by no means restricted to explicitly ethical tasks. Almost every test involving comparisons, for example, required that the subject know those answers that were commonly accepted; "absurd" responses were never given credit. In the bow-knot task, year 7, the examinee was asked to look at a tied bow-knot, and then tie an identical one. In scoring this task, Terman explained that "the usual plain common knot, which precedes the bow-knot proper, must not be omitted if the response is to count as satisfactory, for without this preliminary plain knot a bow-knot will not hold and is of no value."[165] The simple reproduction of the bow-knot's appearance was thus not enough; the intelligent person had also to believe in the importance of making a "proper" bow-knot, one that could "work." Cultural knowledge became objectified in the Stanford-Binet, as social prescriptions were translated into criteria for individual assessments.[166]

Truth and judgment, obedience and order, reason and convention, knowledge and virtue, equality and hierarchy—the Stanford-Binet conjoined these elements and transformed them into a single number, IQ. Stanford-Binet intelligence thus became both differential and universal, a concept that could be applied meaningfully to all people, but that some individuals possessed in greater measure than others, and that was distributed predictably but not equally. For many in the testing community, the Stanford revision set a new standard for methods of assessing general intelligence.[167] In a sophisticated 1916 theoretical analysis of intelligence tests, Arthur Otis had articulated three criteria for a test to be a test of intelligence, including that it "correlate well with general opinion as to what constitutes intelligence." Terman met all of these when developing the Stanford-Binet.[168] Psychologist Walter F. Dearborn observed that Terman's revision "has increased the reliability of this method of estimating intelligence and has led to findings of value and of general psychological interest," while Terman's graduate student Ethel Whitmire found that the Stanford-Binet could be more accurate in determining ability than teachers' judgments.[169] And Goddard, though wary of Terman's earlier efforts at revising the Binet test, waxed enthusiastic over the Stanford-Binet to Abraham Flexner of the Rockefeller Foundation's General Education Board:

> I am bound to say that when Terman's book appeared, with its larger view, the whole subject [of mental testing] seemed to me to be placed in a new light, and I have now come to share fully his enthusiastic belief that with sufficient aid,

> new and more comprehensive methods, that will not be amenable to the criticisms of the old and that will be of the very greatest practical service and would attain general recognition, could be evolved.[170]

Even Robert Yerkes, who had just published his own book outlining his point-scale approach to measuring general intelligence, grudgingly congratulated Terman for doing "a thoroughly good job, scientifically as well as practically considered," and expressed hope that the Stanford revision "should supersede all other forms of the Binet method."[171] Nonetheless, Yerkes quickly published a piece comparing the Point Scale and Binet methods, where he argued forcefully that the Point Scale was superior on both theoretical and technical grounds. The Point Scale, he contended, because it eschewed the Binet age-scale approach for one "based upon the assumption of *developing functions,*" was better grounded with regard to mental development, more easily standardized for different groups, and more fully amenable to statistical analysis.[172] Terman and his collaborators disputed Yerkes' claims vigorously; others in the community observed that the two scales agreed in many of their specific assessments.[173] Moreover, both approaches shared a commitment to the existence of measurable general intelligence.

"By 1917," Leila Zenderland has observed, "professional psychologists were deeply embroiled in a complicated and many-sided debate over the nature and future of intelligence testing."[174] Certainly the disputes between Yerkes and Terman, or Thorndike and the Binet testers, or the existence of many rival versions of the intelligence test all suggest that little had been settled by the outbreak of World War I. Nonetheless, stepping back from the day-to-day skirmishes within psychology, a set of shared assumptions underlying these arguments can also be discerned. Intelligence testing was well on its way to becoming an important, indeed one of the central, fields within the discipline and the commitment to the existence of some sort of global, quantifiable, hierarchical intellectual ability was spreading rapidly.[175] Moreover, in numerous asylums and some schools and even workplaces, intelligence and its tests were beginning to be employed to carry out the practical work of sorting individuals into groups based on their measured level of mental ability.[176]

Worries about the feebleminded and the spread of eugenics within both the professional and lay communities no doubt hastened this turn to intelligence and its tests, as the eugenic concern with the biologically superior and inferior quickly was translated by many into a question of intellectual merit or lack thereof.[177] The eugenics-inspired Race Betterment Foundation, for example, conducted physical and mental perfection contests for children during the 1910s, with mental testing being one of the key methods used to determine a child's superiority, and Goddard carried out an experiment at Ellis Island to see if a modified Binet test could be used to screen out "defective" immigrants.[178] Most telling may have been the reaction to New York State Supreme Court Justice John W. Goff's proclamation in 1916 that Binet test results could not be admitted as evidence of a defendant's mental incapacity because "standardizing the mind is as futile as standardizing electricity." The response to his ruling, at least by the *New York Times,* was quick and stinging.[179] "If such statements as these [by Justice Goff] were not so familiar," the *Times* declared,

> if experience did not prove that though thus progress is delayed it is never stopped—coming from such a source they would be a cause of discouragement, almost of despair. As a matter of demonstrated fact, the Binet-Simon tests, intelligently applied . . . are as trustworthy as the multiplication table. Minds can be standardized, just as electricity is every day and long has been.[180]

Intelligence as a standardized, constant, biological entity that arrayed populations on a continuum from smart to dumb and the tests that made it visible had thus found footholds in American culture. Asylums and schools, especially, began clamoring for information about their charges' cognitive abilities, and proved willing to provide not only subject populations, but positions for full-time psychological experts.[181] The new intelligence, with its obvious practical ramifications, seemed to meet these institutions' needs and thus argued powerfully for expanding psychological science to any domain seeking to sort individuals into a "natural" order. This expansion of psychology to new fields provided a vector for the dissemination of the metric understanding of intelligence to the culture at large.

This was not the only conception of intelligence to achieve prominence in 1910s' America, to be sure. In his popular essay "The Moral Obligation to be Intelligent" (1915), Columbia University literature professor John Erskine emphasized the moral dimensions of the concept, suggesting—contra the British literary tradition of associating intelligence with evil—that every person was called to develop their intellects as well as their characters, so as to make the best, most knowledgeable moral decisions possible.[182] A group of pragmatist intellectuals around John Dewey seized on the term in their 1917 collection of essays, *Creative Intelligence: Essays in the Pragmatic Attitude,* to extol the virtues of creative thought: "the pragmatic theory of intelligence," Dewey observed, "means that the function of mind is to project new and more complex ends—to free experience from routine and from caprice."[183] But in addition to these paeans to intelligence as the source of creativity and individuality in the face of anxiety about mass society, the metric version of intelligence—especially in the guise of the new term "IQ"—was also slowly becoming an important component of the wider public discourse, even though it did not so much celebrate idiosyncrasy as tame it in the name of the mechanical and statistical.

Conclusion

Mental testing's rapid acceptance by a vital segment of the American psychological community had profound consequences for intelligence's status. Before the Binet-Simon scale, psychologists used intelligence principally, if at all, to differentiate groups. Their interest in intelligence per se was low, and they had few means of studying higher mental powers empirically. The Binet test and its offshoots changed this. The test applied the metric version of intelligence to individuals, rendering visible fine shades of intellectual difference. Though the test was initially restricted to children and the feebleminded, testers soon began to use it with a wider variety of subjects. In the process psychologists took with them the vision of intelligence defined by these two populations. For both the feebleminded and children, intelligence as a global description of graded mental power accorded with long-standing social practice, and intelligence tests produced classifications that mimicked that practice's central features.

But for this particular notion of "intelligence" to gain legitimacy, it had first to be constituted as a credible psychological entity. This was not necessarily a straightforward task, as the failure of anthropometric testing demonstrated. Only when psychologists could combine instrument and investigative practices with the appropriate subject populations—those who, by definition, manifested differential intelligence—in a cultural ecology where such distinctions were significant, did they stabilize metric intelligence and transform it into a concept that mattered. Through the lens of the feebleminded and children, intelligence appeared simple and measurable, amenable to the new assessment technology and manifested in every human psyche. The normal, in a sense, became a by-product of the pathological and developmental.

Embodied in the test, this practice of seeing intelligence as a unidimensional, quantifiable object was then readily extended to new populations. When applied to normal adults, the intelligence scale arrayed them along the same simple pattern used for children and the feebleminded, obliterating the heterogeneous features once deemed characteristic of the normal adult mind. At the same time, the new technology expanded professional opportunities for psychological testers, generating a group of professionals whose occupational identities were tied to using intelligence to make sense of the social world. And finally, the simplicity and linearity of the test's concept of intelligence—while not always able to account for individual idiosyncrasies—proved useful to many institutions wishing to sort individuals into practical categories. With such groundwork laid, nothing did more to catapult this version of intelligence into public prominence and to bring the normal firmly within the ambit of the new measuring technology than what happened in 1917, when some psychologists caught the ear of the U.S. Army, and intelligence and its tests were enlisted for service to a nation at war.

Part III

MERIT, MATTER, AND MIND

Six

Out of the Lab and Into the World

INTELLIGENCE GOES TO WAR

ON APRIL 6, 1917, the United States declared war on Germany. American psychologists lost little time in following suit.[1] That day, the Society of Experimentalists, a group of psychologists meeting at Harvard University, proposed several ways in which psychology could be applied to the incipient war effort and drafted a list of suggestions that would broadly guide psychologists' wartime attempts to aid the military. Energized by motives ranging from patriotism to opportunism, American psychologists were extraordinarily active during the war, involved, among other pursuits, in selecting pilots for the new military air wings, maintaining morale among soldiers in camp and overseas, and creating and organizing a new army personnel system.[2] Their most ambitious undertaking, however, lay in the area of mental measurement. Confronting a rapidly expanding military desperate for help in selecting and training inductees, a significant percentage of the American psychological community was mobilized under Harvard psychologist Robert M. Yerkes to administer intelligence examinations to all of the army's new recruits and many of its other personnel.

From a small-scale enterprise limited largely to asylums, some psychology departments, and a few schools, mental testing was transformed during the war into an endeavor that at its height processed 10,000 examinees a day and gained broad public recognition. Adapting psychological testing to military needs, however, significantly altered the nature of the endeavor and the meaning of intelligence. Where before the war, intelligence assessment required expert judgment, and intelligence still carried numerous connotations, by 1919 the measurement of intellect in the United States had been fully routinized and both psychologists and segments of the public had begun to conceive of intelligence as a unitary entity that differed among individuals in degree but not in kind, and was quantifiable on a unidimensional scale. At the same time, the army, while never willing fully to embrace use of the intelligence test, nonetheless came by the early 1920s to incorporate intelligence assessments into many of its recruitment and promotion procedures.

The particular product being "sold" to the American military, intelligence, was only beginning to develop a cultural cachet. Although, as we saw in chapter 5, intelligence's meaning was shifting among American psychologists, that transformation had made only a few inroads into broader public discourse. The story of the U.S. Army's adoption of intelligence is, in part, the story of that transformation, of how notions of intelligence and its tests that had been nurtured largely away from public view came to be disseminated to the larger culture through the intersection of the practical needs of wartime, changing character of American society, and professional ambitions of psychologists. It is also a story of negotiation and accommodation, where the psychologists' desire to persuade the military—a group with its own norms and practices—of their expertise's value forced them to acknowledge the military's needs and mores. However, because army testers were bound by their profession's own standards and practices as well, there were limits to this accommodation. Yerkes caught the tension succinctly when he assured Navy Surgeon General William C. Braisted that "we wish to make our *expert* knowledge *serviceable* to military authorities."[3] On the one hand, as a provider of "expert" knowledge, Yerkes presumed that his claims would carry weight with the public. On the other hand, in desiring to provide "serviceable" knowledge, Yerkes suggested that psychological expertise could be adapted to his client's needs.

From the army's perspective, of course, the importance of knowledge that was *serviceable* could not be overemphasized. Seeking practical solutions to questions of organizing personnel, military leaders required convincing that psychology's "ivory-tower" methods could address their concerns, just as these officials were forced to consider transforming time-honored procedures to cope with the new demands of mass mobilization. As a result, both psychologists and military officials oscillated between embracing and resisting change while trying to negotiate a mutually acceptable modus vivendi and a domain of knowledge valid and useful for both.[4]

Americans were not alone, of course, in facing mobilization for war. The French military too needed to expand rapidly with the outbreak of hostilities in 1914. However, its approach to selecting and sorting troops before and during World War I differed greatly. In the wake of its humiliating defeat in the Franco-Prussian War (1870) and the creation of the Third Republic, the French state had enacted the law of July 27, 1872, which mandated obligatory active or reservist military service for all males aged twenty to forty.[5] Designed to create a citizen army that could be mobilized quickly, the 1872 recruitment law and revisions in 1889, 1905, and 1913 ensured that virtually all French males would receive basic military training and could, if necessary, be quickly reintegrated into the standing army. When combined with the long-standing recourse to elite officer-training schools, this French system of military preparedness faced few of the problems of its American counterpart when war was declared. As an ex-trooper in the French army noted, "August, 1914, was the first time of complete mobilization in the history of the Third Republic, and the system under which the men were gathered back to colours worked smoothly in all its details."[6] Thus, where the American military, largely underprepared for enormous expansion, proved open to offers of help even from academic psychologists, the French army felt little need to turn to outside experts, including those promoting intelligence and its tests, to sort recruits.

The American Army in Peace and War

Before World War I, the American military had had no history of official interest in intelligence as connoting the mind's cognitive powers. Whatever private opinions might have been held, the official language used to describe peers and subordinates rarely mentioned mental capacity. The army's public silence about intelligence is most clearly revealed in its formal regulations, especially those detailing the minimum requirements for recruits and the criteria for officer-performance ratings, their efficiency reports.[7] Up to the war, character, not intelligence, loomed large. Army regulations defining recruit eligibility, for example, specified that candidates had to be between eighteen and thirty-five; citizens or intending to become citizens; literate English speakers; and could not be convicted felons, deserters, insane, or intoxicated.[8] Similarly, prewar officer assessments emphasized the kind of person the officer was, his ability to lead and discipline men, his skill and energy, his judgment, his appearance, and whether he had problems with money or drink.[9] In 1903, for example, Lieutenant Sylvester Bonnaffon III was described as "Resourceful. An excellent officer and of extremely courteous demeanor and gentlemanly bearing." By 1916, then Captain Bonnaffon had become "A 'Very Good' officer. Has good habits. Cheerful disposition + a loyal subordinate."[10] Explicit assessments of an officer's mental capacity were virtually nonexistent.[11]

Such was not the case by the war's end. The army's Committee on Classification of Personnel, headed by Carnegie Institute of Technology psychologist Walter Dill Scott, developed new efficiency report forms mandating numerical ratings in five areas: Physical Qualities, Military Leadership, Character, General Value to the Service, and Intelligence.[12] At the same time, in a program directed by Yerkes, more than 1.75 million soldiers were subjected to intelligence tests—Army Alpha for literates and Army Beta for illiterates—which produced intelligence ratings intended to become part of each examinee's permanent service record.[13] Finally, as reflected in the many official and semiofficial reports that officers had to prepare on one another, during the war, officers increasingly commented on intelligence when assessing subordinates and explaining successes or failures.[14] On July 22, 1918, for example, Brigadier General

L. C. Andrews was described as "an excellent, well-equipped, all around officer, intelligent and competent"; and Lieutenant Colonel A. M. Shipp's brigade commander remarked of Shipp that "while probably lacking the vision ever to become a brilliant officer . . . he has the qualifications and faithfulness to duty to make an excellent officer in command of a unit in a larger organization."[15] As these reports suggest, by Armistice Day intelligence had become a presence in the American military.

The beginnings of an explanation for this change are not difficult to determine. From the standpoint of the human sciences, only as American involvement in the war began had psychology both attained sufficient cultural status and developed appropriate techniques to allow psychological notions of intelligence to reach a wider public.[16] As we saw in chapter 5, American psychology's sporadic interest in intelligence had not borne fruit in any practical sense until the first decades of the twentieth century, when the Binet-Simon and then the Stanford-Binet transformed the field. For most of the preceding period, there would have been few outside pressures for the army to employ the concept of intelligence as psychology understood it.

From a military-history perspective, the turn-of-the-century U.S. Army required few sophisticated techniques to assess its personnel. Small (approximately 6,000 officers and 200,000 soldiers even in March 1917), led by an officer corps trained primarily at West Point, and composed of career soldiers, the prewar army was a relatively intimate organization able to place and rate soldiers based on long-term familiarity and to maintain and propagate a well-entrenched military culture.[17] Although some groups within the military hoped to reform it along more Progressive, "scientific" lines, they had achieved few important successes before 1917. Everything changed, however, with the war. In only a few months, the army grew dramatically—ultimately seventeenfold, with more than 3.5 million soldiers in arms by November 1918, including more than 200,000 officers. With this enormous growth came tremendous stresses.

First, the scale of the increase and the war situation's urgency meant that many peacetime methods for selecting, training, and organizing soldiers no longer applied.[18] Whereas officers had relied on intimate knowledge of their troops and had slowly integrated the few new recruits into an essentially stable military organization, the war brought large numbers of new soldiers needing quickly to be trained, assigned to tasks best suited to their skills, and given leaders—both line and noncommissioned officers (NCOs). As Secretary of War Newton D. Baker noted in 1918:

> We are not getting the men of the same size in the same place, but all sizes in all places. . . . We have no time for men to grow up into those groups evolved by association, but we have to have a selective process by which we will get the

> round men for the round places, the strong men for the strong tasks and the delicate men for the delicate tasks. We have got to evolve a process by which that sort of assortment will take place. . . . [S]ome system of selection of talents which is not affected by immaterial principles or virtues, no matter how splendid, something more scientific than the haphazard choice of men, something more systematic than preference or first impression, is necessary to be devised.[19]

For Baker, as for many contemporary leaders, the "scientific"—equated with the "systematic" and the "objective"—was the key to organizing the great influx of soldiers.

Second, the rapid increase in troop size required a drastically supplemented officer corps. West Point alone could no longer meet the demand, and the consequent temporary appointment of experienced enlisted men and, in particular, inexperienced, usually college-educated members of the middle and upper-middle classes, meant that traditional military culture, already changing some in response to Progressivism, met a host of new influences.[20] Third, especially because of the draft, a much wider variety of Americans entered the military than had at any time since the Civil War. The enormous intellectual, cultural, and linguistic differences arising from a century of immigration and urbanization and from the creation of a large, free African American population presented challenges for which conventionally trained officers were little prepared.[21]

These pressures did not destroy traditional military culture, but they did create space in it for new possibilities, particularly ways of expanding and reshaping existing practices to cope with the new world of the mass army. The story of intelligence's entrance into the American military reveals how a group of scientists, while offering their services to the nation and furthering their own profession, attempted to exploit this niche, and thereby transformed both the knowledge they proffered and their own and their audience's cultures.

Yerkes and Army *a*

Robert Yerkes began his campaign to persuade the military to establish army-wide intelligence testing on April 29, 1917.[22] As chair of the National Research Council subcommittee on the psychological examination of recruits, he presented Surgeon General William C. Gorgas of the Army Medical Corps with his *Plan for the Psychological Examining of Recruits to Eliminate the Mentally Unfit,* completed just three days earlier.[23] Designed to uncover recruits with "intellectual deficiency, psychopathic tendencies, nervous instability, and inadequate self-control," Yerkes proposed administering a ten-minute mental test, individually, to any recruit for whom the recruit's medical or commanding officer felt "special psychological examination is indicated by exceptional or unsatisfactory behavior."[24] Those scoring "inferior" would then receive a battery of tests, with recommendations to the training-camp medical officer about discharge based on the results.

Yerkes' plan adhered closely to American psychology's practices and presumptions. Like the Stanford-Binet and his own Yerkes-Bridges Point Scale, Yerkes' army tests would measure intelligence primarily in order to detect mental deficiency, and required individual administration by trained experts.[25] However, even at this stage features of Yerkes' plan reflected the military's influence as well. Abandoning civilian mental testing's preoccupation with schoolchildren and the feebleminded, Yerkes proposed instead developing an instrument to measure adults. And faced with thousands entering the services daily, Yerkes avoided the painstaking, hour-long examinations of prewar intelligence testing, suggesting instead the creation of an instrument able to sift recruits rapidly, with more time-consuming methods reserved for those (presumably few) of questionable intelligence. Indeed, even Yerkes' focus on weeding out the mentally unfit reflected not just civilian practice, but also his assessment of probable military needs. Canada's wartime experience, Yerkes had learned, had convinced its military of the necessity of eliminating mental defectives from service.[26]

After receiving a tentative expression of interest from Army Surgeon General Gorgas on May 1, 1917—an interest not matched by Navy Surgeon General Braisted—Yerkes solicited funding and then convened his subcommittee at the New Jersey Training School for Feebleminded Girls and Boys in Vineland from May 28 to June 9 and again from June 25 to July 7, 1917.[27] There, Yerkes, Walter V. Bingham, Henry H. Goddard, Thomas H. Haines, Lewis M. Terman, Frederic L. Wells, and Guy M. Whipple—almost all prominent figures in psychological testing—began designing new tests to use with army recruits. Almost immediately the committee members decided to break in several important ways with Yerkes' proposal and common civilian testing procedures. Reconceiving the examination entirely, they envisioned developing several versions of a forty-five-minute multiple-choice group examination, to be administered to all recruits to identify the exceptionally superior as well as the inferior.[28]

In part, these changes reflected the group's diverse concerns. Both Terman and Bingham, for example, were strongly interested in the intelligence spectrum's upper end, and both had been experimenting with the technology of group mental testing for some time. Two other factors were also critical. First, for reasons that are not entirely clear, the committee became convinced that it was important to test every recruit, a decision that

TABLE 6.1
Tests in Army *a* and Their Contribution to Total Score

1.	Following oral directions:	4%
2.	Displaying memory span:	5%
3.	Rearranging sentences:	8%
4.	Solving arithmetical problems:	8%
5.	Answering general information questions:	16%
6.	Giving synonyms or antonyms:	16%
7.	Displaying practical judgment:	4%
8.	Solving number series:	6%
9.	Giving analogies:	16%
10.	Doing number comparisons:	16%

alone required a rapidly administrable and scorable group test. Second, the psychologists' growing appreciation for the military's needs and concerns made them aware of issues—including malingering and cheating—that had not arisen in civilian testing. There had seemed little need for multiple forms to prevent cheating or multiple-choice to allow quick, objective scoring when individually examining children or the feebleminded. In the military, however, things were different, as Yerkes was made aware when the navy rejected his overtures explicitly because malingerers might manipulate psychological methods to avoid their "proper" service.[29] Communications with other military officials confirmed that the fear of malingering was pervasive, and that any new methods the army or navy employed would have to fit their accelerated wartime training schedules.[30]

Having redefined its endeavor, the committee devoted the remainder of its first week to building the new intelligence-measuring instrument, by choosing those suitable for the new instrument from among existing tests and others they could quickly devise. By week's end, after seriously investigating thirteen tests, ten were assembled as the first army intelligence scale, Army *a* (see table 6.1).

Army *a* marked an important break with civilian intelligence-measuring instruments. Although the individual tests were based on commonly used methods, Army *a* transformed them as an overall package by placing a premium on economy of time, security, measurement across a broad spectrum, and adaptability to group administration. Even more significant was the test's elimination of the expert examiner. Previously most mental tests had required a trained professional to administer items, judge answers, and interpret results. But the Vineland committee designed an instrument that substituted technology for professional skill and generated only the most minimal quantitative data—a single number. While the output in some sense measured the same entity, intelligence, as did civilian psychologists, the information produced was enormously reduced and the psychologist's role profoundly altered, even when compared to the Stanford-Binet. With Army *a,* data interpretation rather than behavior observation became the essence of the psychologist's job and the badge of expertise. Given the desire to examine all of the thousands of conscripts entering the military daily, the Vineland committee completed a process—already begun within civilian psychology—of remaking mental testing into an endeavor that subordinated professional judgment to objective determination and statistical manipulation.

The effects of this transformation to fit the needs of the mass army extended down even to the examination's individual elements. Regarding one of the tests excluded from Army *a,* for example, Yerkes noted:

> it was generally agreed that the Trabue type of completion test is a better measure of intelligence than some of the other tests finally accepted, as for example, the number-comparison or memory-for-digits tests. However, the difficulties in securing alternative forms of this test and arranging it for response without writing and objective scoring were too great to be overcome in the time available.[31]

The Trabue test had not failed to meet psychological criteria; it was judged an excellent measure of intelligence. But it also had to meet military criteria; failing them, it was rejected in favor of inferior measures of intelligence. Forced to choose between psychological and military standards, the Vineland psychologists would sacrifice, to a certain degree, even the construct they intended to measure to fit army conditions.

There were limits, however, to how far the committee would push adaptation in this first phase of development. After completing construction of Army *a,* they initiated trial testing on 469 subjects at a variety of institutions with different populations: the Massachusetts School for the Feebleminded, Boston Psychopathic Hospital, Reformatory for Men in Concord, Carnegie Institute of Technology, Fort Benjamin Harrison officers training camp, and Philadelphia Naval Yard, as well as a California high school and a prison.[32] The committee members began analyzing the results when they reconvened two weeks later. Their major task was to assess their scale's validity, i.e., to determine whether Army *a* did indeed measure intelligence. In general, early twentieth-century psychologists employed three methods for validating tests: first, demonstrating that the test adequately sampled the domain being investigated (that a test contained questions of appropriate content and difficulty to demonstrate that the test-taker knew the subject); second, demonstrating that the new test correlated with some already legitimated measuring instrument for the characteristic; and third, demonstrating that the new test correlated with another non-test-linked legitimated evaluation system.[33]

In this first assessment of Army *a,* Yerkes' committee employed primarily the first and second methods, correlating the scores on the trial tests comprising *a* both with scores on the instrument as a whole and with performance on an accepted test of intelligence, generally the Stanford-Binet. The Vineland psychologists accorded greatest significance to the high correlation (about 0.8) of Army *a* with Binet measures, thereby ensuring that *a* would assess the same construct they used in their civilian pursuits, psychological intelligence. Pleased with the results, the committee's only recommendation, other than suggesting some minor adjustments to scoring, was that a much larger trial be conducted, preferably solely on military personnel.[34]

The responses to the new instrument and its approach to testing were many and varied. The military said little, merely approving a further trial of Army *a* at four military institutions. Psychologists, not surprisingly, reacted more vociferously, especially to the "mechanization" of the examining process. While a number applauded the committee, many expressed reservations. Some remarked that the process's routinization disinclined them to participate, because there would be no real demand for their expertise.[35] Others, including H. C. McComas at Princeton, were more troubled, worrying that, without qualitative data, quantitative ratings alone would be insufficient "if our purpose is to supply material which will enable an officer to classify his men."[36] And, finally, some, such as Edward L. Thorndike and James Angell, dismissed the endeavor entirely, convinced that the testing program would be of no real value to the service.[37]

Yerkes had invested too much in the work to agree with Thorndike and Angell. Nonetheless, he did not simply ignore such criticisms. In July 1917 Yerkes issued a new proposal, his "Plan for Psychological Military Service," which formalized the Vineland decisions and added a new feature that marked the next step in Army *a*'s evolution: the correlation of the psychological measurements "with the industrial and military history of the individuals examined." Stressing the need for "immediate military serviceableness" and to have "psychological measurements correlate with the actual performance of soldiers and sailors," Yerkes opened the door to reconstructing the intelligence instruments to better accord with what he presumed would be the military's own standards.[38]

With the aid of a $2,500 grant from the Committee on Furnishing Hospital Units for Nervous and Mental Disorders, a second set of Army *a* trials was carried out on approximately four thousand individuals in four military institutions during July and August 1917. Five committees directed this part of the project: one for each of the four institutions, and a fifth, the Statistical Unit, headed by Thorndike, the eminent psychologist at Teachers College who had been critical of the whole program. Thorndike and associates Arthur S. Otis and L. L. Thurstone had two assignments: first, to weight the contribution of each test in *a* to the overall score, and second, to analyze *a*'s validity as an instrument for measuring intelligence. Thorndike pursued the first even before the data had been compiled. Using "(1) the combined opinions of a dozen psychologists as to the relative weights to be attached to the tests; (2) a rough estimate of the variabilities of the tests; and (3) a rough estimate of their inter-correlations," Thorndike assigned weights ranging from one to three to each of the ten tests.[39] This new scheme increased the contribution to overall score of tests #4-arithmetical problems and #5-information and decreased that of tests #6-synonyms/antonyms, #9-analogies, and #10-number comparisons.

Most of the work for *a*'s developmental phase, however, lay in analyzing the trial results in order to judge and improve the instrument's validity. The Statistical Unit performed three sets of calculations: a rough correlation of results on Army *a* for various groups with expectations about their intelligence, correlations of the tests and the scale as a whole with officers' ratings of examinee intelligence, and intercorrelations of each test with the others in the scale. In two respects this procedure differed markedly from the validation methods used in Vineland.

First, although both the Statistical Unit and the Vineland group analyzed test intercorrelations to check Army *a's* validity, Thorndike and Yerkes disagreed about the optimum standard for intercorrelation. While Yerkes believed that intercorrelations should be high but not too high, probably around 0.5, Thorndike argued that they should be as low as possible, stating that "in proportion as two tests intercorrelate closely, they are repetitive—i.e., are measures of the same fact—and a high weight to each of them will mean an undue weighting of the same fact."[40] This difference reflected their divergent beliefs about intelligence: Yerkes conceived of it as unitary, whereas Thorndike did not.[41]

Second, the Vineland group relied primarily on correlation with a well-established instrument, the Binet scales, to judge validity. Finding that the correlations "were high with outside measures of known value," the committee concluded that Army *a* was satisfactory.[42] The Statistical Unit did otherwise. Acting on Yerkes' recommendation that intelligence rankings be connected to military success, Thorndike's unit divided the test population into three groups—adult defectives, enlisted men, and officer trainees—and determined how each had performed on the test. They found that adult defectives received low scores and enlisted men average scores, while officer trainees achieved superior scores, and that performance on *a* correlated at a moderately high level (about 0.5) with officers' evaluations of their recruits. From this, Thorndike and his associates concluded that Army *a* was a valid intelligence-measuring instrument, able to prophesy reasonably well, as Thorndike put it, "the mental ability which a man will display in the Army."

In the process, however, the Statistical Unit continued shifting Army *a*'s focus from civilian estimations of intelligence toward explicitly military ones. As Thorndike argued to Yerkes:

> The group test is to be used to prophesy the mental ability which a man will display in the Army. *Our best attainable measure of that is the rating for mental ability given to men by their company commanders.* If any one of the ten tests correlates zero with officers' ratings, it deserves zero weight in the composite score used for the prophecy. If it correlates highly it deserves much weight.

For all of his vehemence, and although he recommended some adjustments, Thorndike never changed any of his initial weightings for Army *a,* concluding that "the tests all intercorrelate so closely that the revised weighting would not produce a much better result."[43] Nonetheless, the Statistical Unit's decision to validate Army *a* against officer evaluations, and its willingness to change the instrument to increase *a*'s correlation with this measure indicate the importance they accorded to making Army *a* militarily serviceable.

For Yerkes the importance of a good correlation with officer ratings was essentially rhetorical. Yerkes did not doubt that Army *a* measured intelligence; the initial correlation with Binet scores and his own conception of intelligence as unitary ensured that. What Yerkes needed was evidence of the examination's "serviceability," of intelligence prophesying military performance. Yerkes distinguished between the validity (the "reliability," in his term) of the scale and its serviceableness, turning to the military to develop the latter. Thorndike wanted to go further, emphasizing the composite, context-dependent character of intelligence.[44] For Thorndike, intelligence was heterogeneous; he argued that there was such a thing as army-specific mental ability, and that, therefore, the validity as well as the serviceableness of Army *a* depended on its reflecting the military definition of intelligence. Thus, while Yerkes was content to know that *a* correlated well with officer ratings, Thorndike wanted to rebuild Army *a,* using high correlation with officers' ratings and low intercorrelations with other tests as his criteria. Yerkes, however, could not abide this level of military intrusiveness into psychological affairs, and so Thorndike's suggestion that Army *a* be revamped in light of military definitions of intelligence fell on deaf ears. When Army *a* underwent its final major revision and was transformed into Army Alpha, from December 1917 to January 1918, Yerkes turned to a group of psychologists more in tune with his own views on intelligence and the place of military concerns in the examining program.

From Army *a* to Army Alpha

On August 17, 1917, Yerkes was appointed a major in the Sanitary Corps of the Army Medical Corps, charged with organizing its new Psychological Division. In many respects this marked the beginning of the army's active engagement in the psychological examination of its recruits. Whereas previously contact between Yerkes' psychologists and the military had been informal and the psychologists had operated mainly on the basis of assumptions about the military's wishes, by August, Surgeon General Gorgas and members of the Office of the Chief of Staff and Office of the Adjutant General had begun taking a keen interest in the examining program and what it could accomplish.

Yerkes helped to initiate the army's serious involvement by submitting to the surgeon general and secretary of war a plan for an official army trial of the testing program. Gorgas was suitably impressed, and promoted the proposal to Chief of Staff Peyton C. March by emphasizing the test's ability to detect mental defectives and the fact that its "results correlate highly with officers' judgments of their men."[45] Quickly approved, Yerkes' plan called for administering Army *a* and other measuring instruments to recruits in four cantonments: Forts Devens, Dix, Lee, and Taylor.

Running from September through December, this third trial included more than 65,000 soldiers examined under the same conditions that would obtain if the army adopted the program. The psychologists working in the camps had two major tasks: supervising the administration of Army *a,* and—emblematic of the changes in psychological examining—generating numerous statistical reports (107 in all) analyzing the data in order to suggest practical improvements in the testing procedures vis-à-vis army needs. Preliminary findings indicated that illiteracy was much greater than expected and, significantly for the program's future, that the distribution of intelligence by company varied enormously.[46]

Even before this trial had been completed, Yerkes and his colleagues began receiving feedback from the military. At first the news was not encouraging. Reports from camp psychological examiners indicated indifference or downright hostility from many officers, and in two of the camps psychological examination was significantly delayed. Thus there must have been an enormous sense of relief when the first official evaluations arrived. Major General Adelbert Cronkhite of Camp Lee set the tone in his memo of November 10, 1917. Praising the program, General Cronkhite noted that "the results of the psychological examinations are fully borne out by actual observation of the abilities and the capacity of various officers in the duties assigned to them."[47] Subsequent reports by Colonel Henry A. Shaw of the Medical Corps, Brigadier General H. B. Birmingham of the Surgeon General's Office, and Colonel John J. Bradley of the War College Division of the Office of the Chief of Staff all seized on the close correlation between test results and officer rankings as *the* indicator of the program's value.[48] As Colonel Bradley put it:

> This subject of psychology in its relation to military efficiency is an entirely new one and the War College Division approached it with a good deal of doubt as to its value. A very thorough study of the reports submitted, however, has firmly convinced it that this examination will be of great value in assisting and determining the possibilities of all newly drafted men and all candidates for officers' training camps.[49]

The "reports submitted" were a survey of officers inquiring whether they had found the test ratings to accord with their judgments. Support for the program proved strong: 58 percent of respondents were favorable without qualification, 23 percent favorable with qualification, and only 19 percent in any way unfavorable.[50] The army never wavered in the criteria it used for judging the examining program, consistently asking two questions: Did the psychological ratings match assessments by experienced officers? And, was it worth the time and effort to obtain those ratings? In this first set of assessments the answers were affirmative, and on January 16, 1918, psychological examining was extended to the entire army.

The army's intelligence-testing program was detailed in a memo from Gorgas to Adjutant General John S. Johnston on January 3. This third plan codified the procedures developed and experience gained during the Army *a* trials, scaling them up for army-wide deployment. Materially, it called for a staff of 132 officers, 124 NCOs, and 620 enlisted men to be distributed among thirty-one cantonments; construction of examination buildings in each camp; and establishment of a two-month training school for psychologists at Fort Oglethorpe, Georgia, where both military and psychological training would be emphasized. Procedurally, the plan mandated that individuals receiving the lowest rating, E (very poor), should be individually examined and, if still found mentally inferior, *recommended* for discharge or assignment to a service battalion. And in terms of purpose, it proposed that every enlisted man and newly appointed officer be tested, setting out three goals: to identify the mentally inferior, discover the mentally superior, and reassign men so that "companies and regiments within a given arm of the service may be of approximately equal strength mentally and therefore actively."[51]

Three features of this plan merit particular interest. First, testing new officers as well as enlisted men would eventually pose enormous problems and become the crux of much debate. In practice, the decision meant that those most critical to the military's acceptance of mental testing were as likely to be antagonized as won over by the program, because their own futures could depend on their examination results. Second, by ensuring that psychological examiners received military training and could only make recommendations concerning discharge and placement, the plan highlighted the ambiguous status of psychological examining in the military. Clearly the military's leaders had been convinced that intelligence testing might be of value, but just as clearly they remained unsure, and so sought to limit psychologists' autonomy and remind them of the need to fit into the military system.

Finally, Gorgas's proposal articulated a new purpose for the examining process. Where earlier plans had focused first on eliminating the unfit, and then on identifying the superior, Gorgas added a third: helping commanding officers balance their companies according to intelligence. This broke with standard practice both military and civilian. The army had always had some mechanism for eliminating unwanted soldiers and promoting valuable ones; psychological examining merely contributed to these functions. Balancing companies according to intelligence, however, was new. Before intelligence testing, no information was available that would have differentiated to so fine a degree among the vast majority of enlisted men. But once a numerical evaluation scheme was established, small differences became visible and magnified. The camp psychological examiners' reports noted the phenomenon early: great imbalances existed between companies vis-à-vis the distribution of mentality levels.[52] This "imbalance" was also noticed by company commanders, because they could now *see* the intelligence of their enlisted men expressed numerically and compare their recruits with others. In the ensuing months, officers would consistently cite the possibility of using intelligence ratings to balance companies as a principal value of the program.

With the extension of psychological examining to the entire army, the Psychological Division's next task was to finish analyzing the data from the final trial. Yerkes assigned this duty to a committee headed by Captain Clarence S. Yoakum, from the University of Texas and an officer in the army's Psychological Division; the members included Carl C. Brigham, Margaret V. Cobb, E. S. Jones, Lewis M. Terman, and Guy M. Whipple. Their mandate was to use the results from the full-scale trial to thoroughly revise Army *a,* correcting the defects apparent during the trial.[53]

Yoakum's committee assessed each of Army *a*'s tests, both on its own terms and to determine its contribution to the scale's overall validity. In the main, the committee's statistical findings were similar to those arrived at during prior validations. Like the Thorndike group, Yoakum and his associates determined that Army *a* clearly stratified the examinees into the expected intelligence hierarchy—officers, officer trainees, sergeants, corporals, enlisted men, and the feebleminded—and used this information to discover which tests contributed most to the scores of higher-scoring subjects and which to the scores of lower-scoring ones. Their analyses of various correlations performed for each test also produced results similar to Thorndike's.[54] Although there were some differences with the Vineland data over correlation with the Stanford-Binet, even these discrepancies were relatively insignificant, as both analyses picked the same five tests as best and the same three as worst.

Nonetheless, the instrument that Yoakum's committee produced based on these statistics, Army Alpha, differed significantly from revisions of Army *a* proposed previously. Whereas Thorndike's group, for example, had suggested changes intended to increase correlation with officers' ratings, the major revisions proposed by Yoakum's committee derived from a much different agenda: increasing Alpha's correlation with Stanford-Binet score and reorienting Alpha toward differentiating best in the intelligence scale's higher range. To this end, the Yoakum committee's specific recommendations were to increase the weight of the test exhibiting the highest correlation with Stanford-Binet score, #6-synonyms/antonyms; to drop entirely the two tests that correlated most poorly with Stanford-Binet; and to add harder items to all remaining tests.[55]

Three factors stand out as crucial to explaining why Yoakum's committee sought to change Army Alpha's emphasis. First, the leaders of each evaluation group viewed intelligence differently. Thorndike had long believed in the composite, heterogeneous character of intelligence. Terman, on the other hand—undoubtedly the most influential member of Yoakum's committee—like Yerkes conceived of intelligence as a single entity, differing in degree but not kind and distributed throughout the population. As a result, Terman believed that the individual tests must cohere and that every valid instrument assessing an individual's intelligence should produce essentially the same result.[56] Army Alpha represented this second view. Yoakum and his associates created an instrument embodying a conception of intelligence as universal and unidimensional, distinctly different from what Thorndike's group had envisioned.

Second, differences between Army Alpha and Thorndike's version of Army *a* can be traced to changes in the goals for the testing program. Walter Dill Scott's insertion of intelligence into the criteria for evaluating all army personnel and the desperate need for appropriate NCO and officer-training-school candidates meant that methods for categorizing and assigning all recruits, especially the more able, were of high priority.[57] Yoakum's committee responded to this need by altering the test's character, transforming it from a general measure that discriminated equally well at all intelligence levels, as understood by Thorndike, to one that worked best at the upper ranges, where the distinction between officers and enlisted men was thought to reside.[58]

Finally, differences between Army Alpha and Army *a* arose because Yoakum's group focused principally on the new instrument's adequacy according to professional psychology's standards. Yerkes revealed his own ambivalence about military notions of intelligence in a letter to Thorndike: "Best thanks for your letter of March 18th relative to correlation of army intelligence measurements to intelligence ratings given by officers. I am surprised that the correlation is so high, for in my opinion the officers' ratings are often based upon very peculiar conceptions of what is meant by intelligence."[59] Rather, what Yerkes valued, as he put it succinctly to Terman, were *psychological* measures of intelligence: "Examination *a* is really a remarkable creation of *psychological* intelligence."[60] "Psychological intelligence" was what Yerkes wanted, and Terman and Yoakum complied when they made correlation with Stanford-Binet score the principal criterion for judging Alpha's performance. The Stanford-Binet was *the* measuring rod for intelligence within civilian psychology. Employing it ensured that Army Alpha would seem legitimate to civilian psychologists.

The emphasis Yoakum's committee placed on civilian psychology's norms did not mean, however, that the testing program's military context had had no influence. Indeed, Yoakum's appointment as committee head was explained by "his intimate acquaintance with the conditions and results of examining in the camps."[61] In addition, Yerkes justified the revisions of Army *a* to the surgeon general not in terms of better measurement, but as a means of saving paper and time.[62] What Yoakum's actions indicate, however, is that the military was not the only influence. The psychologists building Army *a* and Army Alpha confronted two sets of pressures. On the one hand, they knew that their primary consumer was the army and that the tests would never be adopted if they failed to meet that organization's needs. On the other hand, these testers were all professionals and thus expected to follow the dictates and practices that defined their community. While these two audiences were not necessarily in conflict, the realities of designing and implementing a new testing program meant that tensions and thus the need for accommodations constantly arose. Different groups of psychologists resolved them differently, but neither Thorndike's group nor Yoakum's could escape entirely the constraints of meeting the demands of these dual audiences.

Army Alpha: Between Psychology and the Military

Revising Army *a* into Army Alpha took Yoakum's committee most of January 1918. For three months psychological examining continued, most probably with Army *a*, at the relatively slow pace of about fourteen thousand recruits per month while new forms were printed, examiners trained, and bureaucratic wars fought.[63] The skirmishing's nominal cause was the Psychological Division's request—in keeping with the surgeon general's January 3 memo—for additional personnel and the construction of examination buildings in each camp.[64] The adjutant general and the Office of the Chief of Staff denied the request in February, provoking a flurry of memos. In the end, camp commanders were instructed to make some existing building available, and the commissioning of additional psychologists was delayed.[65] Though petty, these battles symbolized the continued ambivalence, outside the surgeon general's office, about mental testing's military value. The depths of that uncertainty would be revealed just at the moment of Yerkes' greatest military triumph.

On April 28, 1918, psychological examination of enlisted men, officer-training candidates, and line officers began in earnest. The new forms—Alpha for literates and Beta for illiterates—were introduced, and the testing rate jumped to over 200,000 per month, a level that would continue almost until the war's end. Simultaneously, members of the Chief of Staff's Office expressed concern about the endeavor's real value. Spurred on, probably, by a highly negative report on the testing program from Lieutenant Colonel Lear of the War Plans Division, three full-scale investigations of the program were launched in May.[66] The first, conducted by the Adjutant General's Office, surveyed commanding officers about what benefits they had derived from the psychological work and whether they believed it should be continued.[67] Nearly one hundred responses were received, most fairly skeptical about the test program's value.[68]

The second and third studies—one carried out by Goldthwaite H. Dorr for the assistant secretary of war and the other performed by Colonel R. J. Burt for the chief of staff—mandated personal observation of the program and interviews with those dismissed because of it. Dorr's report, issued June 10, 1918, was generally negative, though it did not recommend eliminating the testing program. Rather, Dorr suggested its transfer to the Committee on Classification of Personnel, its revision with senior officers' input, and its use solely to *aid* in the placement of recruits. Burt's report, delivered June 18, was more favorable. Although he too recommended revisions, especially testing officers only upon their commanders' request, Burt strongly supported the continued examination of all recruits and an increase in testing personnel.[69]

These two reports are most striking, however, not in their differences but in their similarities. They constituted the first extensive investigations of military psychology not conducted by the Surgeon General's Office, and the degree to which Dorr and Burt's findings concurred is notable. Both reported the same three objectives for psychological examining, those Gorgas had articulated in his January 3 memo: eliminating the unfit, identifying the superior, and balancing units by intelligence. Both emphasized that test results should be used solely as *aids* in determining an examinee's fate, that intelligence was only one of many qualities needed by a good soldier, and that the program required close *military* supervision (though Dorr lauded the military training the examiners received). And both noted that officers generally found a high correspondence between their evaluations and the examinations' results. This last point was crucial, as Dorr grudgingly admitted:

> The closer observation I made of the working of all of the psychologic tests, the more impatient I became with their failure through their mechanical character to accurately gauge in certain instances the actual workings of the mind. . . . Nevertheless, I have concluded that in such close observation the defects of this system are unduly magnified and the substantial accuracy of the average result lost sight of. *I have reached this conclusion because of the great weight of testimony of officers who have compared the results of the tests with their own observations of the men tested that there is a striking correspondence in the results of the tests and their own observation as to the mental alertness and agility of the men examined.*

Dorr was not alone in worrying about the examinations' mechanical nature and their failures to accurately gauge a soldier's military value. These fears were widespread and would persist throughout the program's operation. Nonetheless, Dorr supported continuing psychological examining, persuaded by two related factors. Given a situation where "large groups of green men are put into the hands of Company officers, themselves of limited experience," testing seemed to have a practical function.[70] And neither Dorr nor other officials could ignore the support for examining among company commanders. As long as test results matched officer evaluations, the examinations could not be easily dismissed.

The upshot of these three investigations was to maintain the army's ambivalent support for psychological testing in some form, though with varying opinions about specifics.[71] General Order No. 74, issued on August 14, 1918, should have settled the issue.[72] Apparently a victory for testing's critics, G.O. 74—while officially establishing the Psychological Division—sharply limited the program's scope by placing all examining at the discretion of the commanding officer and eliminating individual examinations. It also limited army testing to a single role: helping company commanders balance units mentally. Nonetheless, that month a record 300,000 soldiers were tested, with little diminution subsequently, and in November, Circular No. 65 would urge that intelligence ratings be recorded on *all* enlisted men and officers' qualification cards.[73]

Yerkes' response to these military uncertainties was to redouble his efforts at persuasion, along two lines. As he explained in his *War Diary,* "we should be able to make [psychological examining] more intensely practical and more directly military in appearance."[74] The interest in appearing military had come early in the testing program's history. Directions explaining what the examinees should do, for example, were given as commands—"Attention!" "Go!" "Stop!"—with a premium placed on absolute obedience: "Now, in the Army a man often has to listen to commands and then carry them out exactly. I am going to give *you* some commands to see how well you can carry them out. Listen closely. Ask no questions. Do not watch any other man to see what *he* does."[75] Contrast these instructions with those for the adult level of the Stanford-Binet: "I want to find out how many words you know. Listen, and when I say a word you tell me what it means."[76] Moreover, Yerkes and his colleagues actually joined the army, taking commissions and receiving military training at Fort Oglethorpe.[77]

After the program's rather unenthusiastic reviews, Yerkes repeatedly exhorted his examiners to increase their endeavors' military appearance. He also revised the test's ratings scheme so that they were explained specifically in military terms:

> *What the grades mean.*— All men are classified by the tests as A, B, C+, C, C−, D, DD−, or E as follows:
>
> *A.* *Very superior intelligence.*— High officer type when backed by other necessary qualities.
> *B.* *Superior intelligence.*— Commissioned officer type and splendid sergeant material.
> *C+.* *High average intelligence.*— Good N.C.O. material with occasionally a man worthy of higher rank.
> *C.* *Average intelligence.*— Good private type, with some fair to good N.C.O. material.
> *C−.* *Low average intelligence.*— Ordinary private.
> *D.* *Inferior intelligence.*— Largely illiterate or foreign. Usually fair soldiers, but often slow in learning.
> *DD−.* *Very inferior intelligence,* but considered fit for regular service.
> *E.* *Mental inferiority,* justifying recommendation for Developmental Battalion, special service organizations, rejection or discharge.
>
> *The grades should be consulted.*— (a) In the selection of candidates for officers' training schools; (b) in the selection of all noncommissioned officers; (c) in balancing organizations; (d) in picking men for special detail; (e) in the classification and training of men in Development Battalions; (f) in court cases; (g) in the better understanding of men who are in any way peculiar or exceptional. (h) The tests have also been used effectively in the selection of nurses, Y.M.C.A. personnel, etc.[78]

In all these ways Yerkes and his committee structured the army tests to seem less like foreign objects imposed by ivory-tower academics, and more like familiar techniques and practices developed by army personnel.

In addition to making the examinations more military in appearance, Yerkes also worked to increase the practicality of the instruments. Continuing the theme of serviceability, Yerkes repeatedly sought to constrain those features that the General Staff found most objectionable and to increase the examining process's efficiency. Thus, responding to persistent worries that malingerers were exploiting intelligence testing, the Psychological Division narrowed the range on its lowest score category, level E, so that the number of recruits recommended for discharge would be reduced; the psychologists also stressed that testing's most important function was to indicate the "type of service for which each man or group is fitted."[79] In addition, Yerkes forcefully reminded camp psychologists that one of their primary duties was to persuade camp commanders that the tests had practical value: "practical service is the only justification for the continuation of psychological examining or any other kind of psychological work in the army. . . . [D]emonstrate your usefulness to the officers of your camp and thus command their interest and cooperation."[80] If the response from George F. Arps, chief psychological examiner at Camp Sherman, is any indication, psychologists in the field often strove mightily to meet these expectations. "Re the memo," Arps wrote Yerkes, "I trust you will bear in mind that all scientific niceties were thrown overboard and that I have in this, as in most matters, taken an entirely practical attitude. The Div. Surgeon remarked that it 'cut the critics from under'."[81]

"Cutting the critics from under" was certainly Yerkes' goal. Whether by making the examining program more military, altering the instruments to increase efficiency and their results' palatability, or actively "selling" psychological methods, Psychological Division members worked diligently to persuade the army of their endeavors' value. However, what Yerkes did not suggest changing is as important as what he did. Yerkes made no mention, as Thorndike might have, of revising the scales' guts to produce results even more consonant with army expectations. Partly this was practical. Millions of forms had already been printed by summer 1918, and hundreds of thousands of servicemen examined. Partly, as well, it reflected the fact that commanding officers were already reporting close correspondences between intelligence rankings and their own ratings of soldiers. And partly it stemmed from the army testers' status as experts whose expertise derived from adhering to civilian psychology's standard practices. Thus their turn to correlation with Stanford-Binet as the primary validation measure when transforming Army *a* into Army Alpha, and thus really their desire from the start to induce army leaders to view intelligence as a characteristic of critical concern.

The Army's Response to Alpha

What resulted from this process of accommodation and resistance? Not, as we have seen, the military's easy assimilation of mental testing. While Yerkes was able to convince the army's top echelon to institute examining army-wide, he met with less success in demonstrating to rank-and-file officers and the high command testing's overall utility.[82] In part, Yerkes' problems derived from the army's very organization. Yerkes confronted a structure that either had had no history of interest in his techniques or had developed methods associated with army doctors and neuropsychiatrists little disposed to cede territory to a rival professional group.[83] Thus Yerkes could never demand that his tests alone control elimination of the unfit, officer selection, or other functions; instead, he had to settle for Army Alpha and Army Beta being conceded, at best, an advisory role.

More importantly, the ambivalent reaction to the program can be traced to the range of responses it elicited from officers. Many were persuaded rapidly. Citing, almost invariably, the close correspondence between test results and their own evaluations, these officers generally praised the tests' accuracy and expeditiousness and argued that Alpha would prove especially helpful in dealing with large numbers of unknown recruits needing rapid training and placement.[84] As Captain Norbarue Berkeley noted:

> The report of the psychological examination has been used and much weight given to it in recommending men for the third Officers Training camp, recommending men for appointment as non-commissioned officers and in sorting out the poorest men in the Battery for special instructions and drill.[85]

Other officers, although initially more skeptical, were eventually won over when longer acquaintance with the testers assigned to their units and firsthand experience with the examinations convinced them that intelligence results could aid personnel evaluation without disrupting the unit's functioning or their own command prerogatives.[86]

Nonetheless, many officers, ranging from major generals to first lieutenants, were much less sympathetic. Generally, they advanced one or some combination of three reasons. First, many believed that intelligence tests were unnecessary, that there existed sufficient opportunities to observe one's subordinates early in training to enable time-honored military methods of evaluation and placement. Second, some argued that the examinations simply did not measure intelligence. "This test," as one colonel reported, "is considered more a test of a man's familiarity with general information taking the nature of current topics, etc. rather than a measure of intelligence."[87] Others commented that the tests measured speed and accuracy more than higher mental powers, and that they privileged school learning over military experience.

Finally, a number of officers hostile to the program argued that intelligence alone was not sufficient to make either a good officer or a good enlisted man. As the commanding officer of the 41st Company at Camp Lee graphically put it: "The ability to handle men, the gift of making men follow is seldom shown on paper. The most intelligent clerk can not compete in the gentle art of murder with a two-handed athlete whose ambition does not rise above the rank of sergeant."[88] Other officers stressed bearing, demeanor, loyalty, willingness to follow orders, and especially character when describing the qualities they looked for in a soldier.

This deep division within the officer corps over the test's relevance to military needs precluded effective implementation of Yerkes' program. Although testing continued apace from May 1918 until the Armistice, by war's end no system had been established to ensure even that a soldier's intelligence ranking be included in his personnel file, much less that it be used to determine his duty assignment. Rather, Yerkes found himself constantly struggling to convince company commanders and higher-ups that his tests provided information of value to their immediate concerns.

Nonetheless, the army's intelligence-testing program did have some important effects. More than 1.75 million men were tested, and the program was judged sufficiently valuable for it to be extended to all training camps and maintained even in the face of heavy criticism. Moreover, some individuals' lives were deeply affected by their scores. More than 7,700 recommendations for discharge and 28,000 for transfer were forwarded to army discharge boards based on the army tests.[89] In addition, although there was no official policy on using intelligence test results to select NCO or officer-training-school candidates or to balance units in terms of intelligence, reports demonstrate that many commanding officers relied on these results either to corroborate their own judgments or as a principal selection tool. As Yerkes noted in a memo to the surgeon general, "Lieut. H. T. Moore, Camp Cody, reports that the Personnel Officer has assigned drafted men partly on the basis of psychological ratings, taking care to place an equal number of A, B, C, D, and E men in each organization."[90]

More significantly, a dramatic mental shift occurred: intelligence became a characteristic of consequence in army culture. One indication of this transformation has already been mentioned: officers' growing tendency to assess their subordinates' intelligence in evaluations. The testing program helped teach officers to take cognizance of intelligence, and by war's end many were accustomed to seeing intelligence as a distinct and relevant characteristic of their men, of direct military value. Indeed, intelligence ratings took on enough authority that, as Franz Samelson notes, a technique known as the "passing cull" developed: "unit commanders would hold onto soldiers with high intelligence ratings and try to pass along [to other companies] those with low ratings."[91]

A second indication of intelligence's new place in the military was the army's postwar reliance on mental tests. Not only were intelligence tests administered to recruits for at least a year after the Armistice, but in 1919 the army commissioned a new set of tests for illiterate and non-English-speaking recruits, and it also used intelligence scores to study the relationship between low mentality and criminal or nonmilitary behavior in breaches of discipline.[92]

But perhaps the most telling example of the increased authority accorded intelligence was the November 8, 1918, revision of Special Regulations No. 65, which stated that morons must be unconditionally rejected from the service and defined a moron as "an individual whose mental development is that of a child not over eight years of age, as measured by the Binet-Simon test."[93] Modified in June 1919 to apply to recruits of mental age ten years or less, the decision to create a minimum standard of intelligence, and to do so in the language of psychology using psychological methods represented a significant transformation in the army way of thinking.[94]

Before the war it would have been, almost literally, inconceivable for the army to have promulgated something like S.R. 65. The elimination of morons, in and of itself, was not surprising; the standard recruit medical examination for some time had acknowledged lack of sound understanding as cause for discharge.[95] The turn in S.R. 65 to a precise, test-based standard, numerically defined, however, violated the army's long-standing tradition of personal, subjective evaluations. Things changed over the course of the war. As the military began to understand and accept the new psychological knowledge, and as the authority of that knowledge grew, the military itself became transformed, and intelligence pushed the transformation.[96]

The *Citoyen* as Soldier: France Builds a Republican Army

The situation was quite different for America's ally France. It, too, substantially transformed its military to address the problem of mass mobilization, and faced as well the question of whether to introduce intelligence testing into its recruitment procedures. But throughout the period before the Great War, French military policy was haunted and oriented by France's defeat in the Franco-Prussian War and the consequent desire for *revanche* (revenge). These concerns, combined with the republican commitment to *égalité* for the masses and producing elites via concours and grandes écoles, led virtually all Third Republic governments to maintain a sizable army (with potential for quick expansion) based on universal male conscription and an officer corps trained at elite schools.[97] Mass mobilization, therefore, at least conceptually presented few major logistical difficulties for the French military, nor was it worried about putting into the field a sufficient number of trained officers. Where personnel problems had been encountered, they arose largely out of the duration of a conscript's active service, which was debated and adjusted throughout the period, and the extent of the recruitment, which brought to the fore the question of what to do about physically or mentally unfit inductees. Confronted early, the government put into place a number of procedures—all based on personal assessments performed by civil, medical, and/or military authorities—designed to judge the military fitness of its recruits and conscripts, both before and especially after their entry into the army. Thus, when in 1908 Alfred Binet and Théodore Simon suggested that intelligence testing be experimented with as a way of identifying quickly a certain class of "defective" conscripts or recruits, their proposal was considered against the backdrop of years of experience with other methods of eliminating those deemed unfit.

The Third Republic's first important military legislation was the law of July 27, 1872, which established the groundwork for all subsequent decrees on military recruitment up to World War II. It began by proclaiming that military service was the duty of all French (male) citizens, beginning at age twenty, mandating five years of active duty plus fifteen years of reserves. The July 27 law also set out the conditions for service exemptions. Most were for peacetime, and based either on family status—such as being the eldest of orphans or having a brother killed in active service—or on occupation, including being a teacher or in a religious vocation. In subsequent versions of the law, these exemptions were largely limited or eliminated. Certain other conditions, however, were deemed so debilitating as to necessitate even wartime exclusion. Article 18 specified short stature and too-weak constitution [*complexion*]. And Article 16 stated that "young men whose infirmities render them inappropriate for all active or auxiliary service in the army are exempted from military service."[98] No further explanation was given of the exact infirmities included nor their severity, and subsequent legislation mostly repeated this formulation, without elaboration.[99]

Article 27 placed decisions about such exclusions from service with the *conseil de révision,* a recruitment review committee to be composed of local government and military officials and a doctor, which would hold public hearings on exemption requests.[100] As elaborated in the recruitment laws of 1889 and 1905, the committee could not render any judgment without first hearing the physician's opinion.[101] Medical criteria thus may have played a predominant role in determining who would be excused from service and on what grounds. Certainly this was the view of one contemporary analyst of the recruitment laws, François Roussel, who observed in 1891 that the doctor's advice was to be critical, especially in cases where alleged infirmities could easily be simulated (such as deafness or stuttering), and that the army health committee would determine which infirmities necessitated exemption and which allowed auxiliary service.[102] No mental conditions requiring exemption were explicitly mentioned, though presumably individuals in a mental institution or asylum for the feebleminded were automatically excluded. Still, as an ex-trooper reported in 1914,

> There is as great a percentage of stupid people in France as in any other country; a voluntary army is at liberty to reject fools as undesirable, but the nation with a conscript system must train the fools as well as the wise ones, for, admitting the principle that strength consists in numbers of trained men, then every rifle counts so long as its holder is capable of firing.[103]

Not that even this admittedly low standard meant that anyone and everyone could be a successful recruit. Especially after the passage of the 1905 recruitment law, a number of military doctors, alienists, and others interested in the issue of the quality of the army's enlisted men and conscripts worried about the induction of various categories of the mentally and physically unfit.[104] Convinced that such individuals could do danger to themselves and their fellow soldiers if allowed to remain in the military undetected, they proposed a variety of solutions. Some attention was given to improving the conseil de révision, typically by advising that as much information as possible be collected on the conscript or recruit from local civil and educational leaders and that a psychiatrist or other specialist in mental medicine be included whenever possible.[105] Most of these authors, however, believed that the great speed with which the recruitment boards had to make their decisions militated against anything but the most cursory examination of a potential soldier. Detection of the unfit, they therefore concluded, would have to occur primarily within the military, where the possibility for extensive observation over a significant period of time would allow for careful determinations of a soldier's true level of fitness for duty. As physician Emmanuel Régis pointedly declared, "it is up to the army to determine whether or not the candidate is appropriate for military service."[106] Critical to the process, these authors almost unanimously concluded, was ensuring that experts in mental medicine be involved in the process of assessment, either directly as members of the military review boards or indirectly by mandating that all military doctors receive training in detecting psychological disorders.

There was thus little pressure to introduce fundamentally new methods to detect those unable to carry out military duties. Physicians using tried-

and-true procedures would eliminate the most egregious cases, and familiarity during months of training would allow superiors to remove others unable to cope with military demands. Moreover, as the ex-trooper suggested, the French military's underlying philosophy was to have as many men under arms as possible. With training stripped down to its essentials, and with an approach to warfare that privileged marching and shooting over skill in intricate maneuvers, almost anyone was thought able to fulfill the infantry's demands.[107] Indeed, as one voice for military reform, Captain Hubert Lyautey, declared in a famous 1891 *Revue des deux mondes* article (which got him banished to Indochina for the rest of his career), "in the military schools . . . the individual soldier as he is presented to the students is an automaton; one puts him on the right, on the left, one makes him march, one makes him stop, one dresses him, arms him, places him on a horse."[108] Within such a regime, what value would there have been to knowing a conscript's exact degree of intelligence?

Binet and Simon thought that they had one answer to that question when they suggested that intelligence measurements might help the conseil de révision, by allowing rapid identification of individuals who might have certain mental defects.[109] Invited in 1908 by J. Simonin, professor of legal medicine, to conduct experiments at the military hospital at Val-de-Grâce, Binet and Simon used the 1908 revision of the Binet-Simon scale on eleven subjects, five with less than four months of military experience, five with twenty to twenty-four months of service, and one epileptic considered to be a good soldier. As Simon explained it, this first trial was designed solely "to prepare the method itself," to determine how a scale produced for schoolchildren could be adapted to military needs.[110] According to Simonin, Binet and Simon determined that the soldiers tested at mental ages between twelve and fifteen years, and thus were below the level of normal adults.[111]

Binet and Simon disagreed with Simonin, however, about how to interpret these results. From their perspective, the research was only preliminary, an attempt to generate some data in order to improve their methods for adults and determine exactly how the instrument could best be used. Simonin regarded the experiment as much more definitive. In his view the results suggested that while the Binet-Simon scale might perhaps be of value as an addendum to other methods available to diagnose soldiers once in the military, it could serve few useful purposes during the conseil de révision. As he saw it, conscripts or recruits during this initial screening session were so varied in their emotional states and some were so likely to cheat, that the scale would only be useful in detecting extreme cases of mental debility. Moreover, he maintained that the overriding problem confronting the military lay not in keeping out the weak of mind, but rather those whose morals were degenerate. Thus, Simonin argued for the primacy of the by then standard approach to the problem of the unfit in the military, one that emphasized careful clinical observation over long periods of time, as the only sure way to "establish a serious and unassailable medical verdict."[112]

Binet and Simon, not surprisingly, disputed Simonin's conclusions vigorously, arguing not just that Simonin was relying on preliminary results but that he was rejecting use of the scale on the basis of much broader claims for what the test could accomplish than its creators actually advocated. In their view, the scale was never meant to provide a quick and definitive test for all mental defects. Rather, they saw it as a way of helping the conseil de révision rapidly eliminate the vast majority of individuals whose intellectual level was sufficient for the military (and all agreed that that was not a very high standard—about a mental age of twelve years) and to then identify those few who might need to be examined more thoroughly.[113] Nonetheless, as discussion of the issue at the November 29, 1909, meeting of the Société médico-psychologique suggests, Binet and Simon had little success in convincing their colleagues. With the exception of physician F. Pactet, speaker after speaker both at this gathering and at an earlier meeting of the Congrès des médecins aliénistes et neurologistes de langue française in Nantes adopted positions similar to Simonin's. While few rejected use of the Binet-Simon scale outright, virtually all believed that such mechanical methods alone could never replace careful clinical observation and diagnosis and especially the judgment of expert physicians, particularly those with training in the mental sciences. As the noted alienist Jacques Roubinovitch, an expert on abnormal children, declared, "psychical tests *alone,* of whatever origin they may be—English, German, Belgian, or French—cannot generate an [absolutely] certain diagnosis of mental backwardness [*arriération*]."[114] Rather, the position of most military doctors and alienists was that put forward by Paul Chavigny of Val-de-Grâce: "never forget that diagnosis must remain something medical, exclusively medical, that is to say an object of personal appreciation, of judgment."[115] Given such resistance, it is not surprising that little additional interest was shown in the Binet-Simon or other "mechanical" methods of sorting military conscripts and recruits.

In the selection and promotion of officers, different factors entirely were in play, but ones that also rendered direct measurements of intelligence unnecessary. Individuals typically became officers by one of three routes: some, by direct promotion from the ranks, generally near the end of their careers; noncommissioned officers (*sous-officiers*), by attending the écoles of Saint-Maixent (infantry), Versailles (artillery and engineering), or Saumur (cavalry); and the most elite, by entering the officer corps directly from civilian life via training at the Ecole de Saint-Cyr (infantry and cavalry) or Ecole Polytechnique (artillery and engineering).[116] Chosen in the same way as those entering any of the other grandes écoles, by concours (examinations), the students in these elite schools, though mostly from lower-middle-class or higher backgrounds, could explain their privileged place in terms of the language of republican merit. "What is more democratic and more just," Captain d'Arbeux argued in 1911, "than the examination for Saint-Cyr or Polytechnique? It does not take account of origins, nor of wealth, and the expansion of the middle classes in the lycées has rendered them accessible to all the social classes!"[117] Less conservative commentators were more attuned to the system's limitations, particularly regarding its accessibility to the "lower" social orders, and consequently sought to raise the status of graduates of Saint-Maixent, Versailles, and Saumur.[118] Nonetheless, few advocated abolishing what had become the standard French republican way of parceling out limited social goods: competitive examinations and schools for special training.

The greatest source of prewar debate and controversy, outside of the length of mandatory service, centered not on selection but on officer promotion. Always associated as much with patronage as job performance and merit, promotion became an affair of state and a source of scandal following the Dreyfus Affair. Appalled by the officer corps' conservatism, the radical Socialists under Prime Minister Pierre Marie Waldeck-Rousseau appointed General Louis André minister of war in 1900 and charged him with "republicanizing" the military. A *polytechnicien,* staunch positivist, and disciple of Littré, André reversed the traditional bias in favor of monarchist, Catholic officers and favored those who espoused republican principles and who, at the least, no longer attended mass. His great innovation was to establish a system of promotions "meant to operate with rigid mathematical impartiality," as David Ralston has observed, based on each officer's individual fitness report.[119] Although André was forced to resign in 1904 in the *affaire des fiches,* when it became public that he was supplementing these "impartial numbers" with intelligence on the political and religious outlooks of officers being assessed for promotion, the push to change the officer corps' methods of self-assessment was largely maintained.[120] Convinced that one of the military's prime functions was to educate the male citizenry of France in discipline, order, and good citizenship, the high command insisted that each officer know his men as individuals and serve as a moral example.[121] Thus an officer's character became a key factor determining his success, and for many republican reformers, a major criterion for promotion. With such emphasis on personal character and judgment, mechanical assessments of intellectual ability could have had little place in promotion decisions.

The advent of the so-called *réveil national* in the years just before the war marked a pronounced upswing in French patriotism and support for the army and its distinct culture.[122] Nonetheless, the preceding decade's reforms were left largely intact. Thus, in August 1914 the French military had both the officers and trained soldiers it required to allow for its rapid and largely problem-free growth to a force of over 2 million. Having erected a system founded on the French republican principles of equality tempered by technocratic elitism, France faced none of America's manpower problems when preparing for war. The French male citizenry had already been assigned places in the military structure, and the officer corps had been selected and trained. Even if, after Binet's death in 1911, intelligence and its tests had continued to have strong advocates willing to promote such concepts and instruments to the French military, they are not likely to have stirred much interest. The mass mental testing of soldiers would have seemed superfluous if not ludicrous in a nation fighting for survival largely on its own territory and committed to a military philosophy emphasizing bodies over brains.

Conclusion

Yerkes' program of intelligence assessment thus achieved a modicum of authority in the course of its use by the U.S. Army that was never duplicated in France. Although not all American military officials became convinced of the relevance of intelligence or its measuring instruments to the nation's wartime needs, enough found practical benefits in the testing program for it to have a visible impact on both the army and, as we shall see, postwar civilian society. There were many reasons why the American army proved willing to use intelligence and mental testing in personnel assessment when the French had not. Under Yerkes' leadership, American psychologists strove mightily to fit themselves and their knowledge into the structure of army life. Military training for most members of the Division of Psychology, rapidly administrable and scorable tests with as much army look and feel as possible, close cooperation between psychologists and line officers, and the adaptation of the intelligence construct itself to military culture all contributed to making psychological knowledge and methods palatable to army tastes. More important, however, may have been two other factors.

First, over the course of the war numerous army officers became convinced that intelligence was an influential, if not overriding, characteristic determining military performance. The sudden appearance during the war years of comments about subordinates' intelligence, when contrasted with the absence of such remarks earlier, is striking. Partly, the testing program's very existence may have helped to make intelligence seem both real and noteworthy. The act of measuring, as Norton Wise has observed, imparts value to the object measured, and even more so when prosecuted on such an elaborate scale.[123] Moreover, intelligence's introduction as a category for assessment on the efficiency report form may have forced officers to recognize explicitly what formerly they had noted implicitly. In addition, the draft's wide scope meant that the army was inducting individuals with much more varied backgrounds, linguistic abilities, and competencies than previously. Intelligence may have proven an almost irresistible means of imposing order on this diversity by providing a single explanation, tied to a personal rather than social characteristic, for why certain recruits were less or more able to comprehend army procedures than others. That African Americans and many recent immigrants were routinely assigned to development battalions, created for those deemed too limited intellectually to become combat soldiers, lends this supposition credence. It suggests that army officers, when confronting the training difficulties even of whole classes, tended to see such problems as evidencing an inherent intelligence deficit rather than searching for social, cultural, or other explanations.

Second, the fact that intelligence tests provided information that largely matched officers' judgments became a strong argument in their favor. This feature of the testing program is easily overlooked, but military officials returned to it repeatedly when trying to decide whether to extend or maintain the assessment system. For those who were convinced that intelligence was critical to military success and that war preparation left inadequate time for developing sufficient personal knowledge for assessments, mental testing seemed to provide a useful, reasonably reliable means of making decisions quickly.

Nevertheless, resistance arose at all levels of the army, from recruits who ridiculed testing procedures to major generals who viewed the testers as unwanted interlopers. Any break with tradition, especially in an organization as fundamentally conservative as the American military, was bound to generate hostility, amplified no doubt by the threat that new experts represented to accustomed command prerogatives. And while Progressivism had made some inroads into military culture—accelerated by the entrance of the college-trained into the officer corps—seemingly ivory-tower scientists advising veteran officers about which recruits would make good soldiers was unlikely to be met with universal enthusiasm.

In addition, although both American psychologists and army commanders may have used the same term, "intelligence," there was no guarantee they meant the same thing. Officers had their own ideas about the meaning and value of intelligence, and psychologists could never simply impose their knowledge; rather, they had constantly to try to persuade officers that psychological conceptions of intelligence were better than or at least the same as military notions. Thorndike, with his belief in the contextual nature of intellectual ability, accepted this difference immediately and pushed for adopting military intelligence, as expressed in officer ratings, as the principal standard against which to judge the performance of army measuring instruments. Yerkes and Terman, however, believed strongly that there was only one kind of intelligence, and thus remained firmly convinced that distinctions between psychological and army intelligence resulted solely from military misunderstandings. Their view prevailed in the construction of Army Alpha. Thus, even some of those in the army who regarded intelligence as a concept of value considered the testing program's results—based on psychological intelligence—irrelevant to assessing the intelligence necessary for good soldiers or officers.[124]

The sense of human beings as composed of a complex of traits—mental, moral, and physical—all of which contributed to success or failure constituted perhaps the most significant objection to reducing the evaluation of personnel to intelligence and its tests. While some opposed the army program because of its perceived failure to truly measure intelligence, most military critics emphasized the insufficiency of the measure. For them intelligence was not of overriding importance; it was merely one of a number of attributes, many of which might carry equal or far greater weight, that an officer must evaluate in a soldier. Army personnel needed to be convinced not only of the validity of the psychologists' knowledge, but also of its utility. The company CO at Camp Lee spoke for many when he pointedly noted that combat soldiers needed not intelligence, but high proficiency in "the gentle art of murder." French military officials, if asked to comment on intelligence testing, would undoubtedly have responded in much the same way.

Further, the statistical nature of the intelligence construct propounded by the army examinations struck many officers as a fundamental flaw. Army Alpha, like all intelligence-measuring instruments, was wedded to the law of averages. However constructed, an intelligence test could only be validated by comparing its assessments of a sample population with judgments arrived at in some other manner. Such correlations were never perfect; indeed Thorndike was pleased when Army *a* correlated moderately with officer ratings. Whether these deviations were explained as functions of the probabilistic nature of the intelligence rating or deficiencies in the other methods used to evaluate intelligence, the consequences were clear: there would always be discrepancies between test ratings and personal assessments of individual subordinates. Some officers were not troubled by this. They accepted that the tests' power lay in their aggregate findings and felt that the procedure's occasional inaccuracies could be accommodated.[125] Others, such as Captain Henry H. Burdick, discounted a procedure that selected an obviously unfit person for NCO training and ranked as inferior recruits who, by his lights, would make excellent soldiers.[126] Concerned with assessments at the level of individuals,

Captain Burdick and like-minded officers could not accept an instrument that worked most poorly at that very level.

This distrust of the statistical was amplified when it was suggested that the tests might be turned on officers seeking promotions. Virtually every military evaluation of the testing program recommended that officers either not be examined or be tested only on their superiors' order.[127] Unlike obedience, discipline, loyalty, or knowledge of military procedure, intelligence was not a characteristic amenable to personal control. If found deficient, there was no ready means to correct the problem. For many officers, any evaluation of their intelligence seemed threatening; an assessment not by one's superiors, on the basis of personal knowledge, but by academic psychologists, on the basis of some "objective" scale whose output might be taken to be indisputable, was almost intolerable.

Goldthwaite Dorr, in his evaluation of the testing program, was struck exactly by these tensions in intelligence assessment between the individual and the aggregate. Describing himself as "impatient" with the "mechanical character" of Army Alpha because of its "failure" to assess accurately particular individuals, Dorr nonetheless conceded that by "such close observation . . . the substantial accuracy of the average result [was] lost sight of."[128] Dorr had identified an essential feature of the conception of intelligence and its tests being promulgated by the army psychologists: the products of psychological expertise worked best at the level of the aggregate, as tools of classification, rather than at the level of the individual, as means of exploration. They were artifacts of the modern world, of the mass, anonymous society that prized efficient administration and deemed detailed familiarity impractical. The psychologists' techniques would have had little place in the army if it had remained the intimate organization of the prewar years. Thus it is not surprising that in France, where virtually all the conscripts were known from their obligatory service and reliance on personal judgment by the elite was prized, there were few pressures to adopt mechanisms substituting science for immediate familiarity with recruits.

The prewar applications of intelligence and its tests in the United States were, of course, not restricted to situations demanding quick evaluation of masses of unknown individuals. As we saw in chapter 5, mental tests in America originally served as diagnostic aids in asylums for the feebleminded, where the instruments' ability to guide and embody expertise was particularly appealing.[129] This function's attractions would persist well after the war in situations where tests were used as a first step toward diagnosis of children with possible intellectual difficulties. That would also prove to be their most important postwar use in France. Nonetheless, in America, as we shall see, the military story would ultimately prove more typical.

Seven

Intelligence and the Politics of Merit between the Wars

WRITING IN THE AUGUST 5, 1922, issue of *School and Society,* Henry Holmes, of Harvard University's Graduate School of Education, threw himself into a debate about the nature of American democracy that had been raging almost from the moment the Armistice was declared:

> As a movement for social justice democracy must make real the vision of Lincoln—"a fair chance and an unfettered start in life for every child"; must keep open "the road to talent," which seemed to Napoleon the essence of the matter; must provide genuine equality of opportunity that every man may be able, in the spirit of that superior definition which President Eliot [of Harvard] likes to quote from Louis Pasteur, "to make the most of himself for the common good."[1]

Educators, political thinkers, scientists, public officials, and social commentators alike were puzzling over the nature of human beings and the kind of democracy appropriate for them. They were spurred by a number of factors, including the end of the war, the resumption of the prewar period's labor unrest, the short-lived "Red Scare" of 1919–20, the persistence of eugenics, the "Great Migration" of African Americans northward, the women's suffrage movement, and, most surprisingly of all, certain findings of the U.S. Army mental testing program. Holmes's solution, as he himself acknowledged, was to turn to the tried and true. He sought to elaborate a vision of democracy and merit tied to equality of opportunity and the free play of an individual's talents that could be traced back, as we have seen, at least to the early nineteenth century in America and the Revolution in France.[2] However distant from reality, both France and America had long celebrated the openness of their social structures, maintaining that the recognition of individual achievement was a key factor accounting for their success as democratic republics.

In the war's aftermath, however, Holmes faced changed circumstances as he strove to explain how colleges and universities could contribute to the "movement for social justice" and help fashion a democracy providing "genuine equality of opportunity." As discussed in chapter 6, one by-product of the war for the United States, though not for France, was to bring to the fore a new way of understanding natural inequalities and an individual's capabilities: intelligence. Understood as a unidimensional characteristic varying in degrees, intelligence in this guise referred to an individual's or group's natural mental potential. Where France's military leaders had proven immune to the seductions of intelligence as a way of rationalizing the nation's mobilization efforts, their American counterparts were much less resistant. Ill prepared for the demands of mass warfare, the American military found some value in the new, impersonal and mechanical ways of assessing soldiers that psychologists proffered, especially when these methods were restricted to those, such as new recruits, sufficiently distant in class, rank, and perhaps ethnicity from the officers evaluating the program. This close connection between intelligence, its measuring instruments, and the statistical-mechanical approach to assessing human beings that army testing embodied was one of the program's most important legacies. Thus when Holmes contemplated how higher education could further democracy, he did so cognizant that science had seemingly demonstrated the existence of profound and perhaps ineradicable differences in degrees of individual intelligence. Such differences, he concluded, rendered suspect versions of American democracy promising a playing field level for all because they suggested that merit might be as much a function of biology as of hard work, education, and character.

In France, by contrast, the story is more truncated, as what discussions of intelligence and merit did take place during the 1920s and 1930s rarely ignited worries about the state of French democracy. Before World War I, as we have seen, French psychologists had arrived at little consensus about the nature and meaning of both intelligence and its methods of measurement. Except for the small circle around Alfred Binet, most French psychologists either were uninterested in intelligence or considered it simply one of a number of characteristics that together helped to give definition to an individual. Ignored by the French military during the war, intelligence and its tests did begin to attract some attention from French psychologists during the postwar period, with a few even developing new instruments to replace or employ alongside the Binet-Simon. But the goals of this work were much different in orientation than the most visible of the American projects. Well-established methods of selection were already in place to aid the French educational system in its role as the primary gatekeeper for entrance into the nation's technocratic elite. Thus there was little incentive within administrative circles to employ intelligence testing to sort the population into a hierarchy of merit in order to decide who should receive what types of educational or occupational opportunities. The demographic realities facing the postwar French state reinforced this orientation. The school system had already expanded to include virtually the entire school-age cohort before the war, and the war itself decimated France's adult male population. Thus few French administrators and managers were forced to grapple with the circumstances prevailing in postwar America, where schools and industry had for the first time to sort rapidly ever larger numbers of new or potential entrants. In addition, those in France most marginalized by traditional methods of selection—primarily members of the working class—saw little advantage to using intelligence as a way of sustaining rival claims about merit.

Instead, the French emphasized the diagnostic possibilities of intelligence testing, though generally as only one among an array of assessment methodologies designed to reveal an individual's range of aptitudes. Indeed, when national intelligence testing of schoolchildren was proposed, it was invariably linked to identifying cases of defective intellect and not to serving a broader set of purposes, such as selecting the most able or guiding individuals to appropriate careers. Moreover, French elite culture, as Theodore Porter has shown, placed a premium on judgments rendered by experts and took a dim view of solely quantitative and mechanical solutions to social problems.[3] Thus French psychologists tended to emphasize the complexity, multiplicity, and limited role of intelligence, in and of itself, in shaping an individual's fate. Concerned mostly with employing intelligence to analyze the capacities of individual clients, these psychologists made few claims about differences in intelligence that had political resonance, and certainly none that demanded a radical reassessment of French democracy or the class structure of French society.[4] When these factors are combined with the small size of the professional psychological community and the existence of few occupational positions outside of research centers, it is little wonder that interest in intelligence expanded only modestly beyond its prewar niches and never took on enormous general significance for the French public.

The French approach to understanding intelligence as an aid to constructing individual profiles at the hands of a trained expert certainly had its proponents across the Atlantic. Indeed, a number of American psychologists worried about treating intelligence as a singular entity that alone was accorded a preponderant influence in determining an individual's—or a group's—fate. Instead they argued for more complicated assessments that took account of a host of factors, a position that Holmes, among other commentators, shared. However, for most psychologists and laypeople alike, the appeal of the IQ version of intelligence was nearly irresistible. Embedded in an easily disseminated technology of display, the mental test, IQ determinations made visible fine grades of intellectual difference, distinctions that psychological research tied to class and occupational hierarchies. By these means singular, differential intelligence took on a reality for most Americans that proved difficult to dispute.[5]

The army's wartime testing program—and especially the extraordinary publicity some of its findings received—was crucial, helping to transform an endeavor that had existed mainly on the margins (in academic psychology laboratories and asylums) to one that seemed relevant to many aspects of American life. The moment was especially opportune, just as Americans were beginning to confront aspects of mass society for the first time. In the face of tremendous expansion in the secondary and higher education systems, rapid growth of urban centers, and decades of Catholic and Jewish immigration from eastern and central Europe and of African American migration from south to north, educators, industrialists, administrators, and politicians, among others, were desperate for new methods to manage this burgeoning multitude. The new objective and efficient intelligence tests, seemingly able to reveal an individual's hidden biological potential, beckoned as one solution, and suddenly it seemed that everyone was either being tested or advocating the use of intelligence tests to address some social or administrative problem. Discussions of intelligence and its tests broke out nationwide, the IQ conception of intelligence becoming a powerful cultural resource for marking and explaining difference.[6]

In response, especially when testing first burst on the national scene during the early 1920s, Holmes and many other educators, administrators, and political leaders grappled with the question of differences in intelligence and what they might mean for American democracy. Some turned to language reminiscent of Thomas Jefferson's about the natural aristocracy, and called for the dedication of higher education to producing "an aristocracy of brains"—possessing in the intelligence test, they believed, a means of at last being able to scientifically determine who merited special training, the natural aristocrats. According to its proponents, intelligence was the principal factor explaining why some individuals were at the top of the social and occupational hierarchies and others were at the bottom, and thus why America was fundamentally a meritocracy. It was also a means, certain marginalized groups such as Jewish Americans claimed, to challenge traditional allocations of resources on the basis of exclusionary class and ethno-racial networks. In this guise, intelligence testing had an enormous range of proposed uses during the 1920s and 1930s: as a method for school systems to sort their students and place them in distinct educational tracks, for universities to choose among applicants, for industries and the government to select for entry-level white-collar jobs, for the immigration service to decide on an individual basis whom to admit into the country, and even for voting, jury duty, police promotions, and automobile licensing.

Rarely, however, was this interpretation of the meaning and purview of intelligence allowed to go unchallenged. Others, including most of those worried about the racial implications of the new science, struck more cautionary notes, seeking to temper, if not outright contravene, the proposals for how education—and society—should be transformed to reflect the "fact" that the citizenry was naturally stratified according to level of intellect. In general these critics—ranging from William Bagley to Horace Mann Bond to Otto Klineberg—emphasized the power of education to improve the mental abilities of everyone and underscored the limited contributions that intelligence alone, however understood, made to determining an individual's success in school, business, or life. Objecting strongly to placing great weight on any single attribute or performance on a solitary test, they argued for a more holistic approach to merit, one that took account of the many different characteristics an individual manifested and that might even see intelligence itself as multivalent and open to improvement.

Consequently, throughout the interwar period the status of intelligence and its tests was under constant negotiation, though by the start of World War II a rather complicated settlement had emerged. Intelligence as a singular, measurable entity was accorded greatest weight as a constituent of merit when there was a perceived need to sort rapidly a relatively undifferentiated mass, such as in the expanding educational systems, and least when individual assessments emphasizing complex, multidimensional analyses were most desired, as was the case in most management-level promotion decisions. While few Americans would declare outright that intelligence was of no consequence in questions of assessing merit, their conclusions about what it was, whether one thing or many, and how strongly it should be weighted varied tremendously depending on the particular determination to be made and, often, their judgment of how they or their group would fare if intelligence measurements were relied on to open opportunities otherwise likely to be denied them.

By the outbreak of World War II, therefore, the ways in which the concept of intelligence was understood and routinely deployed in France and America differed substantially. In the United States, with its highly decentralized educational system, heterogeneous population, and culture that viewed claims to expert authority skeptically, many professionals and members of the public embraced intelligence as naturalized entity able to provide an otherwise unachievable uniformity and standardization to assessments of merit. Moreover, by seeming unbiased, measurements of intelligence could be represented as according with fundamental notions of equality and justice.[7] In France, by contrast, centralized institutions and methods of standardization to deal with the most and least able long preceded the development of the intelligence scale. At the same time, interest in providing complex portraits especially of members of the intellectual and social elite for the purposes of occupational counseling and self-understanding remained strong. United in the project of understanding a fundamental feature of the mind, French and American psychologists, in the end, created distinct versions of intelligence tied to the specifics of their social, intellectual, political, and cultural contexts. As a result, the links between quantitative assessments of intelligence and determinations of merit that Americans had come, by the 1940s, to see as natural had few parallels in France, where the seemingly unproblematic way to select an elite was through competitive examinations judged by experts within a class structure where upward mobility was rare.

Vive la différence: Intelligence after Binet

World War I, which brought metric intelligence to national prominence in America, did not, as has been suggested, have the same effect in France. Because the French military never turned to intelligence testing to assess its personnel, the war provided neither great publicity to intelligence's values nor a group of trained testers looking for new domains for their expertise. Nonetheless, the end of hostilities did allow French psychology to emerge from over half a decade of neglect, and after the war, intelligence was one of the elements of human psychology that received renewed attention. In contrast to the American story, however, intelligence in France retained its prewar connections with individual variations and pathology. The new directions that French psychologists explored in the postwar period largely accentuated these associations, as psychologists entered peacetime profoundly shaken by the death and destruction of mass, "scientific" warfare, with its ruthless devaluing of the individual and personal.

When the French psychological community regrouped with the war's end, it in part revived prewar laboratory-oriented experimentation: Benjamin Bourdon, Marcel Foucault, and Henri Piéron, among others, continued to produce large-sample, instrument-based investigations of basic mental processes. But what mainly marked postwar French psychological research, in contradistinction to the American turn to mass testing, was the resurgence of intensive clinical studies of individual subjects.[8] Reflecting disappointment with statistical methods and the reductions of mental phenomena to a single number, as well as continued fascination with extraordinary individuals, French psychology celebrated the mind's complexity, the intellect's multivalency, and the individual's irreducible singularity.[9] At the same time, when actively using intelligence as an analytical category, be it for educational, occupational, or legal purposes, French psychologists most often employed intelligence measurements as just one way of assessing individuals, with some even warning that reliance on single measures for life-shaping decisions would be foolhardy and unscientific. Resolved not to place too much weight on any single criterion, they advocated an approach that put a premium on expert holistic judgments.

As we saw in chapter 4, French psychology had struggled long before the 1920s to understand what intelligence was and how best to define it.

Alfred Binet wrestled with these questions in his various explorations of intelligence, providing different answers depending on the specific focus of his research. When performing his craniometric investigations, Binet assessed intelligence by conducting single measurements of a single characteristic; when working with Théodore Simon on the metric scale, Binet adopted a pluralistic method combining performances on numerous tasks; and when undertaking his intensive individual case studies, Binet employed a variety of investigative methods, concluding that entirely different types of intelligence can be present. Each approach had proponents among his fellow prewar psychologists, some of whom also suggested—adopting a position similar to the American Edward L. Thorndike's—that intelligence had to be assessed contextually because it was a collection of discrete abilities. Others simply ignored the question of intelligence entirely.

In the postwar period, these debates revived, reflecting the community's rival camps. Some, François Parot has argued, came together around Ignace Meyerson, who in the 1920s became both the lead editor of the *Journal de Psychologie normale et pathologique*—one of France's two principal psychology journals—and general secretary of the Société de Psychologie.[10] Meyerson sought to develop a psychology of the whole person that crossed the divide between individual and society and aimed to integrate the human sciences into a comprehensive left-oriented project. Particularly influenced by Durkheimian sociology, Meyerson and his circle focused on a synthetic psychology that was "developmental, objective, [and] historical," interrogating and integrating the connections between psychological functions and social environment.[11] They paid little direct attention to the concept of intelligence, and even less to testing and measurement.

A few psychologists outside Meyerson's camp continued to hew relatively closely to Binet's metric-scale vision of psychology and to speak of intelligence as a singular entity open to straightforward measurement. Simon, not surprisingly, was in the forefront of this group, sticking doggedly with the scale's 1911 version and refusing even to make improvements he knew were necessary.[12] Others, such as the Swiss psychologist Alice Descoeudres, adopted the Binet-Simon scale as a model, but not one that had to be treated inflexibly.[13] For all in this group, intelligence remained something universal and quantifiable, as well as an entity whose practical significance lay in its ability to allow differentiations and comparisons. "Imagine the interest," Descoeudres enthused,

> in a precise scale of development, rectified, perfected finally, in order to compare not only different children to each other, not only one child with himself to measure how he has developed, but to *compare the sexes,* children of different *social milieus* . . . and finally—by translating the tests, adapting them to other tongues—children of diverse *races* and *tongues.*[14]

Marcel Foucault, a professor of psychology at Montpellier known mostly for his work in psychophysics, saw similar possibilities for intelligence measurement, though with an instrument more inspired by the Binet-Simon scale than derived from it. His 1933 volume *La mesure de l'intelligence chez les écoliers* described a new intelligence-measuring instrument, one that he hoped users would find "analogous to a thermometer."[15] Announcing that he wanted to measure children's intelligence "as directly as possible," Foucault developed five tests—"usage of things," construction of "genre-species" categorizations, identification of "opposites," formulation of "part and whole" relationships, and making of "analogies"—that he used to determine an individual's overall mental level. The advantage of this approach, Foucault explained, was that "this method measures not only intellectual level and development, but also aptitude, and finally, can be applied to schoolchildren of all ages." However, like the Binet-Simon, Foucault's test was predicated on a unitary conception of intelligence. Even when he spoke of intelligence's different modalities—either logical or intuitive—he suggested that these were only "two ways of applying intelligence"; intelligence in and of itself could be measured as a singular and integrated entity.[16]

Skeptical about such a vision of intelligence as unitary and integrated was the other major figure in interwar French psychology, Henri Piéron, editor of *L'Année psychologique* and founder of the Institut de psychologie. Although Piéron and others who worked with him or shared his perspective agreed with Foucault that intelligence was open to quantitative analysis, they were more impressed with intelligence's complicated, ultimately almost idiosyncratic nature than its holistic quality.[17] Like Binet in his study of his daughters, Piéron emphasized the variety of forms and modalities in which intelligence should be assessed. "It appears to be necessary," Piéron explained in 1931,

> no longer to consider one "intelligence" but several aspects or multiple forms of it; therefore, in order to characterize an individual, it would be necessary to trace his profile in as detailed a manner as possible, giving place to the principal types of problems which he is called upon to solve in ordinary life, distinguishing at the same time the phases of comprehension, invention or criticism which intervene unequally according to the manner in which the problems are actually presented by various professional activities.[18]

Operationalizing this approach, Piéron and his wife Marguerite produced a series of individual case studies of intelligence whose principal product was a set of graphs representing a participant's intelligence in four modes—numerical, verbal, logical, and general—and along three axes, comprehension, critical ability, and invention.[19] While the Piérons did not reject measurement, and indeed used the Binet-Simon scale as an assessment tool, their aim was to produce not a single measure, but an individual profile charting intelligence's strengths and weaknesses along multiple dimensions. "Though we always employ the same word, intelligence, for the aptitude to solve problems," Henri Piéron noted in 1927, "it is still necessary to understand that under this term the mental action may be quite different, depending on the nature of the problems to be solved."[20]

For Piéron, in other words, intelligence was context specific, a sentiment echoed by Benjamin Bourdon in 1926 when he argued that even genius-level proficiency in one area did not ensure ability in other domains, and by Jean-Marie Lahy in 1935, who asserted that "subjects of equal global value" still differ in which aspects of their intelligence predominate.[21] Lahy, in fact, undercut the strictly biological understanding of intelligence still vital by insisting that intelligence was as much a product of social context—in his studies especially of occupational class—as of heredity and physiology, and therefore open to a wide array of influences and expressions.[22] The Swiss educational psychologist Edouard Claparède, although a former associate of Binet's and deeply influenced by him, basically concurred with Piéron, Bourdon, and Lahy. Conceiving of intelligence as composed of several distinct operations, Claparède declared that even two people with identical scores on Binet's or Terman's intelligence tests "are not identical; their aptitudes are not the same," concluding that "perhaps it would be better to avoid here the term intelligence, and speak only of global aptitude."[23] This rejection of the 1911 Binet-Simon vision of intelligence in favor of the clinical sense was pushed to its limits by one of Piéron's students, Jeanne Monnin, who in 1934 underscored the highly complicated nature of intelligence and the mistake in trying to reduce its measure to a single number:

> Intelligence is complex; it presents differences in quantity and quality in each individual. . . . Practically, it is improper to characterize someone on the basis of the notion of a general level of intelligence. . . . The analytical method must be preferred in all cases where it is necessary to know the possibilities for success in different domains, or to predict success in a specific area of activity.[24]

A conclusion at odds with the view of intelligence that came to predominate across the Atlantic, it captured well the position of many interwar French psychologists, who believed that intelligence perhaps could be characterized, but certainly not simply and straightforwardly assessed.[25]

Nevertheless, as William Schneider has demonstrated, single measures of intelligence conducted on a large scale did attract some state attention. Mass testing became a serious public issue first in the mid-1930s, with the Popular Front's rise to power and its desire for a nationwide survey of schoolchildren to determine the percentage of abnormal children requiring special educational services.[26] Postponed with the leftists' defeat in the late 1930s, the proposal was resuscitated by the Vichy government in 1940—spurred by its own interest in the health of the family and nation—but only came to fruition in 1954 under the Fourth Republic, with the report "The Intellectual Level of School-Age Children."[27] Even at this point, the role for intelligence measurements remained sharply circumscribed: associated with detecting such pathologies as idiocy and imbecility, they were seen as ways of categorizing intellectual deficits of particular concern to French educators and administrators.

The Practical Career of Intelligence in France

The state's occasional forays into mass intelligence assessment notwithstanding, in interwar France the major areas in which intelligence determination had practical consequences lay in educational decision making, individual counseling, and occupational placement. With the exception, to a certain degree, of detections of mental pathology in schoolchildren, however, French practitioners in general considered determinations of intelligence alone as insufficient for individual assessments. Rarely concerned with mass categorization and quick classification, French psychologists and administrators instead emphasized the multiple aspects of intelligence and the multiple traits influencing individual success. Intelligence thereby was domesticated in such a way as to enhance opportunities for the application of expert judgment in interpreting assessment results and to minimize the possibility that intelligence measurements could be used as independent checks on established selection procedures.

Certainly this was the case for French education. Binet never succeeded in making the Binet-Simon intelligence scale a mandatory part of the procedure for examining schoolchildren suspected of being arriéré and assigning them to an educational track—normal, classes de perfectionnement, écoles de perfectionnement, or asylums. Nonetheless he did convince many individual school administrators and private practitioners that intelligence testing might aid their determinations. After the war most psychologists referred to this function as one of Binet's legacies, and believed that testing's practical value for such work was beyond question. As even a thorough skeptic about intelligence measurement, Albert Challand, conceded in his doctoral dissertation in pedagogy, "the Binet series, for example, is a useful expedient for discovering the *arriérés;* its success in this domain is completely legitimate."[28] Psychologist Henri Delacroix, although critical of some of the Binet-Simon scale's tests, shared Challand's enthusiasm, as did Descoeudres and the psychiatrist Georges Heuyer, who employed the scale extensively in his studies of juvenile delinquents.[29] In all of these cases, however, the role of the scale was advisory: an individual's Binet-Simon score served as one piece of information that a psychological or educational expert would take into account when determining an individual's intellectual status.

Much the same can be said of the uses of intelligence and its tests at the educational spectrum's other end. Although the concours continued to be higher education's prime gatekeeper, the growth of *écoles secondaires* as more advanced versions, and rivals, of the higher primary schools (*écoles primaires supérieurs*) opened space for new selection methods. In the *Revue pédagogique* in 1925, R. Duthil, a professor at Nancy's Ecole Primaire Supérieure, applauded the decision of some school directors to turn to intelligence measurement in response to a September 1924 ministerial directive advising principals to "take full account of the aptitudes of the children." Duthil, an enthusiast for testing, argued for its value in terms reminiscent of those employed by U.S. Army officers during World War I: "Employed in combination with other tests of academic knowledge and of special aptitudes," Duthil observed, " . . . it [the intelligence test] furnishes a remarkable prognosticator and permits, from the first days of return to school (*rentrée*), knowing the children and anticipating what they will accomplish (*donneront*) a year later."[30] As such, he argued, the tests could supplement the methods already used to predict which students would benefit most from the academic opportunities provided by the écoles secondaires.

Several French psychologists, including Foucault and Piéron, were also impressed with the ways intelligence testing could contribute to choosing students for advanced training. Echoing Duthil, Piéron distinguished between assessing a student's knowledge, the function of *les examens,* and predicting "if he was going to profit from a new education," the purpose of *les tests.* Piéron emphasized the tests' impartiality and objectivity, and, like his American counterparts, the assurance of justice they could thereby provide: "The important thing is that one be able to choose a certain number of candidates and that it have the air of being just."[31] Virtually no one in France, however, suggested that intelligence measurement alone could or should serve as the sole vehicle for making such momentous determinations, nor that they should supersede expert judgments. As Foucault, one of testing's strong proponents, remarked:

> I do not want to say that access to secondary instruction must be determined, in an exclusive fashion, by the result of one, or even several measures of intelligence: this [selection] should be done, on the grounds of equity and reason,

> only by the concordance of a plurality of indications. The results reported from the primary level studies constitute one of these indications, the appreciation of the intelligence and the moral qualities of the children by their teachers who know them is another: the measure of intelligence could be usefully joined to the information coming from the preceding sources.[32]

In addition to aiding in selection, advocates of testing argued that one of the great values of les tests—when used in conjunction with other assessment methods—was that they could reveal the diverse individual aptitudes teachers would encounter in the classroom. Thus the anthropologist and psychologist Gustave Le Bon suggested in 1918 that "one of the most important roles for professors [is] to diagnose the real aptitudes of a student and to direct him toward the studies for which he has a natural disposition."[33] After the war, Duthil, the Piérons, and Foucault all saw diversity of aptitudes among students as a primary reason why schools should adopt the testing technology. The Piérons, along with physiologist Henri Laugier, for example, suggested in 1934 that their method of creating and interpreting individual intelligence profiles could "furnish for each child an analytical profile which characterizes him [so as to] place him in a homogeneous group for each aptitude analyzed."[34] This possibility of creating homogeneous groupings among students already selected rather than of actually aiding in selection, in fact, was one of the main ways in which interwar French psychologists sought to "sell" intelligence and its tests to the educational system. Their success is unclear, however, as schools routinely sorted students within the classroom according to daily performance, and thus may have found aptitude profiling of little practical value, especially as the law mandated the same curriculum for all "normal" students.

Intelligence and its tests followed a similar path in the field of occupational placement and career guidance (*orientation professionnelle*).[35] The Piérons, Lahy, Laugier, and Edouard Toulouse pioneered an approach founded on using tests to develop comprehensive individual profiles that could make clear a person's aptitudes and provide occupational guidance. The Binet-Simon scale and its notion of intelligence was an important element in their repertoire of investigative tools, though, again, as only one among many such instruments and modalities of observation.[36] Schneider has pointed out that support for the establishment of orientation professionnelle spread broadly in interwar France, with centers being opened in a number of cities during the early twenties and the Institut national d'orientation professionnelle (INOP) established in Paris in 1928 by Piéron, Laugier, and educator Jules Fontegne.[37] Although there were clear differences in emphases in each of these leaders' work, all shared a commitment to intelligence's multiplicity and its limited usefulness, in and of itself, as a criterion for orienting individuals toward appropriate careers. "Here are, for example . . ." Lahy remarked,

> the psychological profiles of two subjects having the same index of intelligence as measured by our test, 32 correct responses. One imagines easily how these two individuals are different in life. The one . . . has become an excellent railroad switchman, and the other has been obliged to retire from the switching office because of the numerous accidents he caused.[38]

By underscoring the insufficiencies as well as contributions of single measures of intelligence, French psychotechnicians (as they sometimes called themselves) made clear that assessing individuals could not be reduced to the mechanical output of a single instrument.[39] In this they differed significantly from the most enthusiastic of their American counterparts, who routinely advocated using intelligence assessments alone to make quick decisions about appropriate career choices, and proved little interested in individualized personal profiles, at least when proposing assessment programs for the masses.[40] Nonetheless, French promoters of intelligence testing could not escape some of the same criticisms that bedeviled the American testers. Worried that overreliance on test results might perniciously suggest that destiny was fixed from birth, those skeptical about intelligence assessments' value celebrated instead the possibility of development and change. Challand, for example, insisted on the unique capabilities of every individual and denounced those who assumed that test results revealed the limits of an individual's abilities. "One does not allow the first person who comes along," Challand scathingly observed, "on the pretext that he is a psychologist, to decide in a few minutes whether one is or is not an acceptable sample of humanity, and to settle definitively the possibilities that one might have for success in one's career."[41]

The response of French psychotechnicians to such criticisms was largely to agree with Challand. The entire project of those interested in intelligence and its tests, as we have seen, was to argue that good psychological science dictated the production of comprehensive individual profiles, with interpretations generated out of expert judgment and geared to the specifics of the context in question, be it occupational advice, educational placement, or what have you. Little interested in remaking the social composition of the upper levels of French education or culture, French psychologists by and large sought to integrate their methods into existing procedures for choosing and justifying the technocratic elite. Thus, for all of American culture's supposed celebration of individuality, ironically French psychologists proved more interested in incorporating diversity and difference into their theorization and instantiation of notions of intelligence and practices of measurement, albeit in ways that largely supported the status quo.

"The Thin Red Line": Intelligence and the Problem of Democracy

In America, as we saw in chapter 6, the exigencies of coping with mass mobilization during World War I opened the U.S. Army to establishing the nation's first large-scale intelligence-testing program, for the purpose of classifying and sorting new recruits. In the aftermath of the war, the status of intelligence and its tests was dramatically transformed. Where French psychologists slowly had to reconstitute their profession after the war's devastations, American psychologists exited the conflict with a community, if anything, dramatically strengthened by its wartime service and with a product, the intelligence test, that had become part of the experiences of hundreds of thousands of adult males. The publicity generated by the testing program alone placed intelligence assessment on the national stage, advertised as a seemingly objective way of sorting groups into the naturally superior and inferior.[42]

Moreover, results from that testing—particularly the widely reported finding that the average American soldier had the mental age of a thirteen-year-old—were quickly incorporated into a series of debates that erupted in the early postwar period about American democracy's nature and health. Couched largely in terms of worries about the population's biological fitness for democratic governance, pessimists about democracy in particular seized on the Army data as clear indication that the republic was in peril because of the inferior intelligence of the vast majority of its population. Could a democracy, some wondered, really provide the same kinds of citizenship to all its adult members if there were significant individual differences in the ability to be, or even choose, a good leader? Spokesmen for those more sanguine about democracy quickly responded, in part by attacking the data itself and the testing mechanisms from which it was derived, and in part by developing more fundamental challenges to the notion being promulgated of intelligence as a unidimensional biological characteristic with unparalleled implications for an individual's position in society. Although neither side completely prevailed during the early 1920s, in the process of arguing, intelligence itself became a topic freighted with enormous consequence, especially with regard to determining which individuals or groups merited what sorts of educational, occupational, and even citizenship opportunities.

Intelligence became an issue in these democracy debates largely for two reasons. First, at a general level, although the United States had survived

World War I relatively unscathed, seeming to have made good on its pledge to make the world safe for democracy, nonetheless, in the immediate postwar period many Americans shared their European counterparts' concerns about the future. Eugenics, the "Red Scare," and fears of "race suicide," along with the emergence of post-Victorian values grouped around efficiency, order, and control all reflected a cultural moment in which many middle-class Americans felt disoriented.[43] Massive social transformations, America's growing international entanglements, and immigration, both of southern and eastern Europeans (predominantly Catholics and Jews) to America and African Americans to the north, had troubled members of the old elite, especially, even before the war; after, some feared that the Armistice marked not the triumph of civilization but another moment in its precipitous decline.[44]

Analyses psychologists performed on the Army's World War I mental-testing data—trumpeted first in Clarence Yoakum and Robert Yerkes' *Army Mental Tests* (1920) and with greatest effect in Carl C. Brigham's *A Study of American Intelligence* (1923)—contributed to this unease, as they seemed to legitimate both the optimism and anxieties of middle-class Americans.[45] Although buoyed by these studies' "proof" that individuals of northern-European descent were distinctly superior in intelligence to all other groups, many white Americans—already fearful about "reds," immigrants, workers, and other "threats" from within—were nonetheless unsettled on learning that a significant percentage of adult American males had been discovered to be feebleminded or worse.[46] Notions of a nation in biological and cultural peril abounded, reflected not only in the vogue for eugenics but also in the Immigration Act of 1924, which sought virtually to eliminate the immigration of southern and eastern Europeans in part on the grounds of their biological unfitness, and in Supreme Court Justice Oliver Wendell Holmes's famous opinion in *Buck v. Bell* (1927) upholding enforced sterilization of the feebleminded.

Second, a more immediate cause was the startling news thrown into this cauldron in 1921 from the Army examination program that the average mental age of the American soldier was under thirteen.[47] For many, this simply confirmed their worst suspicions, and they seized on this "fact" as a golden opportunity to decry publicly the state of the American republic, initiating intense debates in the culture at large and especially within education over intelligence's implications for democracy. Echoing long-standing cultural worries about national degeneration, for example, Cornelia James Cannon, wife of noted Harvard physiologist Walter B. Cannon, opined in the pages of the *Atlantic Monthly* in February 1922 that "the lower grade man is material unusable in a democracy."[48] George B. Cutten, in his inaugural address as president of Colgate University, followed suit in October, suggesting that "we have never had a true democracy, and the low level of the intelligence of the people will not permit of our having one."[49] Perhaps most inflammatory were Boston lawyer Lothrop Stoddard's claims in *The Revolt Against Civilization* (1922), in which the testing data formed part of his eugenicist and frankly racist portrait of civilization under siege:

> Against these assaults of inferiority; against the cleverly led legions of the degenerate and the backward; where can civilization look for its champions? Where but in the slender ranks of the racially superior—those "A" and "B" stocks which, in America for example, we know to-day [because of the World War I Army testing data] constitute barely 13½ per cent of the population? It is this "thin red line" of rich, untainted blood which stands between us and barbarism and chaos. There alone lies our hope. Let us not deceive ourselves by prating about "government," "education," "democracy": our laws, our constitutions, our very sacred books, are in the last analysis mere paper barriers, which will hold only so long as there stands behind them men and women with the intelligence to understand and the character to maintain them.[50]

Like his friend Madison Grant, whose own racial call to arms, *The Passing of the Great Race* (1916), had been a best seller, Stoddard wove together fears of degeneracy, miscegenation, race war, and primitive savagery with visions of corporeal and racial purity, social Darwinist renderings of evolution, and skepticism about education and democracy, all to render vivid the image of a beleaguered aristocracy of red-blooded intellect—the "A" and "B" men—upon whose powers to repress and procreate rested civilization's future.[51] Florid though his account surely was, it drew on the popular pronouncements of much more distinguished professional scholars. In fact, one of Stoddard's chief sources was Scottish-born Harvard psychology professor William McDougall, whose *Is America Safe for Democracy?* (1921) also used the army testing data to buttress dire conclusions about the biological warrant for democratic politics.

Lacing his tract with the new psychological knowledge—derived especially from mental testing and eugenics research—McDougall argued that civilization was becoming ever more precarious. The increasing complexity of modern industrial urban life, he suggested, meant that "the demand for A and B men steadily increases," while the supply inexorably diminished. Without intervention, McDougall concluded, disaster must result. McDougall's account was concerned above all with race, understood in ethno-national terms as much as in broad color-based distinctions. He sought to preserve the presumed apex of humanity, Nordic or northern European stock, from degeneration from within, symbolized by the procreative menace of the feebleminded, and degradation from without, symbolized by the specter of miscegenation. Throughout his text, data on white/"colored" group-level differences in IQ—derived from at times tortured interpretations of the army testing results—loomed large as the touchstone for arguments about the inferiority of non-Nordics, anchoring his conclusion that only vigorous eugenics policies could make America again ready for robust, white-dominated democracy.[52]

McDougall and Stoddard's intertwining of modernity's perils and potentials around notions of civilization's fragility and racial preservation (through careful tending of its most meritorious biological specimens), Daniel J. Kevles has noted, articulated common hopes and worries among segments of the white middle class.[53] Campaigns for prohibition, crusades against urban vice, and calls for immigration restriction, not to mention the imposition of various forms of segregation and quota systems, all marked the early twentieth century, as middle-class Americans looked to "reform" movements to regulate the masses, exclude the "undesirable," and bring order to the republic. Among those in the forefront of championing such efforts were America's growing community of social scientists—including sociologists, anthropologists, political scientists, and psychologists—who by the late nineteenth century had become virtually obsessed with the nation's biological status and its implications for the social order's health. While some of these concerns subsided after the war, for many social scientists, science's contributions to America's military success simply strengthened their sense that major social problems were now ready to be tackled. Scarcely six months after the Armistice, for example, Joseph Kinmont Hart, a University of Chicago education Ph.D. teaching at Reed College, wondered about the fate of American democracy and the contributions social science could make to sustaining it.[54] Fearing that some would learn from the war to associate science with destruction and so consider it inimical to the civilian social order, Hart produced an impassioned plea for a different view of science's postwar role. Inspired by Frederick Jackson Turner's arguments about the frontier's place in sustaining American democracy, Hart argued that what westward expansion had been for America, scientific exploration could become.[55] This new frontier, however, resided not in the American landscape, but in its inhabitants. As such, the new social sciences would be critically important. Their domain was human beings in all their complexity and the social sciences' peacetime duty would be no less than to act as savior of democracy itself:

> Without science there can be no democracy, but only old prejudicial social forms. . . . In the future, all crucial action of a social nature must be determined by scientific investigation, rather than by customs, and men must be brave enough to fight for these things, even to the losing of their—jobs![56]

Hart's grand ambitions for the social sciences were not unprecedented. Progressives had championed applying scientific methodologies to social problems since the late nineteenth century, and with a modicum of success.[57] Practitioners of disciplines as diverse as economics, sociology, anthropology, social work, political science, and psychology challenged what often seemed to them sentimentalized Victorian conceptions of society and the individual and sought to replace those notions with objectively determined facts made visible through the cold light of science.[58] In many respects, Hart simply repackaged these Progressive commonplaces. He went further, perhaps, only in the implications he drew about science's place in the modern polity. Spurred by the experience of the war, with its vast marshaling of materiel and manpower and its heady application of expertise to managing the economy, Hart imagined a sociopolitical order not just open to science's authority, but subservient to and transformed by it.[59] In Hart's postwar America, science was to be the final arbiter, dispassionately settling social questions and dispelling prejudices through objectivity and impartial reason.[60] Democracy required no less, Hart contended, for only science—particularly social science—could legitimately establish the boundaries of a true democracy's operation.[61]

An Aristocracy of Brains?

Although Hart's hyperbolic plea that scientists risk even their careers to establish the social authority of science may have fallen on deaf ears, his vision of the place of the social sciences in the postwar world did not. Psychologists in particular returned from their military duties determined to carry the gospel of science and mental testing to the public at large, and thereby to enhance American democracy, not to mention their own professional opportunities.[62] Although the army's response to the psychologists' efforts had been mixed at best, the publicity the program received as well as the fact of almost 2 million recruits being exposed to intelligence and its tests helped to transform testing's status in the postwar period.[63] Finding employment primarily in academe, public education, and industry, psychologists constituted a powerful interest group whose livelihoods were linked to promoting intelligence and its importance. And what they disseminated above all was the sense of measurement of general intelligence as the key to addressing a range of social issues because it would allow, they believed, the replacement of subjective and unreliable assessments with objective determinations of individual capability and merit. Thereby, true equality of opportunity could be provided, because each person could be trained or selected for the role for which he or she was best suited, regardless of gender, class, or race/ethnicity.[64]

One of the earliest examples of this attempt to re-engineer society according to the dictates of an intelligence-based concept of merit was the creation in 1919 of the National Intelligence Tests (NIT), a joint product of onetime rivals but wartime colleagues Terman and Yerkes (along with Melvin E. Haggerty, Edward L. Thorndike, and Guy M. Whipple).[65] With $25,000 from the Rockefeller Foundation's General Education Board, Yerkes et al. met over the course of 1919 to design the NIT, modeled on Army Alpha but modified for a school-age population.[66] Completed by winter, the NIT was a collection of group-administered, multiple-choice instruments able to rank the entire American school population on a single scale.[67] Once adopted, its creators contended, American public education could be transformed. No longer need students of varying abilities be grouped in the same classroom; no longer need all students be subjected to the same curriculum; and no longer need every student be prepared for the same future. Rather, as B. R. Buckingham put it in 1921 in an editorial in the *Journal of Educational Research:*

> Our educational and intelligence tests permit us to ascertain the capacities of pupils far more accurately than ever before. Thus, the teacher becomes a guide and director. . . . Instead of prescribing the same treatment for all, he will become the expert diagnostician. On the basis of mental ability he will reclassify children, and because of their special abilities, he will further subdivide them.[68]

Where once the single-room schoolhouse had symbolized the commitment to universal primary education, now American psychologists celebrated the modern substitute: the multi-tracked school, in which the results produced by impersonal assessment mechanisms could be translated into objective systems of classification and separate educational destinies.[69] Equal treatment, they argued, required no less, because science had proven that children were more different than the same, and less susceptible to molding than sieving. "From this day forward," A. E. Winship wrote Terman on the publication of Terman's *Intelligence of School Children,* "it will be a crime against childhood, against humanity, to continue the 'course of study for all children' in any school. I am as glad to have lived to see this day educationally as to have seen the eleventh of November 1919 for democracy."[70]

Yerkes, in fact, went so far as to suggest that children be grouped by intelligence level as early as kindergarten, and after fifth grade be sent off onto distinct educational tracks: professional, for high-intelligence group-A children; industrial, for medium-intelligence group-B children; and manual, for low-intelligence group-C children.[71] Educational efficiency and democratic ideals themselves required, these psychologists believed, that the schools be transformed according to science's dictates. "I believe that the real meaning of democracy," University of Michigan psychologist Guy M. Whipple noted in 1922, "is properly safeguarded in the notion of 'equity of opportunity,' and if any nation is destined to perish it is that one which fails to provide the best possible educational training for those of its rising generation that show promise of educational leadership."[72] Yerkes concurred, condemning the privilege embedded in older methods of selection by contending that equality of opportunity was the sole true form of democratic education and that only ability grouping would allow "the free intermingling of children of the various [class] strata in any given intelligence section."[73]

Needless to say, this interpretation of American education, merit, and democracy—reorganization on the basis of natural inequality objectively determined or "educational determinism" in William C. Bagley's provocative characterization—was controversial at best. Some Americans rejected it out of hand, skeptical about the pretensions of experts, comfortable with time-honored educational structures and pedagogical approaches, and content with existing systems of reward and exclusion. For them, the common school was a potent icon of American democracy and its commitment to equality as uniform treatment of all, with success or failure a matter of personal initiative and, perhaps, family connection; any scientific attempts at restructuring threatened to undermine adherence to basic social values.[74] Others, more self-consciously modern and liberal, however, could not slough off the claims of science or the cult of opportunity so easily. Bagley, Walter Lippmann, John Dewey, and other Progressives were as committed as the psychologists to objective methods and reforming democratic institutions according to empirical fact, and shared as well the belief that opportunity and access underlay the American concept of equality. While they found various features of the psychologists' educational and political vision troubling, they could not simply discount it. Rather, in publications ranging from the *Saturday Evening Post* and *Atlantic Monthly* to *School and Society* and the *Journal of Educational Research,* they challenged specific factual claims and interpretations, so that psychological findings might be domesticated within their own conceptions of democracy, scientific objectivity, and merit.

On February 27, 1922, William Bagley of Teachers College, Columbia University, fired the opening salvo. Addressing the Society of College Teachers of Education, Bagley impassionedly denounced a number of the psychologists' arguments, especially as applied to basic education.[75] Rejecting what he termed belief in "educational determinism," or the primacy of innate mental ability, and the theory of aristocracy he believed it implied, Bagley instead celebrated education's power to expand the common man's intelligence and championed basic education for all. "If education is to save civilization," Bagley declared, "it must lift the common man . . . *to new levels of thinking and feeling.*"[76] In Bagley's view, education contributed at least as much to an individual's intelligence as all other factors combined. While he did not reject the concept of general intelligence, or the value of intelligence testing for particular purposes, he did strongly denounce the vision of a society ruled by an intellectual elite chosen virtually from birth. Only mass education, American style, Bagley proclaimed, could truly open society to the voices and efforts of all.[77]

A few months later, Bagley's cause was joined by Walter Lippmann. In a series of six articles in the *New Republic,* Lippmann took Stoddard and the mental testing community to task for shoddy procedures in generating and interpreting the army test results.[78] Lippmann argued that the figure of thirteen years for the army recruits' average mental age was on the face of it absurd (because such a large sample as the data on all army recruits must itself define normal adult intelligence); that IQ tests were mechanisms for classifying groups, not for producing absolute measurements; that predictions about school performance had little relevance to success in life; that intelligence was ill defined within psychology; and that there was little evidence that intelligence tests measured an innate trait. He then concluded that however useful IQ examinations were for specific classifications in specific settings, such as helping schools create more homogeneous classrooms, they failed to measure anything like pure intelligence, while according mental testers inordinate social power. "If the intelligence test," he warned, "really measured the unchangeable hereditary capacity of human beings, as so many assert, it would inevitably evolve from an administrative convenience into a basis for hereditary caste."[79] The following year, in the *Century Magazine,* Lippmann attacked the army testing data McDougall had used to establish the intellectual inferiority of "coloreds" by pointing out that there were extreme regional variations in IQ regardless of race and that these variations corresponded to the local school systems' quality. Passionate in his defenses of the "Negro's" intellectual ability, the value of education, and the possibilities of a democracy open to all, Lippmann also revealed the depth of his animosity toward his opponents' views:[80] "I hate the impudence of a claim that in fifty minutes you can judge and classify a human being's predestined fitness in life. I hate the pretentiousness of that claim. I hate the abuse of scientific method which it involves. I hate the sense of superiority which it imposes."[81]

Lippmann's observations struck a nerve with many readers, among them John Dewey, who responded to Lippmann by questioning not the existence of individual differences but rather their limitation to any single construct, even intelligence.[82] In his own reflections in the *New Republic,* Dewey explained that democracy's essence was radical individuality, the belief that each person encompassed a unique set of attributes and that education should allow those talents to flourish. Mental testing was ill conceived, he continued, because it tried to hammer complicated human beings into simple administrative boxes, thereby producing a society at odds with the goals of true "civilization." The fetish for numbers, statistics, and quantitative categories, Dewey observed, derived from "our mechanical, industrialized civilization" and produced a "reverence for mediocrity, for submergence of individuality in mass ideals and creeds" inimical to both true education and true democracy.[83] Dewey rejected notions of superiority and inferiority, whether racial or individual, on the grounds that, while morally equal, human beings were otherwise incommensurable: each had to be appreciated in his or her own unique way.[84]

These attacks on the testing community and its instruments did not go unchallenged; sarcastic responses from Terman and more nuanced replies from other psychologists soon appeared in popular and professional journals.[85] Terman turned first to Bagley. In the *Journal of Educational Research,* Terman argued that Bagley's refusal to concede the significance of differences in individual mental endowment was a denial of scientific truth and a return to superstition, and that Bagley was actually imperiling democracy.[86] Terman preached the need to adapt the curriculum to each child's needs, and not vice versa, as most efficient for both child and society. He then lambasted Bagley for failing to understand how intelligence testing could aid in producing a truly egalitarian democracy, one in which inherited privilege was minimized and opportunity could flourish through identifying and training those who merited it most—the very able—regardless of class.

Terman was even more dismissive of Lippmann, suggesting in his *New Republic* rejoinder that Lippmann lacked the expertise to judge psychologists' work and that Lippmann's understanding of intelligence was laughably naive.[87] Characterizing (or caricaturing) Lippmann as asserting that "the essential thing about a democracy is not equality of opportunity, as some foolish persons think, but equality of mental endowment," Terman again celebrated the use of mental tests to "sift the schools for superior talent . . . in whatever stratum of society it may be found."[88] Terman swept aside most of Lippmann's technical criticisms of the army tests, although he did attempt to explain the controversial thirteen-year average mental age by conceding that the professional community disagreed over the exact point where adult intelligence began.[89] Most significantly, where Lippmann had insisted that the number of high-grade "A" and "B" men was a function of the time allotted to complete the test and that more would have scored well if additional minutes had been added, Terman countered that timing had little effect on a person's overall ranking, and that more time would simply have increased the number of correct responses required to be classified in the A or B group, without changing its meaning. Terman saw the proportion of most intelligent as fixed, and used relative test performance to identify them; Lippmann, conversely, considered measurement of absolute level of performance critical, and so maintained that the most intelligent were those who exhibited proficiency to a certain level (i.e., those who got a particular number of questions right), and not simply the top "N" percent. Both conceded that mental tests could reveal intellectual merit, but understood that merit in decidedly different ways.

For the next two or three years, insults continued to fly and positions continued to be argued, with each commentator wrestling with the implications innate intelligence differences held for basic education and, by extension, American democracy. The debate even extended to the more rarefied realm of higher education, where the new demographic pressures of the postwar period also initiated a wholesale examination of what the purpose of collegiate-level education in a democracy should be. On the one side stood such figures as President Cutten of Colgate and President Ernest M. Hopkins of Dartmouth College, not to mention Henry Holmes, who contended that the vast increase in applicants to America's colleges and universities meant that rigorous methods of selection would be necessary. In no other way, they believed, could higher education remain true to its calling of training, as Hopkins put it, the "aristocracy of brains, to whom increasingly the opportunities of higher education ought to be restricted."[90] In their view, "nature's inexorable law is inequality," proven most recently by the army mental testing program, which showed "only 13½ per cent of the population able to get through college well."[91] Although Cutten and Hopkins argued, like the mental testers, that this intellectual aristocracy was open to members of all classes, they also believed that the gap between the classes was growing ever greater, as the intellectually more able rose to their "natural" social position. Inevitably, Cutten concluded, "the effect of mental tests upon the problem of democracy" will be that "they will result in a caste system as rigid as that of India, but on a rational and just basis," one where even the right to vote would be limited.[92]

Other educational leaders, including President Alexander Meiklejohn of Amherst College, Melvin Rigg of Kenyon College, and U. S. Commissioner of Education John J. Tigert were not so impressed with this celebration of the aristocracy of brains.[93] Tigert complained that even on practical grounds, "colleges have not reached the point where the heads of the various institutions can diagnose the abilities nor the intellect of the prospective student" and rejected reliance on selection methods such as intelligence tests, while the Institute for Public Service suggested that the problem with higher education was not too many going to college, but, because of inadequate finances, too few.[94] Meiklejohn also voiced reservations about the reliability of the selection methods, though his main target was the vision of limiting college, not to mention suffrage or other fundamental citizenship rights, to only a few. Instead, he argued for a system of providing broad-based opportunities for education so that "each man in the measure of his own capacity, can be joined together in the attempt to build up a social order in which each individual may find some adequate expression of himself."[95]

In the end, however, despite the intensity of these debates and the popular press's continuing interest in mental testing throughout the 1920s and 1930s, after 1925 few authors addressed specifically the implications of intelligence differences for democracy.[96] The immediate impetus of the postwar moment and the anxieties it had stirred up passed, making questions about the nature of democracy seem less urgent. In addition, by the mid 1920s testing itself had become part of the cultural landscape: intelligence assessments were routinely being proposed, used, and contested as aids in, for example, hiring decisions, asylum admissions, capacity determinations, and educational tracking placements. Not surprisingly, focus in the press shifted largely to the tool, and articles about the use, or misuse, of intelligence tests predominated. Discussions of such "technical" issues could still carry a political valence, as Lippmann's criticisms of the army testing program make clear, but the more general worry that intelligence differences might cause fundamental problems for American democracy subsided, eased perhaps by the middle class's seeming success in containing the more radical demands of the masses in the immediate postwar period.

The position of intelligence in American society had become more secure, and the debates over testing and democracy can be seen as part of that transformation. By forging connections between testing results, educational structures, merit, and possibilities for democracy, both supporters and critics of intelligence testing created a setting in which such knowledge about human nature had become a matter of consequence with distinctly political dimensions. "The facts of biology and psychology," F. H. Hankins noted in his 1923 article "Individual Differences and Democratic Theory" in the *Political Science Quarterly*, "are forcing a conscious recognition of inequality, of differences in origin and destiny, of differences in the rôle played in the life and fortune of one's nation."[97] From a concept of only limited cultural purview before the war, by the mid-1920s, intelligence was becoming an established way of talking or worrying about biological differences at the level of individuals as well as groups, by providing a language for discussing and a means of assessing the relative superiority/inferiority of whoever was at issue.[98] Whether the initial public response was to embrace or despise their work, psychologists had credibly represented testing as a project whose goal was to determine objectively individuals' intrinsic intellectual merit so that social resources could be allocated efficiently and perhaps justly. Thereafter, simply ignoring such a project proved difficult. Rather, sorting out rival claims about, and agendas for, testing pushed public institutions and psychologists themselves, as we shall see, to make responses and concessions, thereby leaving the links between intelligence and merit, while undeniably present, never fully resolved or stabilized.

Intelligence and Interwar American Culture

"Can't we Americans," an anonymous author rather whimsically wondered in the November 24, 1923, *Saturday Evening Post,* "erect a tariff wall to protect a new infant industry and start to manufacture a commodity which ought to be in great demand not only among the young but among grown-ups? Why not have quantity production of intelligence?"[99] Why not, indeed? While most psychologists would doubtless have disputed our author's suggestion that intelligence could be manufactured, they would certainly have agreed that the 1920s and 1930s were marked by enormous interest in this "infant industry" and intense demand for the identification of those possessing high and low degrees of this much sought-after product. All the tensions surrounding the intelligence-democracy conundrum notwithstanding, at the practical level the business of intelligence assessment boomed during the interwar period.[100] Intelligence became a kind of mass-marketed commodity, in some ways no different from the vast range of consumer goods flooding American stores and reshaping American culture.[101] Within psychology the measurement of intellect was one of the largest and most successful of the discipline's fields; outside, intelligence testing became a lasting feature of the American educational system and, for a while, a significant practice of the modern corporation and many governmental agencies, including the military.[102]

American psychologists, not surprisingly, stood in the forefront of the effort to spread the regime of intelligence to American culture during the interwar period.[103] The wartime testing program had created a virtual army of advocates, trained by the military, who were ready to offer their services in peacetime as psychological testers.[104] While not all shared Terman and Yerkes' enthusiasm for intelligence assessment as the key to society's rational reorganization, most were convinced that intelligence constituted a fundamental human characteristic, one that could be measured and used, in one form or another, to diagnose mental defect or help guide selection of the right persons for the task or training appropriate for them.[105] This vision of what intelligence assessment could provide became more readily disseminated once intelligence itself was commodified, packaged into the numerous standardized tests developed during the period, sold by private companies—most notably the Psychological Corporation, C. H. Stoelting Company, Houghton Mifflin, and World Book

Company—and administered by psychologists occupying new positions in education and industry. Such practices and technologies allowed intelligence to permeate the culture and become part of the everyday experiences of millions, from World War I veterans to schoolchildren to job applicants.[106]

In another 1923 *Saturday Evening Post* article, for example, its author, Elizabeth Frazer, made clear how thoroughly intelligence tests had become part of the corporate landscape, as symbolic of modernity as the companies' new skyscrapers: "On the sixth floor of the massive business building more than a score of stories high, whose slim, spirelike campanile of granite cleaving the sky dominated that entire business section . . . they were giving the fresh applicants for jobs the new modern mental tests." Frazer described the scene in detail: how clerical applicants to this "gilt-edged conservative" company were screened twice before being led off to the final, and most important, assessment, the mental test. The goal, as the examiner put it, was "to shut out the slow, stodgy, stupid and defective brains," because, as she continued, "we require a high-grade girl in both mental capacity and character."[107]

The interest in character was nothing new, but the focus on mental capacity as revealed by tests certainly was. Merging the language of scientific management—"high-grade"—with that of intelligence assessment, the examiner revealed that notions of intelligence and its degrees had permeated her understanding of her role and the criteria by which she would choose applicants. She now had a language that allowed her to go beyond the traditional tripartite division of human intelligence—genius, normal, idiot—and to speak of "high-grade" minds as well as those that were "slow" or "defective," gradations she could use to construct a scale of merit relative to the positions she was trying to fill. In this respect, the examiner was not much different from Frazer herself. Reporting on the issue of women in the workplace, Frazer compared the worlds of white-collar work and the factory, concluding that they were fundamentally different, with intelligence being a principal factor keeping them separate. "I began to see a big mental gulf," Frazer explained, "between the factory worker and the white-collar girl. I had already discovered there was a social gulf . . . but here was a breach of brains."[108]

By the mid-1920s, this "breach of brains" was no laughing matter in American culture. Intelligence beckoned as a pragmatic way to address a number of social problems. The transformations, from urbanization to corporate capitalism's rise, that were producing elements of mass society in the United States had concomitantly opened space for new methods to regulate what seemed to many a nation of immigrants and strangers. When confronted with hundreds of thousands of largely unknown recruits, for example, as we saw in chapter 6, the army had turned to intelligence assessments as one way of rapidly classifying and sorting this multitude. After the war, the swelling school systems followed suit, seeing in ability grouping a means of efficiently organizing students and providing large-scale education for the tenfold increase in student population the high schools alone had experienced since the 1890s.[109] Within the government, proposals to exclude unwanted immigrants, sterilize the unfit, or justify practices of segregation or voter restriction all on the basis of "defective" intelligence were rife. At the same time, growing corporate bureaucracies found in intelligence both an explanation of what white-collar labor contributed to production and a criterion to help discriminate among applicants for desk work and sales positions.[110] As Frazer succinctly observed: "Sheer brawn, youth, quickness no longer count all. It needs something else to get by. And that something is gray matter. Brains."[111] Moreover, even some of those individuals highly critical of aspects of America's existing social structure pushed for the adoption of intelligence testing and the regime of "brains," seeing objective assessment as a powerful means of undermining traditional exclusionary practices based on ethnicity or gender in the name of individual competency and merit. Thus psychologist Henry Garrett, for one, conducted a number of studies in the interwar period dedicated to demonstrating the intellectual superiority of Jewish Americans as revealed by intelligence examinations.[112]

During the 1920s and 1930s, this belief in the overriding importance of "brains" underwrote an enormous variety of attempts—mostly orchestrated by psychologists, school administrators, and the newly developing profession of personnel managers—to employ intelligence tests to sort the population. Surveys conducted in the mid-1920s found that 85.6 percent of cities responding were already employing intelligence tests in the schools for classification of students, with ability grouping being the most important application, especially in elementary schools, and the method used typically some combination of IQ measurement and teacher judgment.[113] As psychologist Frank N. Freeman, no overzealous advocate of testing, explained in 1924: "The fundamental basis for the practice of homogeneous grouping is the psychological fact of extreme differences in the capacity of children to do school work."[114] Newspapers and magazines throughout the period were full of reports about, or discussions of, the decision of this school district or that university to turn to intelligence testing, as well as celebrations of educational systems rearranged, collegiate admissions improved, and new prodigies discovered through testing—such as Elizabeth Benson of Los Angeles, who became a minor celebrity in the mid-1920s when, as an eight-year-old in 1922 her IQ was measured at 202, the highest on record.[115]

The U.S. Army, for all its hesitancy about testing, continued to employ psychologists immediately after the war and would periodically drop and then reinstate intelligence assessments for various kinds of recruits throughout the interwar period, mounting a major testing effort again with the onset of World War II.[116] Indeed, in 1929, according to the *New York Times,* the army's adjutant general, Major General C. H. Bridges, even declared that the use of intelligence tests to sift recruits had helped reduce desertions.[117] Other governmental agencies—federal, state, and local—turned to intelligence as a means of determining who did and did not merit being employed, promoted, or permitted some licensed activity. Both the federal and New York State Civil Service examinations, for example, from early in the period included a mentality test as one of their methods for selecting various kinds of potential government workers.[118] A number of large urban police departments experimented with testing (including Los Angeles, Philadelphia, and New York), especially for those under consideration for promotion to detective; intelligence testing was used in New York State for first-time voters unable to prove they had graduated from the eighth grade; and various schemes were proposed to require mental tests for marriage, driver's licenses ("Court Seeks Curb on Moron Drivers," a *New York Times* headline proclaimed), and jury duty, all to rather mixed receptions.[119]

Perhaps the most significant exploration of intelligence assessments' value and usefulness outside education was undertaken in the business world.[120] At first, many industrial psychologists and personnel managers saw the World War I army testing results—especially the discovery of a strong correlation between intelligence level and occupation—as a clear demonstration of mental tests' potential to make the process of personnel selection more efficient and cost effective, and numerous firms experimented with the possibilities of selection, placement, and promotion aided by measurements of general intelligence.[121] As J. P. Lamb, employment manager with Cheney Brothers, a silk manufacturer in Connecticut employing five thousand, noted in 1919:

> modern industry demands less waste in relation to the human factor. A selection of employees to accomplish this purpose and obtain increased efficiency, accompanied by contentment, compels more care and method in the study of human characteristics in relation to placement. An accurate intelligence rating is necessary to this purpose.[122]

In 1922 psychologist Clarence Yoakum informed readers of the business magazine *Forbes* about the success of one New York City company in using intelligence tests to pick executives, and studies in the 1930s demonstrated "a definite and consistent relationship between intelligence scores and advancement in clerical work."[123] For those who championed testing, its most important use in business, as Donald A. Laird noted in *The*

Psychology of Selecting Employees (1937), was to establish the minimum and often maximum amounts of intelligence required to perform a given job successfully.[124] Even skeptics about testing admitted that there was value in eliminating those with defective intellects from the selection pool, and many agreed as well that certain executive positions did require greater than average general intelligence.[125]

For all of this interest, however, business leaders and the applied psychologists who worked with them overall proved much less enthused about assessing individual intelligence than their brethren in education. Like the French psychologists developing orientation professionnelle, American industrial psychologists were most likely to emphasize the multiplicity of factors that went into business success, whether as a cashier or company executive, and to insist that no single characteristic determined how individuals would, in practice, perform. Some, as has been suggested, did believe that an individual's intelligence was an important, if not the preeminent, feature that needed to be assessed. Many other psychologists and personnel managers, however, strongly disagreed. Whether or not they accepted the existence of general intelligence, these critics by and large subscribed to Edward L. Thorndike's conception of intelligence and argued that intelligence as it was relevant to business was multiple and contextual. Industrial psychologist Henry C. Link, for example, argued as early as 1923 that "we cannot say that a man's general intelligence fits him for any particular occupation," and then concluded that "the practical value of any intelligence test must be specifically established, for specific occupations, under specific conditions."[126] By 1928 Link was even more explicit, declaring that "there is no such thing as general intelligence, and that, if there were, it would be of little use to us in employment work because we are interested in specific abilities or *kinds* of intelligence and not in degrees of intelligence *per se,*" a position wholly sanctioned by Morris Viteles in his 1932 textbook on industrial psychology.[127]

The corporate records of the Pennsylvania Railroad (PRR) reveal a similar skepticism: while the PRR established a committee to look into intelligence testing in 1934, the company as a whole remained doubtful that tests would prove of any real value to them. "As applied to the Railroad," F. W. Hankins opined, "there are a great many angles in a 'psychological test' that would be of little benefit to the man who is responsible for the selection of men in various capacities. . . . My personal view on the selection of men is that it should be handled by contact of the employing officer, following the workmanship of the men and weeding out those who do not perform." However, as another railroad official wrote in response: "I take no exception to what Mr. Hankins has to say relative to the intelligence test. At the same time, we have got to admit that we make mistakes in our present way of promotion. We get men without 'aptitude' pretty well along in years before we discover what we might have observed—through a proper test—earlier."[128]

This ambivalence, as we shall see, marked much of the response to mental measurement during the interwar period, within not only the culture at large but even the discipline of psychology itself. Certainly the proliferation of intelligence testing in schools, government, and industry, coupled with the growth of white-collar bureaucracies where "brain work" was assumed to be the essence of such workers' occupational identities, meant that determinations of intelligence could and often did profoundly affect an individual's future. At the same time, a range of criticisms were articulated, exceptions established, and limitations proposed that served to cast doubt on the value of intelligence assessments as methods of determining merit, whether individual or group. Claims about intelligence became a part of American culture during the 1920s and 1930s, but remained always controversial, points of debate as much as ways of making and justifying decisions.

Controversies, Resistances, Accommodations

As has already been suggested, the success of mental testers in promoting intelligence to the public as a product that ought to be "in great demand" did not prevent serious disagreements from erupting over the concept and its instruments. Four issues proved particularly vexing, both within psychology and for the public: how to define intelligence, whether it was one thing or many, what the relative importance of nature and nurture were in its development, and what weight it should be accorded in various decision-making situations where merit was at issue. None of these questions, as we shall see, was ever fully resolved; rather, they constituted critical sites of dispute that persisted throughout the interwar period and, indeed, the rest of the century.

Most embarrassingly, throughout the 1920s and 1930s the mental-testing community proved completely unable to define the nature of intelligence in a way that enjoyed wide support. A 1921 symposium on intelligence and its measurement is a case in point. The thirteen psychologists participating offered eight distinct definitions of intelligence, including ability to learn, ability to give correct responses, ability to think abstractly, and behavior that brings advantage.[129] A 1925 collection of definitions gathered by psychologist Henry H. Goddard provided a similar range.[130] Thereafter, definitions continued to be routinely proposed and defended, as psychologists sought to fill a problem at the center of the burgeoning field of intelligence assessment that Lippmann had identified in his first arguments with Terman.

In 1940, Frank N. Freeman, in the lead article for *Intelligence—Its Nature and Nurture,* still found the community troubled over how to define its fundamental concept.[131] His own solution was to turn to the philosophy of instrumentalism, and to argue that intelligence should be defined simply as whatever it was that intelligence tests test. Circular though that definition may sound, throughout the 1920s and 1930s, partly in response to criticisms of testing, various mental testers came to define intelligence in much the same way.[132] Not that many remained fully satisfied with such an approach. No sooner had Freeman equated intelligence with its tests, for example, then he defined it anew. "Intelligence," he went on to note, ". . . is the ability to learn new acts or to perform new acts that are functionally useful."[133] By no means a novel way of characterizing the concept, what is revealing is that intelligence, to Freeman and most other psychologists, was more than a score on a test. Intelligence was a thing—real, concrete, and independent of its measurement—even if not directly perceptible and even if many had to resort to roundabout ways of defining it scientifically.

Psychologists were perturbed by more about intelligence than its definition. One aspect of intelligence's nature that provoked especially heated debate, as we have seen, was whether the term referred to a single ability or a composite of different skills. The English psychologist Charles Spearman and his fellow travelers, including Terman and Yerkes, formed one end of the spectrum of opinions with their commitment to the existence of the general intelligence factor *g,* and *s,* a collection of specific factors that influenced but were secondary to *g.*[134] On the other end stood Thorndike and a few others, who continued to insist that the mind was composed of vastly numerous, intrinsically independent abilities.[135] And in the middle, arguing that the primary abilities were multiple though few in number, stood psychologists such as L. L. Thurstone and the Scot Godfrey Thomson.[136] The strains these opposing tendencies produced could have split the testers into warring factions. Debates between the Spearman, Thorndike, and Thurstone camps certainly took on this quality, as they developed ever more elaborate statistical justifications for their respective positions.[137] Nonetheless, for all the intensity of their arguments and pressure to redefine their pursuits, no irreparable breach developed. Psychologists of all persuasions shared the same perception of the proper form of a mental test and the appropriate statistical methods for its construction, and the same journals continued to publish all three factions' articles.

In the 1920s, for example, after vociferously criticizing his colleagues for failing to design instruments based on a well-developed theory of cognition,

Spearman used his own ideas about *g* and *s* to construct a theory-validated test.[138] His instrument was composed of four principal item types—analogies, antonyms and similarities, identities, and completions—selected as those most heavily *g* loaded. Ironically, although Spearman's theory of cognition was unquestionably novel, his test was not. All four of the item types he chose were commonly employed by other psychologists—including Thorndike, Thurstone, and Terman. Despite the controversy over theories of intelligence, these four item types, along with vocabulary knowledge, information, reading comprehension, and arithmetic, formed the core of a pool of question types routinely relied on by all parties in test construction. When Princeton professor Carl Brigham created the Scholastic Aptitude Test in 1926, for example, he drew on these elements while trying to produce a college-entrance intelligence test valid anywhere in the country.[139]

The testing community also shared a set of statistical techniques for measuring tests' reliability and validity. Methods such as correlation, regression, partial correlation, normal distribution, and Probable Error—largely derived from the work of Francis Galton, Karl Pearson, and Charles Spearman—became standard assessment tools for mental testers.[140] Two statistical assumptions dominated. First, distribution of intellectual ability had to fit the normal (bell-shaped) curve; that is, the number of very able and least able had to be about the same, and most individuals had to group around the population average. Second, a test's validity was thought best assessed by using a particular correlation coefficient (typically the Pearson product-moment correlation coefficient or some variant thereof) to determine how closely the results on the test matched some other, already legitimated method for assessing intelligence.[141] Both of these statistical assumptions were treated as black boxes by most psychologists: they were used routinely and rarely questioned. A valid instrument was one that produced scores that internally fell into a normal distribution and correlated well externally with some other standard of intelligence, be it teachers' evaluations, marks in school, or performance on some other standardized test. Even the most statistically informed mental testers—including Truman Kelly, Thorndike, Thurstone, Haggerty, and Spearman—worked mostly to elaborate or refine these concepts rather than to replace them with new ones.

The third issue that generated controversy within the discipline (and outside of it) was over the characterization of intelligence as a biological potential genetically determined from birth. Especially potent at the level of groups, where this use of intelligence, as we saw in chapter 3, had proven integral to the development of scientific racism, the biological basis for an individual's degree of intelligence had been promoted by most of the first generation of American testers as well as by their allies in the eugenics movement. At first, group-level studies of the intelligence of African Americans and immigrants—though not women—all seemed to produce results similar to Brigham's 1923 analysis of the army testing data: that northern Europeans were superior in intelligence and other groups distinctly inferior, and that these differences were permanent, a product of nature rather than environment.[142] However, as early as the 1910s, questions were raised in the scientific literature about the "nature" interpretation of intelligence, most significantly in the research of the anthropologist Franz Boas on migration and changes in skull size among native peoples of northwest North America.[143]

With enthusiasm for wholly biological and especially eugenical explanations of social phenomena waning in America by the end of the 1920s and the culture concept gaining sway, a number of psychologists joined Boas in advancing more decidedly environmentalist interpretations of IQ at the level of race and ethnicity.[144] In 1930, Brigham dramatically recanted his 1923 *Study of American Intelligence,* in which he had argued for the existence of a biological hierarchy of European groups (Nordic, Alpine, Mediterranean).[145] At about the same time, Boas's student Otto Klineberg undertook research on the mean IQs of these European peoples and demonstrated that Brigham's initial findings had been the result of specific environmental conditions and not of underlying biological differences.[146] Klineberg went on to challenge assertions about the innate intellectual inferiority of African Americans, demonstrating that African American migration to northern cities produced IQ gains that could best be explained in terms of the different educational environments of the North and South.[147]

Throughout the period, African American psychologists and intellectuals were among the earliest and most vigorous in attempting to discredit the nature interpretation of intelligence differences, especially as promulgated at the group level. W.E.B. Du Bois, for example, while accepting distinctions in individual intelligence in his famous 1903 essay extolling the importance of the "talented tenth," nonetheless resoundingly rejected claims about the inferiority of the "Negroes" as a group.[148] During the interwar period, a number of African American psychologists and sociologists—including Horace Mann Bond, Charles S. Johnson, Howard H. Long, and J. St. Clair Price—generated elaborate critiques of the intelligence-testing regime. In addition to producing numerous studies revealing the strong effect of environmental changes on IQ scores, many also undertook exacting examinations seeking to cast doubt on research showing that intelligence levels were largely products of nature.[149] Nonetheless, even most African American researchers couched their objections principally as problems with procedures and interpretations, and not with the notion of an intelligence test itself. "It is not with Intelligence Tests that we have any quarrel," Bond observed. "[I]n many ways they do represent a fundamental advance in the methodology of the century. It is solely with certain methods of interpreting the results of these tests that we, as scientific investigators, must differ."[150] Concerned with establishing their scientific bona fides, it may not seem surprising that many African American scholars tempered their criticisms of testing. In addition, like many of their Jewish American colleagues, they may have suspected that objective assessment methods, for all their problems, provided protections to marginalized groups that more subjective procedures associated with quotas and segregation simply did not.

For the most part, by the 1940s, belief in the hereditarian interpretation of intelligence at the level of groups had given way to one version or another of Du Bois's position: while there might be even profound natural differences in intelligence among individuals within groups, differences between groups were almost entirely a product of environmental influences and could be reduced with better education, improved living conditions, and other sociocultural changes.[151] At the level of individuals, however, the issue was much more contentious. Although the work of Bond, Long, Klineberg, Melville J. Herskovits, and others had done much to suggest that an individual's IQ measurement might improve if he or she relocated to a markedly better environment (such as moving from a poor rural southern area to a more economically and educationally advantaged urban northern one), in the main, belief in the constancy and biological character of individual intelligence was widespread in the psychological community. As Hamilton Cravens has convincingly shown, the one group of researchers who argued consistently during the interwar period for a much more environmentalist understanding of the development of an individual's intelligence—those associated with the Iowa Child Welfare Research Station—were relentlessly criticized and largely marginalized by Terman and his followers for what they suggested were the Iowa Station researchers' naive views on intelligence and shoddy statistical methodology.[152]

For most psychologists, too many studies had already established a number of critical facts: that, by school age at the latest, IQ measurements remained fairly constant; that education could not make feebleminded individuals normal; that IQ scores, like other natural phenomena, were distributed according to the bell-shaped curve; that correlations of identical-twin intelligence scores were much higher than those for fraternal twins; and that intelligence levels seemed to run in families.[153] By the time the National Society for the Study of Education published its two-volume *Intelligence—Its Nature and Nurture* in 1940, virtually all researchers had concluded that both biology and culture had profound effects on the development of an individual's general intelligence, with nature establishing the potential that nurture might or might not realize.[154] While thus conceding the environment an important role, most psychologists still accepted not only the reality of intelligence as a heritable characteristic, but the value of measuring it and making social decisions based upon it.

In the culture at large, however, the public was much less sure about how much assessments of intelligence should count in determining individual merit. Unquestionably throughout the period, there were many individuals and groups who were wholehearted enthusiasts for using intelligence assessments, especially in schools, as objective and fair ways of sorting the population. On September 14, 1922, for example, the *Los Angeles Times* proclaimed that "Mentality Tests for Children Prove Value," while a reader of the *Washington Post* wrote in September 1924 to "express appreciation of intelligence tests given in the schools," and the *New York Times* reported on October 12, 1930, that educators at Columbia University had rallied to defend intelligence tests and the university's use of them, tests similar to those pictured in the *Washington Post* on March 10, 1940, as part of "making the school system run smoothly."[155] Ernest Greenwood, vice president of the Board of Education for the District of Columbia, wrote a series of articles on "grading human beings" for the *Independent* in 1925, in which he began by painting himself as a skeptic but concluded by declaring that "the importance of these scientific developments in methods of testing, examining, and grading human beings . . . cannot be overestimated."[156]

Nonetheless, almost from the moment that testing went public, others echoed the concerns and criticisms of Bagley, Lippmann, and Dewey. An "Anxious Parent" writing to the *Washington Post* on September 12, 1924, voiced opposition to the introduction of intelligence testing into the public schools because "the idea of promoting one group more rapidly than another is un-American" and declared that "giving every child an equal chance is the true and tried American way."[157] Only a month later the *Christian Science Monitor* reported on the Public School Protective League of California, which vowed to oppose mental testing of school children on the grounds that "such tests do not and cannot determine efficiently or honestly the ability of a child." A similar position was adopted by the Illinois State Federation of Labor in 1927, when it rejected intelligence tests and "the philosophy that would teach to our children that nature has placed them on levels either high or low from which there is no escape," and by Rollo G. Reynolds, principal of the Horace Mann School (a part of Teachers College), in November 1930 when decrying the segregation of "bright" and "dumb" pupils.[158] Critics were particularly worried about four aspects of intelligence and its tests: that the tests did not truly measure native intelligence, that they would create invidious distinctions not in keeping with a democratic culture, that intelligence was more complicated and multiple than IQ suggested, and that intelligence alone was a poor predictor of future success.

Some members of the public took such critiques to heart, concluding that assessments of intelligence were of little value and should be straightforwardly rejected.[159] Most, however, found it difficult to discount entirely the tests and their results, no matter how controversial they might be, because of a combination of the tests' administrative usefulness, their ability to make decision-making procedures more objective and accountable, and their ideological tie to notions of merit. Neither Bagley nor Lippmann, for example, suggested that mental testing lacked value or that its use in schools should be precluded. Indeed, Lippmann was careful to specify ways in which intelligence tests could contribute to school and society, and even Dewey conceded that there were certain practical situations where classification and tests were appropriate.[160] Many others skeptical about some of the claims for intelligence sought, in one form or another, to imagine possible accommodations with testing, most typically by attempting to redefine some central feature of the intelligence construct or how it would be used. Thus, in a September 13, 1921, article "Mental Tests in the Schools," the *New York Times* explained that "as regards the individual, their [mental tests'] significance is limited . . . but 'statistically,' as regards large groups, the results are of great value."[161] Reynolds adopted a similar approach to intelligence tests in a series of articles he wrote for the *New York Times* in March 1931, arguing that test results alone were too limited and should be balanced with teacher assessments and analyses of the individual's actual accomplishments, that tests could not reveal an individual's special abilities, and that many factors other than intelligence influence success or failure.[162]

Angelo Patri, author of the "Our Children" column for the *Washington Post,* revealed particularly clearly the ambivalence that many Americans felt about these new instruments. On February 2, 1922, he urged parents not to worry about assessments of their children's intelligence but rather to "welcome such a test. It is a fairly accurate measurement of the children's condition." Immediately after, however, he also advised that "the teacher's judgment is to be given consideration" and that "whatever you do, don't look upon an I.Q. as fixing your child's position in relation to his life work or his mates. Nothing is ever fixed that concerns a child."[163] This pattern of simultaneously welcoming and yet seeking to limit the ramifications of intelligence testing was reflected in psychologist Herbert S. Langfeld's declaration in 1926 that mental testing was fine within limits, but should never be used to compare those of different backgrounds, and in Frank Hill's article for *New York Times* readers in 1934, in which he described how IQ tests work and then explained that such tests assess certain aspects of intelligence, but not others, and that there are important non-intelligence factors—"patience, determination, imaginative power"—the tests simply cannot measure.[164] In all these cases, and many others, the goal seemed to be less to deny all value to mental testing than to find ways to constrain both how it was used and what it meant.[165]

Indeed, consider another attempt to rein in the ambitions of the testers and their claims. In the November and December 1923 issues of *America: A Catholic Review of the Week*, Austin G. Schmidt, S.J., wrote a series of articles on intelligence tests in which he sought to prove to Catholic educators "that mental tests, when used properly, can be of signal service in school work; and to point out the chief limitations of tests, and the precautions that should be observed in their administration."[166] The key phrase here was "used properly." Schmidt sought to preserve the substance of the tests while stripping away any meanings that he and his Catholic brethren found objectionable. In particular, this meant disavowing claims that intelligence was a unitary entity or innate capacity, and that it was the principal factor determining success in life. Repeatedly Father Schmidt argued that standard intelligence tests measured only one kind of intelligence—which he dubbed the FCA, or *Facultas Comprehensiva Academica*—and that worldly success depended as much on physical, social, moral, and environmental factors as on this type of intelligence.

In theory, then, the movement of intelligence and its tests from psychology to Catholic education would require a radical redefinition of both concept and instrument. But in practice it is not clear that that is what transpired. For Schmidt's text also can be viewed as a primer on intelligence and its tests for a community otherwise hostile toward both. Schmidt explained in accessible language what IQ was, how quantitative estimates of intelligence were arrived at, and what the statistical terms "correlation" and "reliability" meant. More importantly, Schmidt seemed to reintroduce, via practice, many of the features of testing and intelligence that he had rejected in theory:

> [Intelligence] tests do the following things. They enable us to identify pupils who could do much better, and to discover what is holding them back. They prevent injustice to pupils of whom we might expect too much. They make it possible for us to put pupils—especially those who are just entering the school and with whom we are unfamiliar—in sections of approximately equal ability.[167]

In short, tests did for Schmidt everything they did for all except their most rabid proponents within psychology. Consider Schmidt's proposal to rename intelligence tests FCA tests, advanced on three occasions in his series of essays.[168] Each time, he almost immediately abandoned the term and returned to the word "intelligence." "*Facultas comprehensiva academica*" hardly trips off the tongue, to be sure, even for a Jesuit; nonetheless that alone does not explain his reluctance to eliminate "intelligence" from his vocabulary. Schmidt believed in the value of intelligence tests, and such tests had certain characteristics: they produced a single number characterizing each examinee; they promised that this number would relate to the examinee's mental equipment and success in life; and they were called tests of "intelligence." This was the reality of the tests, and Schmidt could no more escape the name, linked to the everyday language of his audience, than he could the approach to human nature built into the tests. For all of the qualifications that he tried to erect around intelligence and its instruments, Schmidt accepted the promise of the technology: that it could make visible an examinee's latent ability, and do so in relatively fine gradations, all the while corresponding to some basic conception of fairness to the students.

Like Bagley, Lippmann, Dewey, and thousands of other Americans, therefore, Father Schmidt too found it difficult to reject completely the world of intelligence he was so assiduously working to constrain. Given the place of scientism within American intellectual culture, mental testing's immunity to full-scale rejection is hardly surprising.[169] Nonetheless, what is equally striking is that the IQ version of intelligence in particular generated such strong opposition. Lay people as well as professionals could see the implications of the construct psychologists laid before them as one solution to the problem of merit, and they were equivocal in their response. Scientific authority did not simply triumph; rather, the psychologists and their critics reached a complicated set of accommodations. Psychologists gained extensive powers to categorize and manage those deemed marginal, especially the feebleminded, and institutions often carried out coercive practices such as sterilizations based on test results. Administrators established ability-grouped, multi-tracked schools throughout the United States, and mental-test results were often a key criterion determining placements. During the 1920s and 1930s, a number of companies and government agencies used intelligence tests as aids in employment and placement decisions, especially in hiring entry-level white-collar workers. And many colleges and universities incorporated intelligence testing into their admissions and advising processes.

The critics of testing, however, also achieved some important results: no state or private agency ever established a system of testing, classifying, and then preparing children for career trajectories based solely on intelligence measures. Universities rarely used mental tests alone to decide admissions. Except for brief moments, public officials never turned to intelligence tests as important gatekeepers even for immigration or access to voting. And many Americans—from recruits making snide remarks about army testing to fundamentalists celebrating Christian over secular values to Deweyites committed to radical individuality to parents sure that good upbringing trumped native propensities—continued to embrace more complicated understandings of merit, including ones that celebrated characteristics other than brains. In addition, the criticisms of testing propounded by Lippmann, Dewey, Bagley, et al. helped create a rhetoric for expressing doubts about psychological instruments, especially in regard to the meaning of statistical findings for individual cases. More generally, the broader social vision of the "determinists" was largely rejected. Their highly rationalized and hyper-efficient "brave new world"—in which each citizen would be slotted into his or her occupation through the objective determinations of psychological experts—won few adherents.[170]

Michael Sokal and Franz Samelson have argued that intelligence testing reached its zenith in America in the early to mid-1920s, and then, as a result of widespread criticisms, underwent a period of retrenchment.[171] While it may be true that intelligence testing, except in education, declined in the late 1920s and 1930s, then stabilizing in the form it has maintained to the present day, that trajectory cannot be applied to intelligence itself. Once postwar intelligence tests publicly revealed (or produced) small differences in individual intelligence, the reification of those differences was difficult for most people to avoid. They became part of one's consciousness, so much so that in 1927 the *New York Times* could print the following, only half in jest:

> **Browsing Among Brows**
>
> Once more an English educator has risen to demand consideration for the mental fate of the great army of human beings who rank between the high and low brows. . . . Americans, having read so much about psychological tests and ratings, and about the effects of internal secretions on personality and of certain foods on brain power, must be disappointed to see that no sliding scale has been offered to measure the exact height of the individual brow so as to ascertain just where its possessor fits in the social order. How is it possible to decide the precise amount of intellectual diversion to which he can be subjected unless the brow rating is known in advance? The old problem still awaits solution.[172]

As "Browsing Among Brows" suggests, the concept of intelligence reified by such instruments as Army Alpha, the NIT, and the SAT had become an everyday feature of the American intellectual landscape during the interwar period. Intelligence tests took psychologists' ideas and made them "real" for the mass of Americans; these notions then developed a cultural life of their own, sustained even when the tests themselves receded somewhat in popularity and importance. Intelligence became a way in which Americans could argue about who merited what social goods, be it particular educational opportunities, job promotions, or rights to licensed activities. While intelligence was never unequivocally accepted as a criterion of merit, neither could it be completely rejected. Rather, claims for, or arguments against, the use of intelligence assessments were constantly open to contestation. Americans acculturated intelligence during the interwar period by acknowledging at the same time both its persuasive power and its deep limitations as a way of understanding merit.

Conclusion

By 1940 rather different versions of intelligence and its tests had achieved places in the American and French intellectual and social worlds. The extraordinary career of intelligence in interwar America derived largely from the popular conviction that intelligence might be a critical determinant of an individual's place in society and success in life. Consonant with the rationalizing imperative for order, cooperation, and efficiency widespread among America's managerial and professional classes, intelligence conceived of as a simple, linear, biological, and hereditary characteristic provided one means of explaining not only why certain individuals were at the top of the social hierarchy and others were not, but of justifying this arrangement within the language of meritocratic democracy.[173] It also provided the terms for assigning ethno-racial or class groups to distinct social strata or certain kinds of occupations or lives, as well as for vigorously contesting such designations on the basis of measures seemingly independent of traditional systems of privilege.

Intelligence and its tests thus came to be closely linked with analyses of American democracy and how equality, citizenship, and merit should be understood in light of the tests' findings. But why? Why should a new scientific procedure, intelligence testing, and the data it generated have come to play such significant roles in American culture? And why did they take on such different functions in France? The connections Americans forged between intelligence and merit, as we have seen, had few direct French parallels. For French psychologists, measurements of intelligence were one part of assessing and understanding an individual's capabilities and deficits. No less obsessed with merit than Americans, and even more entranced by technocratic solutions to social problems, republicans in France nonetheless felt little need to stabilize their ways of determining merit by recourse to natural objects such as intelligence. Rather, given a political vision in which the state's active intervention was deemed critical to maintaining the nation and its citizens, French republicans looked to the government to select and mold the next generation of leaders. The free play of "natural" talents, for them, was less significant than meeting the nation's needs through a system of training open to all, though with key moments of gatekeeping accomplished via expert judgment and not mechanical objectivity. Too multidimensional and underinstitutionalized to be used readily as a counterweight to existing methods of assessing merit, intelligence in France—when not deployed to sort the pathological—thus served primarily as a way of understanding the complex skills and abilities of discrete, mostly elite individuals and providing insights that might aid, but could not determine, career choices and life goals.

To their American promoters, by contrast, intelligence measurements promised simultaneously to reveal one of an individual's fundamental characteristics and to allow seemingly objective and neutral decisions about that person. As a universal though differentially distributed human feature, intelligence seemed perfectly suited to facilitating comparisons across regional, gender, and ethno-racial divides. American advocates of the IQ version of intelligence—ranging from psychologists to educators to members of certain historically marginalized groups—used these characteristics to promote a vision of the social order founded on belief in human differences, equality of opportunity, and the need for efficiency. To critics, however, the vogue of intelligence threatened to undercut American democracy by naturalizing a social hierarchy and substituting the norms of mental testers for those of the nation; it threatened, as well, to challenge established ways of distributing privileges. These critics too celebrated democracy, equality, and accountability, but their democracy emphasized human malleability, a common cultural heritage, and social mobility. Their critiques of the tests were partly technical and partly more fundamental, contending that any technology threatening the American ideal of the liberal, self-directing citizen was intrinsically problematic.

The clash over the nature of intelligence and the ways to assess it, from this perspective, was a struggle over who should have authority to define what was equal, democratic, and fair.[174] It was a political argument, as well as an argument over how to pursue politics in an age of human science.[175] If completely victorious, psychologists might well have attained just such extensive powers to decide who would and would not advance as Lippmann, for one, had dreaded. What made this possibility so real was that psychologists connected their new concept and instruments with long-standing concerns about how to balance merit and democracy, seeking almost invisibly to naturalize particular definitions of merit and determinations of who should have access to what social goods. The shift in focus of the cultural conversation about intelligence from its implications for democracy to the technicalities of testing itself is one indication of the testers' success. Yet for all its proponents' efforts, critics never allowed intelligence tests to become completely transparent instruments unproblematically linking observer and observed. Bagley, Lippmann, and Dewey—not to mention the countless "commentators" who subverted, poked fun at, or rejected the tests and their results—continually fought to keep the tests themselves visible, and to suggest that these instruments' legitimate applications were at best highly circumscribed. Unable or unwilling to deny that social scientific knowledge was essential to managing twentieth-century America, critics instead sought to define limits, to domesticate intelligence and complicate its links with notions of merit, without abandoning entirely the transparencies and accountabilities objective mental measurement could provide.

What emerged, therefore, in America by the start of World War II was a situation in which intelligence had neither fully triumphed nor been completely rejected as a criterion of merit. Rather, it came to serve simultaneously as a means by which older methods of asessment could be challenged, a way of scientifically rendering and legitimating selection decisions, and a site where determinations could be contested on both technical and ideological grounds. As such, while intelligence testing achieved a modicum of cultural importance in the United States, it never evolved into the means of establishing a "hereditary caste" that Lippmann and other critics had feared. Where French culture tamed the ambitions of its practitioners of human science by limiting the scope of their findings vis-à-vis issues of merit, American culture achieved a similar result by maintaining both the authority of the scientists and that of their critics. The measure of merit thus became in each nation not only a site of contestation, but a way of bounding larger struggles over the problem of inequality in a democratic republic.

Epilogue

"IQ TESTS: THOSE scores invented me." So observed the noted twentieth-century New York intellectual and short story writer Harold Brodkey in his autobiographical "A Story in an Almost Classical Mode." First published in the *New Yorker* in 1973, Brodkey's article described his childhood in 1940s America, recalling "a decisive piece of destiny." It was 1943 and he was thirteen:

> I was supposed to have a good mind—that supposition was a somewhat mysterious and even unlikely thing. . . . I composed no symphonies, did not write poetry or perform feats of mathematical wizardry. . . . But I did well in school and seemed to be peculiarly able to learn what the teacher said—I never mastered a subject, though—and there was the idiotic testimony of those peculiar witnesses, IQ tests: those scores invented me.
>
> Those scores were a decisive piece of destiny in that they affected the way people treated you and regarded you; they determined your authority; and if you spoke oddly, they argued in favor of your sanity. But it was as easy to say and there was much evidence that I was stupid, in every way or in some ways or, as my mother said in exasperation, "in the ways that count."[1]

As Brodkey's mother recognized, what was at issue here was precisely to define "the ways that count." In America in the 1940s, IQ tests were what helped determine what counted, and why Brodkey had access to authority and destinies not open to most of his peers. Today, Americans still live enough in the culture of IQ—witness the furor over the *Bell Curve* or affirmative action, as well as the persistent debates over the SAT—that Brodkey's reminiscence retains its salience.[2] Talk of multiple intelligences or "emotional IQ" notwithstanding, the meaning of intelligence dominant in the United States in the 1940s is still alive today: intelligence as a measurable, unidimensional, biological, and heritable entity that some people have more of than others, and that can be used to justify why such individuals are accorded opportunities—"destinies" in Brodkey's language—that others are denied.[3] Intelligence tests of various sorts have maintained the role they gained during the interwar period, and help to shape the educational and professional possibilities, not to mention personal self-images, of millions. They continue to provoke, as well, profound skepticism little different from Brodkey's mother's or that of such interwar critics as Walter Lippmann or William Bagley. And all within a culture celebrating, at the same time, commitments to both equality and merit.

In France, too, tests were, and remain, a critical determinant of an individual's future. Sociologist Pierre Bourdieu, no doubt influenced by his own experience with the French educational system during the 1940s, wrote extensively about how the highly competitive achievement examinations that determine entrance to the lycées and grandes écoles are part of the process of creating the French technocratic elite.[4] From peasant roots, his glittering career in French academia was made possible through examinations that earned him entrance to Lycée Louis-le-Grand in Paris and then the Ecole Normale Supérieur, and was assured with his successful performance on the agrégation in philosophy at the end of his time at Ecole Normale. The French culture of testing, as we have seen, had long been in place by the time Bourdieu negotiated it so successfully, and only some very recent moves toward instituting affirmative action procedures have marked a serious challenge to this French approach to integrating equality and merit. However, in contradistinction to Brodkey's experience in America, from Bourdieu's perspective this elite's hallmark was not that they were endowed with extraordinary mental capacity—that they had high IQs—but that they were formed into a particular *habitus,* a way of being that defined everything from what they wore to what they read and how they thought.[5] Similar in believing that commitments to equality could be combined with the resort to tests to measure merit, therefore, the French and Americans have differed profoundly in how they have constructed their systems of merit and what they presumed those tests meant about intelligence's nature.

In France intelligence was associated with an elite class and the social and educational experiences that brought it into being. While no less committed than Americans to the notion that individual capacities strongly shape one's destiny, the French have emphasized to this day the importance of training, of using the state's resources to produce its technocratic elite. They have also viewed that elite as imbued with sets of characteristics only in the crudest sense open to direct quantification and measurement.[6] Indeed, the purpose of the various concours has been less to measure individual contestants than to establish a threshold, sorting those allowed to continue on from those eliminated from further training. In the United States, on the other hand, while intelligence has also been considered an individual possession, the role of institutions in developing and transforming it has, in many respects, been minimized. Certainly Americans have retained from the earliest days of the republic a strong commitment to basic education; nevertheless, they have also typically conceived of intelligence as an intrinsic characteristic, existing independently of the contexts in which it has been developed and assessed and able to explain why an individual deserves access to opportunities, often regardless even of his or her actual accomplishments.

At one point it was possible that America and France would adopt similar methods for constructing their elites. In 1787, Thomas Jefferson proposed a system of schooling and culling of students for ever more advanced studies paralleling the approach to education born out of the French Revolution.[7] However in a society deeply committed to decentralization and local control, this plan remained stillborn. In its stead, the United States has experimented with a variety of methods and ideologies for selecting those appropriate for special training or advancement and justifying the advantages such individuals then received. As we have seen, in the nineteenth century Americans relied primarily on a language of talents, understanding human capabilities along the lines of the Common Sense philosophy as multiple, varied, and products principally of training. Under the pressures of immigration, urbanization, and industrialization in the aftermath of the Civil War, however, new ways of conceptualizing human nature became pervasive. With the emergence of elements of a mass society in America and the desire among many middle-class white Americans to anchor social and racial differences in biology, numerous American educational, industrial, and governmental institutions turned to methods of selection and organization for which they could claim the highest legitimacy possible—scientific objectivity. In this context, unidimensional intelligence and its measures proved for many to be an alluring means of guiding or justifying their determinations, seeming to allow impartial decisions to be made on the basis of individual merit. By 1940, therefore, intelligence had become something real and available to do important work in both France and America, even though the ways in which merit was understood and intelligence acculturated in each nation diverged sharply.

The post–World War II era did not disturb these patterns so much as reinforce them. The French continued to regard examinations as the key to maintaining a system of merit consistent with democratic principles, and intelligence as an attribute whose primary value lay in helping to characterize individual aptitudes rather than sorting multitudes. At the same time, Americans turned in some respects even more enthusiastically to intelligence as a way of recognizing and justifying its meritocracy while still celebrating opportunity for all. Central to this project, as Nicholas Lemann has shown, was the widespread adoption of the SAT by American universities in the postwar period as a key selection admissions criterion.[8] Especially important for entrance into elite institutions, the SAT was represented by its enthusiasts as critical to opening the most prestigious colleges and universities and the advantages they could provide to any applicants with sufficient ability, regardless of social background or schooling. The equivalence this established between testing and some combination of fairness and efficiency proved so enduring that even the 2001 proposal by the president of the University of California system, Richard C. Atkinson, to abolish the SAT/ACT requirement for undergraduate admission applicants, though receiving wide support, was quietly shelved once the College Board approved addition of a writing assessment and some other changes in the SAT.[9]

At the level of individuals, intelligence in America has largely retained the authority it achieved during the interwar period, reflected not only in the continued reliance on tests such as the SAT but also in the ongoing search for physiological markers of intelligence, whether through DNA sequencing or PET and fMRI scanning.[10] From the moment tests became important factors in how Americans adjudicated questions of merit, however, they also spawned a range of skeptics who proved adept at mounting challenges to the truth claims advanced for both concept and instruments. This has been particularly the case for pronouncements regarding the intellectual status of groups, where arguments for the mental inferiority of African Americans or southern Europeans or women have been met with sustained and often effective critiques. In mounting such opposition, critics have helped define a kind of middle ground, where the authority and even veracity of intelligence and its tests, though not insignificant, has nonetheless most typically been left open to doubt and at times serious dispute.

As just one illustration, in 1969 the Russell Sage Foundation, one of the grandes dames of American social science research, published a survey of student opinions about intelligence. Based on the responses of over nine thousand high school students—varied in terms of class, race, age, religion, gender, kind of school attended, and so forth—the study provides a revealing portrait of how intelligence as a concept was understood and, to a degree, internalized by teenage Americans.[11] Five findings stand out. The first is that African American students ("Negroes" as they were called in the survey) were more likely to believe that intelligence was mostly inborn than white American students, 31 percent to 14 percent, and Jewish students more than Protestant or Catholic students (20 percent to 14 percent and 15 percent). The second is that "intelligence" was ranked by about 92 percent of respondents as either "extremely important" or "important" to have, just behind "good health" (about 95 percent) and ahead of "drive to get ahead" (about 87 percent) and "good marks" (about 80 percent). Third, when asked what they believed intelligence tests measure, a little over 60 percent responded that they measure mostly or only learned knowledge, with around 90 percent declaring that an average person's intelligence score could be improved given a good diet, education, and the like. Fourth, out of a list of nine possible areas where intelligence tests might be used to help make decisions, only one—whether to place children in special education classes—received close to 50 percent support.

Finally, when asked about the accuracy of the tests, almost 75 percent declared that such tests were very or somewhat accurate, although only about 45 percent found them accurate when measuring their own intelligence (about 46 percent believed the tests underestimated their ability).[12]

What is striking about these pieces of data, examined together, is how discordant they are. While at one moment intelligence was described as a characteristic of immense value to one's future, at the next, intelligence tests were accorded little credence even as *aids* in most decision-making situations. Opinions about the innate nature of intelligence similarly seem to tell stories pointing in opposite directions. Although no group believed very strongly that intelligence was mostly an inborn trait, African Americans and Jewish Americans—whose intellectual capacity the scientific literature and general culture had framed largely in biological terms well into the twentieth century—responded slightly more in keeping with the inborn position than did their white Protestant or Catholic counterparts, whose intelligence had been routinely represented as including elements of both the nature and nurture interpretations of the origin of mental abilities.[13] Finally, although there may have been a broadly shared confidence that the tests, on the whole, were reasonably accurate, when the results came to press down on the individual him- or herself, they were understood by virtually everyone as open to improvement, a function as much of effort and good diet as genetic endowment. Well before the proliferation of standardized-test coaching courses, therefore, few considered one's measured intelligence as set from birth.

Thus, rather than simply accepting or rejecting the knowledge that the psychometricians were peddling, American high school students of all races and religions, however unwittingly, appropriated and transformed it. Accorded sufficient authority to help individuals make sense of their lives, psychological science was at other moments also readily ignored. The ambiguous status of intelligence and its tests that such attitudes reveal, from this perspective, may be seen to be as much strategic as contradictory, less the unexamined impulses of a confused public than a practical, if not always deliberate, accommodation to expert knowledge within a democratic culture. It may, in other words, represent a solution—a way of balancing the demands of equality and merit in a democracy—rather than simply an error in interpreting scientific claims.

Telling the story of the equivocal and contestable, and yet nonetheless powerful, nature of measures of merit has been one of this study's principal goals. Brodkey's tale of the IQ test "inventing him" or Bourdieu's of the peasant boy made good through the concours are complicated once we take account of their own skepticism about whether such measures really revealed their intelligence in "the ways that count." *The Measure of Merit* has tried to suggest why different professional and lay groups—psychologists and administrators, political philosophers and social commentators, and children and adults whose intelligence or ability was being probed—assessed the implications of (in)equality and the relationship between mental ability, merit, and tests differently in various times and locations. By emphasizing the contingency and cultural specificity of the places and ways in which Americans and the French have measured merit and understood intelligence, the book has argued not that such methods and concepts are illegitimate, but that they represent only possible, not necessary, solutions to the problems of allocating scarce social resources and comprehending human nature.[14]

One of the attractions of merit-based selections in both republics has been the promise they seem to provide of meting out rewards on the basis of individual accomplishments (or potential) that plausibly suggest superiority vis-à-vis the decision at issue. In so doing, such allocations could avoid reliance on such "artificial" criteria as family, wealth, and connections, and thereby seem just.[15] One key to making this turn to merit work, however, has been the availability of a scale of comparison that would allow individuals to be assessed and ranked so that the meritorious could be discovered and chosen. As we have seen, developing such scales that seem both workable and socially legitimate has been no easy task in either nation. Individuals in both societies have been too readily characterized as intricate combinations of skills, traits, and beliefs to make the construction of linear rankings a straightforward endeavor. And yet, without the ability to create such unilinear measures that could rank all relevant parties on a single scale, how could unambiguous assessments of relative merit be rendered? Both nations, in a sense, found at least one solution by creating measures that aggregated in a single factor what had been, or still were, seen to be separate abilities and talents.

The French used the Revolutionary moment, with its demands for the elimination of aristocratic privilege and its glorification of republican equality, to construct an educational system whose ruling philosophy of fairness through uniform treatment of all was able to override worries about homogenizing the individual. Performance on the various concours was key. Whatever different skill sets were brought into play by individuals competing in the examinations, the result was aggregated into a single ranking of all contestants, with those not performing to sufficiently high standards eliminated. The most meritorious were, by definition, those who survived, scoring at or near the top. Americans, by contrast, at first were content to valorize the multiplicity of an individual's talents and to celebrate the triumph of merit through the successful play of those talents in the marketplace. By the early twentieth century, however, the pressures of organizing rapidly expanding school systems and business enterprises convinced many that efficient selection methods based on assessment of solitary characteristics was of critical importance. One preferred solution, as we have seen, was to turn to intelligence, a seemingly "natural" attribute that represented a collection of (perhaps) diverse abilities and yet could be represented as a singular entity open to quantification and able to array entire populations in a simple rank order. While the loss in terms of registering an individual's complexities has been great, the gains in terms of administrative efficiency, consistency of diagnosis, and seeming transparency in decision making have also seemed significant to a number of governmental and private institutions and, at moments, to the public at large.

Facing such conflicting demands—that the system of merit simultaneously assess the individuals under consideration as holistically as possible, make decisions on objective criteria, provide transparency and accountability, and establish efficient selection methods—it is an open question whether other approaches to merit than those that the French and Americans developed could, or should, have been adopted. Certainly, numerous alternative or additional criteria of merit have been proposed, and often employed, from assessments of character to measurements of personality to itemizations of achievements to, most recently, development of new measures such as Robert J. Sternberg's "Successful Intelligence."[16] Nonetheless, unless these multiple criteria are somehow combined into a single assessment that can then be used to compare one individual with another, decision makers still face the problem of selecting among individuals on the basis of contrasting profiles with no clear criteria for determining which, in the given situation, is the best. In the end, some sort of ranking must be produced, and distinctions, however invidious, drawn, at least for the purposes of the particular decision to be made.

Where the Americans have gone further than the French, perhaps, at least when employing intelligence assessments to measure merit, is to suggest that such evaluations are not temporary and situational, but permanent, reflective of a natural inequality that helps define the individual or group being assessed. As we have seen, this turn to nature as a way of justifying inequality has deep roots, going back to the republic's very founding and becoming more ingrained particularly when many Americans seized on biology to justify various practices of racial exclusion and domination. Yet this recourse to nature has often been regarded as controversial, if not utterly suspect. From the inception of mental testing early in the twentieth century, for example, psychologists not only chose test items so that the scale would stratify test takers by age, but also so that the overall intelligence means for boys and girls would be approximately the same.[17] In so doing, they may have been reflecting, among other factors, the success of the nineteenth-century struggles to establish women's mental equality with men in shaping their generation's beliefs about women's intellect—beliefs that may then have been reinforced by the presence of large numbers of women in the new profession of psychology. As we saw in the last chapter, the thoroughgoing dismantlement by Klineberg, Bond, and other scholars of the "selective migration" explanation for the World War I army testing data showing that northern African Americans were superior mentally to southern African Americans did much to undermine claims based on this data about African American intellectual inferiority. Moreover, during the 1930s the strong hereditarianism of the early decades of the twentieth century yielded somewhat to a more environmentalist interpretation of intelligence, one stressing the many ways in which changes in social context could alter, often dramatically, individual IQs and in which group-level differences seemed closely correlated with quality of education and other economic and cultural resources.[18]

After World War II, and in response to Nazi eugenics policies, a series of UNESCO conferences on race pointedly denounced not only the concept of race itself as a biologically meaningless one, but also the existence of natural intergoup differences in mental abilities.[19] This rejection of hereditarian explanations for racial and group differences was virtually unanimous in America during the 1950s and 1960s, until controversy erupted again in 1969 in relation to the work of Arthur R. Jensen, professor of education at Berkeley, and supporters such as psychologists Richard J. Herrnstein and Hans J. Eysenck.[20] Questioning the basis of Head Start and other programs promising to improve poor children's academic futures by enriching their preschool years, Jensen et al. contended that environmentalist claims about intelligence were overstated and that significant divergences in native abilities exist at both the individual and group level. Once again, scholars could be heard claiming that social stratification was largely the inevitable and just consequence of intelligence differences. Such arguments were reiterated in 1994 by Herrnstein and Charles Murray in *The Bell Curve*—in which they argued that the U.S.'s socioeconomic (and racial) stratification was meritocratic, a reflection of disparities in innate intelligence levels—and continue to attract support from a range of mostly politically conservative scientists and social critics.[21]

In each instance, objections to such views, by both experts and (generally politically liberal) public intellectuals, have been immediate and intense, and often paralleled criticisms of testing advanced by Walter Lippmann et al. in the early 1920s. During the 1960s, for example, biologists and psychologists including Richard C. Lewontin, Stephen Jay Gould, and Leon Kamin joined with New Left college students and other social critics to organize opposition, at both the technical and policy levels, to the hereditarian interpretation of intelligence being advanced by Jensen and Herrnstein. A similar reaction occurred in the 1990s in response to *The Bell Curve*. In each case, charges of cultural bias, faulty methodology, and inadequate definition of the construct being tested that had long been leveled against intelligence measurement were raised anew by these opponents.[22] African American intellectuals and organizations were, and have been, particularly vigilant, committed to dismantling any biological warrant for policies that seemed to reintroduce notions of "separate but equal."[23]

At the center of the firestorm, clearly, lay the question of whether differences in test performance reflected natural inequalities in ability, or environmental factors such as disparities in education and family income, or cultural biases built into the instruments. There has not been, and likely never will be, any firm resolution of these issues, either in popular or scientific discourse. On the one hand, while discredited for groups, at the level of individuals it has seemed almost self-evident that people are born different and that those initial natural differences might in later life be translated into particular sets of strengths and weaknesses, be they for running fast, making music, writing poetry, or solving equations. The possibility that nature might simply be abandoned when addressing questions of merit thus seems dubious; it becomes even less plausible once the enormous practical advantages of appeals to nature are recognized, particularly for liberal democracies seeking to justify inequalities on objective grounds. On the other hand, it has seemed just as indisputable that almost no skill develops without experience and training, and that improvements in diet and education have alone wrought considerable changes in individuals, both in body and mind. Moreover, our certainty about what is inborn versus learned is constantly prone to destabilization. Some earlier mobilizations of the language of natural inequality in the name of science, for example—such as the eighteenth-century claim that persons born unable to hear and speak were, like idiots or imbeciles, of limited intellectual ability—now carry little weight, potent reminders that the authority of nature is less an absolute than an accomplishment and thus always, at least potentially, open to reconsideration.[24]

And so nature seems both indispensable and suspect when linked to systems of merit. The French have managed this problem largely by ignoring it, presuming that the best succeed in competitions and not interrogating too closely that superiority's origins. With nature marginalized, contestations over the merit system, when they occur, have tended to focus on whether all French citizens, especially those of the working class, actually do have the same opportunity to succeed and be selected for advanced training. Americans, by contrast, at least when discussing intelligence, have worried the problem, debating both the nature and nurture positions, and are seemingly resigned to leaving the issue in flux. And perhaps, in the U.S. context, that is not such a bad resolution. If measures of merit must exist—and at least some are probably unavoidable—then maintaining a sense of their contingency and contestability may be one good way of employing them without also being trapped by them and by the structures of inequality they seek to justify. Embracing an agonistic approach, whether along the French or American lines, may also better reflect the complicated and dynamic ways in which scientific knowledge and social order not only interact and continually reconstitute each other, but also shape the lives of the individuals living in their midst.

Notes

Preface and Acknowledgments

1. Linda Greenhouse, "Justices Back Affirmative action by 5 to 4, but Wider Vote Bans a Racial Point System," *New York Times,* June 24, 2003, p. A1; and "Excerpts from Justices' Opinions on Michigan Affirmative Action Cases," *New York Times,* June 24, 2003, p. A24.
2. See, for example, Steven A. Holmes, "Leveling the Playing Field, but for Whom?" *New York Times,* July 1, 2001, Section 4, pp. 4, 6; Leon Botstein, "The Merit Myth," *New York Times,* January 14, 2003, p. A27; and "Excerpts from Arguments Before the Supreme Court on Affirmative Action," *New York Times,* April 2, 2003, p. A14.
3. Burton Bollag, "French Court Upholds Landmark Program of Affirmative Action in College Admissions," *Chronicle of Higher Education,* November 7, 2003.
4. The classic example is the work of Michel Foucault; see *The Archaeology of Knowledge,* trans. A. M. Sheridan Smith (New York: Harper & Row, 1976); *The Order of Things: An Archaeology of the Human Sciences,* trans. Alan Sheridan (New York: Vintage, 1973); and *Discipline and Punish: The Birth of the Prison,* trans. Alan Sheridan (New York: Vintage, 1979).

Introduction

1. Steve Shapin and Barry Barnes, however, argue that even the old elite used meritocratic rhetoric to justify their social position. See "Head and Hand: Rhetorical Resources in British Pedagogical Writing, 1770–1850," *Oxford Review of Education* 2 (1976): 231–54.
2. See especially Kurt Danziger, *Naming the Mind: How Psychology Found Its Language* (London: Sage, 1997), chap. 5. Also important are Danziger, *Constructing the Subject: Historical Origins of Psychological Research* (Cambridge: Cambridge University Press, 1990); Carl N. Degler, *In Search of Human Nature: The Decline and Revival of Darwinism in American Social Thought* (New York: Oxford University Press, 1991); Stephen Jay Gould, *The Mismeasure of Man* (New York: W. W. Norton, 1981); Elizabeth Lunbeck, *The Psychiatric Persuasion: Knowledge, Gender, and Power in Modern America* (Princeton: Princeton University Press, 1994); Theodore M. Porter, *The Rise of Statistical Thinking, 1820–1900* (Princeton: Princeton University Press, 1986), chaps. 4–5; Nikolas Rose, *Governing the Soul: The Shaping of the Private Self* (London: Routledge, 1990); Rose, *The Psychological Complex: Psychology, Politics and Society in England, 1869–1939* (London: Routledge & Kegan Paul, 1985); Gillian Sutherland, *Ability, Merit and Measurement* (Oxford: Oxford University Press, 1984); Adrian Wooldridge, *Measuring the Mind: Education and Psychology in England, c. 1860–*

1990 (Cambridge: Cambridge University Press, 1994); and Leila Zenderland, *Measuring Minds: Henry Herbert Goddard and the Origins of American Mental Testing* (Cambridge: Cambridge University Press, 1998).

3. Theodore M. Porter, *Trust in Numbers: The Pursuit of Objectivity in Science and Public Life* (Princeton: Princeton University Press, 1995), chap. 6.

4. This phrase is adapted from Barbara J. Fields, "Ideology and Race in American History," in *Region, Race, and Reconstruction: Essays in Honor of C. Vann Woodward,* eds. J. Morgan Kousser and James M. McPherson (New York: Oxford University Press, 1982), pp. 143–77, quote on p. 152. See also Michel Foucault, *Discipline and Punish: The Birth of the Prison,* trans. Alan Sheridan (New York: Vintage, 1979).

5. Sheila Jasanoff, "Ordering Knowledge, Ordering Society," in *States of Knowledge: The Co-Production of Science and Social Order* (London: Routledge, 2004), pp. 13–45; and Jasanoff, *The Fifth Branch: Science Advisers as Policymakers* (Cambridge: Harvard University Press, 1990). See also Yaron Ezrahi, *The Descent of Icarus: Science and the Transformation of Contemporary Democracy* (Cambridge: Harvard University Press, 1990).

Chapter 1

1. John Adams to Thomas Jefferson, 9 July 1813, in *The Adams-Jefferson Letters,* ed. Lester J. Cappon (Chapel Hill: University of North Carolina Press, 1988), p. 352; 15 November 1813, p. 400.

2. For an earlier consideration of rule by a natural aristocracy, though expressed with less enthusiasm, see Melancton Smith, "Speech before the New York Ratifying Convention; June 21, 1788," in *The Debate on the Constitution: Federalist and Antifederalist Speeches, Articles, and Letters during the Struggle over Ratification,* ed. Bernard Bailyn, vol. II (New York: Library of America, 1993), pp. 760–63. On the place of the "natural aristocracy" in late eighteenth-century politics, see esp. Saul Cornell, *The Other Founders: Anti-Federalism & the Dissenting Tradition in America, 1788–1828* (Chapel Hill: University of North Carolina Press, 1999), pp. 68–80, 96–120; Gary Kornblith and John M. Murrin, "The Making and Unmaking of an American Ruling Class," in *Beyond the American Revolution: Explorations in the History of American Radicalism,* ed. Alfred F. Young (DeKalb: Northern Illinois University Press, 1993), pp. 27–79; Edmund S. Morgan, *Inventing the People: The Rise of Popular Sovereignty in England and America* (New York: W. W. Norton, 1988), pp. 239–62; Alan Taylor, "From Fathers to Friends of the People: Political Personas in the Early Republic," *Journal of the Early Republic* 11 (1991): 465–91; David Waldstreicher, *In the Midst of Perpetual Fetes: The Making of American Nationalism, 1776–1820* (Chapel Hill: University of North Carolina Press, 1997), esp. pp. 67–107; Gordon S. Wood, *The Creation of the American Republic, 1776–1787* (New York: W. W. Norton, 1969), esp. pp. 471–518; and Wood, "Interests and Disinterestedness in the Making of the Constitution," in *Beyond Confederation: Origins of the Constitution and American National Identity,* eds. Richard Beeman, Stephen Botein, and Edward C. Carter II (Chapel Hill: University of North Carolina Press, 1987), pp. 69–109.

3. For an excellent account of one aspect of the transatlantic character of this conversation, see Joyce Appleby, "America as a Model for the Radical French Reformers of 1789," *William and Mary Quarterly,* 3rd ser., 28 (1971): 267–86.

4. I am here taking much more literally the function of the word "natural" than does Waldstreicher in *In the Midst of Perpetual Fetes*. In this, Waldstreicher follows such anti-Federalists as the Federal Farmer and Brutus, who argued against the existence of any true "natural aristocracy."

5. On Jefferson and slavery, see Alexander O. Boulton, "The American Paradox: Jeffersonian Equality and Racial Science," *American Quarterly* 47 (1995): 467–92; Winthrop D. Jordan, *White over Black: American Attitudes toward the Negro, 1550–1812* (New York: W. W. Norton, 1977), esp. chap. 12; Larry R. Morrison, "'Nearer to the Brute Creation': The Scientific Defense of American Slavery before 1830," *Southern Studies* 19 (1980): 228–42; and Peter S. Onuf, "'To Declare Them a Free and Independent People': Race, Slavery, and National Identity in Jefferson's Thought," *Journal of the Early Republic* 18 (1998): 1–46.

6. Daniel T. Rodgers, *Contested Truths: Keywords in American Politics since Independence* (New York: Basic Books, 1987), pp. 45–79. On the concept of "natural rights" in Anglo-American discourse, see also Orlando Patterson, "Freedom, Slavery, and the Modern Construction of Rights," in *Historical Change and Human Rights,* ed. Olwen Hufton (New York: Basic Books, 1995), pp. 131–78; Morton G. White, *The Philosophy of the American Revolution* (New York: Oxford University Press, 1978); and Michael P. Zuckert, *Natural Rights and the New Republicanism* (Princeton: Princeton University Press, 1994). On the gendering of the natural rights, see Ruth H. Bloch, "The Gendered Meanings of Virtue in Revolutionary America," *Signs: Journal of Women in Culture and Society* 13 (1987): 37–58; and Rosemarie Zagarri, "The Rights of Man and Woman in Post-Revolutionary America," *William and Mary Quarterly,* 3rd ser., 55 (1998): 203–30.

7. Wood elaborates on the inherent tension in republicanism between "equality of opportunity" and "equality of condition" in *Creation of the American Republic,* esp. pp. 70–75. See also White, *Philosophy of the American Revolution,* pp. 9–141; and Gordon S. Wood, *The Radicalism of the American Revolution* (New York: Knopf, 1992), pp. 180–86. For examples of meritocratic projects, see particularly Ken Alder, "French Engineers Become Professionals: or, How Meritocracy Made Knowledge Objective," in *The Sciences in Enlightened Europe,* eds. William Clark, Jan Golinski, and Simon Schaffer (Chicago: University of Chicago Press, 1999), pp. 94–125. Also valuable are David D. Bien, "The Army in the French Enlightenment: Reform, Reaction and Revolution," *Past and Present* 85 (1979): 68–98; and Jay M. Smith, *The Culture of Merit: Nobility, Royal Service, and the Making of Absolute Monarchy in France, 1600–1789* (Ann Arbor: University of Michigan Press, 1996).

8. John Adams, *Defence of the Constitutions of Government of the United States* (1787), in *The Works of John Adams,* vol. 4, ed. Charles Francis Adams (Boston: Little, Brown, 1850–56), pp. 391–98, quote on p. 391.

9. For an example of this outlook, see Jefferson's system of education, outlined in *Notes on the State of Virginia* (1781–85), in *The Portable Thomas Jefferson,* ed. Merrill D. Peterson (New York: Penguin Books, 1977), pp. 193–98.

10. For a thoughtful discussion of attempts to establish a ruling elite in America, see Kornblith and Murrin, "Making and Unmaking of an American Ruling Class." Morton White has argued in *The Philosophy of the American Revolution* that notions of human inequality were fundamental to the philosophies of John Locke and Jean Jacques Burlamaqui, and thus were written into such foundational American documents as the Declaration of Independence. White, *Philosophy of the American Revolution,* pp. 9–141.

11. See Adams, *Defence of the Constitutions,* pp. 391–98; and on his sense of the value of the natural aristocracy in the preservation of liberty, John Adams to Samuel Adams, 18 October 1790, in *The Works of John Adams,* vol. 6 (Boston: Little, Brown, 1850–56), pp. 416–20.

12. Isaac Newton, *The Principia: Mathematical Principles of Natural Philosophy,* trans. I. Bernard Cohen and Anne Whitman (1687; reprint, Berkeley: University of California Press, 1999); and Adam Smith, *An Inquiry into the Nature and Causes of the Wealth of Nations,* vol. 1, bk. 1, chap. 2 (1776; reprint, New York: Modern Library, 1994). On notions of "nature" in the Enlightenment, see esp. David Carrithers, "The Enlightenment Science of Society," in *Inventing Human Science: Eighteenth-Century Domains,* eds. Christopher Fox, Roy S. Porter, and Robert Wokler (Berkeley: University of California Press, 1995), pp. 232–70; Peter Gay, *The Enlightenment, an Interpretation,* vol. 2, *The Science of Freedom* (New York: Knopf, 1969); Henry F. May, *The Enlightenment in America* (New York: Oxford University Press, 1976); Roy Porter, *Enlightenment Britain and the Creation of the Modern World* (London: Penguin Press, 2000), esp. chaps. 6 and 13; Roger Smith, "The Language of Human Nature," in *Inventing Human Science,* pp. 88–111; and White, *Philosophy of the American Revolution.*

13. As historical scholarship for over the last quarter century has shown, a plethora of political languages flourished during the late eighteenth century in the Atlantic world. The historiography of the American colonial/early national period has been particularly concerned with exhuming and nuancing the discourses of republicanism and liberalism, at times to the exclusion of all others. For an overview of this historiography, see Daniel T. Rodgers, "Republicanism: The Career of a Concept," *Journal of American History* 79 (1992): 11–38; and Robert E. Shalhope, "Republicanism and Early American Historiography," *William and Mary Quarterly,* 3rd ser., 39 (1982): 334–56. Critical analyses include Isaac Kramnick, *Republicanism or Bourgeois Radicalism: Political Ideology in Late Eighteenth-Century England and America* (Ithaca: Cornell University Press, 1990); Anthony Pagden, ed., *The Languages of Political Theory in Early-Modern Europe* (Cambridge: Cambridge University Press, 1987); J.G.A. Pocock, *The Machiavellian Moment: Florentine Political Thought and the Atlantic Republican Tradition* (Princeton: Princeton University Press, 1975); Pocock, *Virtue, Commerce, and History: Essays on Political Thought and History, Chiefly in the Eighteenth Century* (Cambridge: Cambridge University Press, 1985); Quentin Skinner, *The Foundations of Modern Political Thought,* 2 vols. (Cambridge: Cambridge University Press, 1978); Martin van Gelderen and Quentin Skinner, eds., *Republicanism: A Shared European Heritage,* 2 vols. (Cambridge: Cambridge University Press, 2002); and Sheldon Wolin, *Politics and Vision: Continuity and Innovation in Western Political Thought* (Boston: Little, Brown, 1960), chaps. 9 and 10.

14. For an extended discussion of the political ramifications of "talents," see Wood, *Radicalism of the American Revolution,* esp. pp. 229–43, 271–86. Also relevant are Morgan, *Inventing the People,* pp. 288–303; White, *Philosophy of the American Revolution;* and Wood, *Creation of the American Republic,* pp. 506–18. On the limitations of the equality of virtue, esp. for women, see Bloch, "Gendered Meanings of Virtue"; Jan Lewis, "The Republican Wife: Virtue and Seduction in the Early Republic," *William and Mary Quarterly,* 3rd ser., 44 (1987): 689–710; and Marisa Linton, "Virtue Rewarded? Women and the Politics of Virtue in 18th-Century France," *History of European Ideas* 26 (2000): 35–49, 51–65.

15. See Mary Wollstonecraft, *A Vindication of the Rights of Woman* (1792; reprint, London: Penguin Books, 1992), p. 92. On other projects to meld science and politics in the late eighteenth century, see Carrithers, "Enlightenment Science of Society"; Judith N. Shklar, "Alexander Hamilton and the Language of Political Science," in *Languages of Political Theory in Early-Modern Europe,* pp. 339–55; and Wood, *Radicalism of the American Revolution,* pp. 145–68.

16. See, for example, Thomas Paine, *Rights of Man* (1792; reprint, New York: Penguin Books, 1985), pp. 125, 140, 172–76.

17. See Wood, *Creation of the American Republic,* p. 480; and more generally pp. 478–518, 532–36, 596–602. On the people as the rhetorical foundation of a republic, see Morgan, *Inventing the People;* Wood, *Radicalism of the American Revolution,* pp. 11–42; Pocock, *Machiavellian Moment,* chaps. 11–15; Skinner, *Foundations of Modern Political Thought,* vol. 1, chaps. 1–2, 4–6, and vol. 2, chap. 9; and the various essays in vol. 1, pts. 1 and 2, of van Gelderen and Skinner, *Republicanism.*

18. *The Declaration of the Rights of Man and Citizen* (1789), in *The Portable Age of Reason Reader,* ed. Crane Brinton (New York: Viking, 1956), p. 201 (emphasis added).

19. See, for example, Paine, *Rights of Man.*

20. Gordon Wood observes that "equality was thus not directly conceived of by most Americans in 1776, including even a devout republican like Samuel Adams, as a social leveling." Wood, *Creation of the American Republic,* p. 70; see also pp. 475–83. Jean Starobinski notes much the same of Rousseau: "he [Rousseau] does not ask for social leveling but simply for a proportioning of civic inequality to the natural inequality of talents." Jean Starobinski, *Jean-Jacques Rousseau: Transparency and Obstruction,* trans. Arthur Goldhammer (Chicago: University of Chicago Press, 1988), p. 302. The lack of interest in radical egalitarianism is most clear in the assessment of other races, women, children, etc.

21. On this point, see particularly Joan Scott's trenchant analysis of equality and difference in the French Revolution in "French Feminists and the Rights of 'Man': Olympe de Gouges's Declarations," *History Workshop* 28 (1989): 1–21.

22. There is an extensive literature on the concept of "virtue" in the early modern period. See particularly Bloch, "Gendered Meanings of Virtue in Revolutionary America"; Carol Blum, *Rousseau and the Republic of Virtue: The Language of Politics in the French Revolution* (Ithaca: Cornell University Press, 1986); James T. Kloppenberg, "The Virtues of Liberalism: Christianity, Republicanism, and Ethics in Early American Political Discourse," *Journal of American History* 74 (1987): 9–33; Mona Ozuf, "La Révolution française et l'idée de l'homme

nouveau," in *The French Revolution and the Creation of Modern Political Culture,* vol. 2, *The Political Culture of the French Revolution,* ed. Colin Lucas (Oxford: Pergamon Press, 1988), pp. 213–32; Pocock, *Machiavellian Moment;* and Wood, *Radicalism of the American Revolution,* pp. 95–109, 213–25.

23. For more, see Wood, *Radicalism of the American Revolution,* esp. pp. 229–43, 271–86.

24. For a discussion of the historiography, see Carson, "Differentiating a Republican Citizenry." Particularly valuable are Daniel Walker Howe, *Making the American Self: Jonathan Edwards to Abraham Lincoln* (Cambridge: Harvard University Press, 1997), esp. pp. 48–103; Mary Poovey, *A History of the Modern Fact: Problems of Knowledge in the Sciences of Wealth and Society* (Chicago: University of Chicago Press, 1998); and Simon Schaffer, "States of Mind: Enlightenment and Natural Philosophy," in *The Language of Psyche: Mind and Body in Enlightenment Thought,* ed. George S. Rousseau (Berkeley: University of California Press, 1990), pp. 233–90.

25. John Locke, *An Essay Concerning Human Understanding,* 2 vols. (1690; reprint, New York: Dover, 1959).

26. See John Carson, "Minding Matter/Mattering Mind: Knowledge and the Subject in Nineteenth-Century Psychology," *Studies in the History and Philosophy of the Biological and Biomedical Sciences* 30 (1999): 345–76. More generally on the history of the human sciences during this period, see Marina Frasca-Spada, "The Science and Conversation of Human Nature," in *Sciences in Enlightened Europe,* pp. 218–45; Gary Hatfield, "Remaking the Science of Mind: Psychology as Natural Science," in *Inventing Human Science,* pp. 184–230; Sergio Moravia, "The Enlightenment and the Sciences of Man," *History of Science* 18 (1980): 247–68; Graham Richards, *Mental Machinery: The Origins and Consequences of Psychological Ideas, 1600–1850* (Baltimore: Johns Hopkins University Press, 1992); and P. B. Wood, "The Natural History of Man in the Scottish Enlightenment," *History of Science* 27 (1989): 89–123.

27. Locke, vol. 1, bk. 2, chap. 1, sec. 2.

28. Ibid., intro., sec. 1, p. 25; bk. 1, chap. 1, sect. 1, p. 38. John O'Neal makes a similar point in *The Authority of Experience: Sensationist Theory in the French Enlightenment* (University Park: Pennsylvania State University Press, 1996), chap. 1.

29. On this point, see Christopher F. Goodey, "John Locke's Idiots in the Natural History of Mind," *History of Psychiatry* 5 (1994): 215–50. In *Some Thoughts on Education* (1693), Locke argued for the fundamental similarity of most human minds, seeing only a few exceptional individuals as naturally better or worse than the majority of humankind. Morton White, however, has argued that Locke relied on the existence of individual variations in either the capacity or willingness to reason to accommodate both claims to universal equality and the privileges of reason among the elite. See White, *Philosophy of the American Revolution,* pp. 9–36, 61–78.

30. On Locke and the popularity of his ideas in Britain, see Porter, *Enlightenment Britain,* pp. 60–71.

31. David Hartley, *Observations on Man. His Frame, His Duty, and His Expectations* (London: Leake, Frederick, Hitch, and Austin, 1749). On associa-

tionism, see Hatfield, "Remaking the Science of Mind"; Richards, *Mental Machinery;* Roger Smith, "The Background of Physiological Psychology in Natural Philosophy," *History of Science* 11 (1973): 75–123; and Robert M. Young, *Mind, Brain, and Adaptation in the Nineteenth Century: Cerebral Localization and Its Biological Context from Gall to Ferrier* (New York: Oxford University Press, 1990).

32. Hartley, *Observations on Man,* vol. 1, p. 360.

33. For a valuable discussion of eighteenth-century philosophical speculation about statues being brought to life, see Julia V. Douthwaite, *The Wild Girl, Natural Man, and the Monster: Dangerous Experiments in the Age of Enlightenment* (Chicago: University of Chicago Press, 2002), chap. 2.

34. "Judgment, reflexion, desires, passions, &c. are only sensations differently transformed." Condillac, *Treatise on Sensations,* pp. xxxi–xxxii.

35. O'Neal makes a similar point in *Authority of Experience,* pp. 58–59.

36. Condillac, *Treatise on Sensations,* pp. 45, 239.

37. See Francis Hutcheson, *An Essay on the Nature and Conduct of the Passions and Affections . . .* (1742; reprint, Gainesville: University of Florida Press, 1969); Thomas Reid, *Essays on the Active Powers of Man* (1768), and *Essays on the Intellectual Powers of Man* (1785), both in *Philosophical Works* (Hildesheim: Georg Olms Verlag, 1983); Dugald Stewart, *Elements of the Philosophy of the Human Mind* (1792; reprint, Boston: J. Munroe, 1859); and Stewart, *The Philosophy of the Active and Moral Powers of the Mind* (1828; reprint, Boston: Phillips, Sampson, 1859). On the development of the Scottish science of man as an alternative language of civic morality, see Nicholas Phillipson, "The Scottish Enlightenment," in *The Enlightenment in National Context,* eds. Roy Porter and Mikuláš Teich (Cambridge: Cambridge University Press, 1981), pp. 19–40.

38. For general background on the Common Sense philosophy, see Frank M. Albrecht, "A Reappraisal of Faculty Psychology," *Journal of the History of the Behavioral Sciences* 6 (1970): 36–40; Garland P. Brooks, "The Faculty Psychology of Thomas Reid," *Journal of the History of the Behavioral Sciences* 12 (1976): 65–77; Gladys Bryson, *Man and Society: The Scottish Inquiry of the Eighteenth Century* (Princeton: Princeton University Press, 1945); Selwyn A. Grave, *The Scottish Philosophy of Common Sense* (Oxford: Oxford University Press, 1960); White, *Philosophy of the American Revolution,* pp. 97–131; and Garry Wills, *Inventing America: Jefferson's Declaration of Independence* (New York: Vintage, 1979).

39. See, for example, Reid, *Essays on the Active Powers of Man,* essays 1–2; also Brooks, "Faculty Psychology of Thomas Reid."

40. For some of the gender implications of the spread of the Common Sense philosophy in the United States, see Bloch, "Gendered Meanings of Virtue."

41. Jean-Jacques Rousseau, *Discourse on the Origin and Foundations of Inequality among Men [1755]* in *Rousseau's Political Writings,* eds. Alan Ritter and Julia Conaway Bondanella (New York: W. W. Norton, 1988).

42. For an important analysis of the *Discourse on Inequality,* see Elizabeth R. Wingrove, *Rousseau's Republican Romance* (Princeton: Princeton University Press, 2000), esp. pp. 24–57. Also of value are Blum, *Rousseau and the Republic of Virtue*; and Judith Shklar, *Men and Citizens: A Study of Rousseau's Social Theory* (Cambridge: Cambridge University Press, 1969). Wingrove concurs that the

place of nature in Rousseau's account is of critical importance, and her comments particularly on this section on Rousseau have been invaluable in shaping aspects of the argument I present here.

43. Rousseau, *On the Social Contract* (1762) in *Rousseau's Political Writings,* p. 85.

44. Ibid., *Discourse on Inequality,* pp. 8–9.

45. Ibid., pp. 31–32. On this point, see Steinbrügge's discussion of the *Discourse* in Lieselotte Steinbrügge, *The Moral Sex: Woman's Nature in the French Enlightenment,* trans. Pamela E. Selwyn (New York: Oxford University Press, 1995), chap. 5.

46. Rousseau, *Discourse on Inequality,* pp. 41–42.

47. See, for example, Jean-Jacques Rousseau, *Emile, or On Education,* trans. Allan Bloom (1762; reprint, New York: Basic Books, 1979), p. 125.

48. See, for example, Richard Price, *Additional Observations on the Nature and Value of Civil Liberty, and the War with America* (1777), in *Political Writings,* ed. D. O. Thomas (Cambridge: Cambridge University Press, 1991), pp. 85–88.

49. Michel Foucault has elaborated on the consequences of this point at length in *Discipline and Punish: The Birth of the Prison,* trans. Alan Sheridan (New York: Vintage Books, 1979). For excellent analyses of the implications of Rousseau's philosophy for women, see Steinbrügge, *Moral Sex,* and Wingrove, *Rousseau's Republican Romance.*

50. See Pocock, *Machiavellian Moment,* chap. 3.

51. For Rousseau's reiteration of this position, see *Emile,* p. 193.

52. I am indebted to my colleague Dena Goodman for this point. For more on the culture of the philosophes, see Dena Goodman, *The Republic of Letters: A Cultural History of the French Enlightenment* (Ithaca: Cornell University Press, 1994).

53. See Rousseau, *Emile,* p. 339. It was also a criterion that, notoriously, Rousseau believed would serve to remove women from any claims to governance. See Blum, *Rousseau and the Republic of Virtue,* esp. pp. 119–27; Steinbrügge, *Moral Sex,* esp. chap. 5; and Wingrove, *Rousseau's Republican Romance.*

54. Claude Adrien Helvétius, *De l'esprit; or Essays on the Mind and Its Several Faculties* (1758; reprint, New York: Burt Franklin, 1970); and Helvétius, *De l'homme, de ses facultés intellectuelles, et de son éducation* (1772–73), 2 vols., trans. W. Hooper (London: Albion Press, 1810).

55. On Helvétius's adoption of sensationism, see O'Neal, *Authority of Experience,* pp. 84–86.

56. For an excellent discussion of Helvétius's philosophy and how it changed between *De l'esprit* and *De l'homme,* see C. Kiernan, "Helvétius and a Science of Ethics," *Studies on Voltaire and the Eighteenth Century* 60 (1968): 229–43.

57. On this point, see O'Neal, *Authority of Experience,* pp. 94, 177.

58. Helvétius, *Essays on Mind,* p. 359.

59. Ibid., p. 498.

60. See particularly Kiernan, "Helvétius and a Science of Ethics."

61. See O'Neal, *Authority of Experience,* pp. 174–75.

62. David W. Smith, *Helvétius: A Study in Persecution* (Oxford: Clarendon Press, 1965), p. 1.

63. See ibid., p. 13.

64. Jean-François LaHarpe, *Réfutation du livre de l'Esprit, prononcée au Lycée Républicain, dans les séances des 26 et 29 mars et des 3 et 5 avril* (Paris: Migneret, An 5–1797), p. 3.

65. For the details of the story of the response to *De l'esprit,* I have relied on the excellent account provided by Smith in *Helvétius,* esp. pp. 1–3.

66. See Henry Vyverberg, *Human Nature, Cultural Diversity, and the French Enlightenment* (New York: Oxford University Press, 1989), pp. 58–61.

67. John Adams to Thomas Jefferson, 13 July 1813, in *Adams-Jefferson Letters,* p. 355.

68. On reactions to *De l'esprit* among philosophes, see Smith, *Helvétius,* pp. 157–71.

69. Denis Diderot, "Refutation of the Work of Helvétius Entitled *On Man*" (1773–76) in *Diderot's Selected Writings,* ed. Lester G. Crocker (New York: Macmillan, 1966), pp. 293, 292, 291.

70. François Marie Arouet de Voltaire, "D 18431. Voltaire to Prince Dmitry Alekseevich Golitsuin. 19 Juin 1773," in *The Complete Works of Voltaire,* vol. 124, *Correspondence,* vol. 40 (Oxfordshire: The Voltaire Foundation, 1975), p. 29.

71. William Godwin, *Enquiry Concerning Political Justice and Its Influence on Morals and Happiness,* ed. F.E.L. Priestley (1793; reprint, Toronto: University of Toronto Press, 1946), vol. 1, p. 147.

72. On Paine, see Eric Foner, *Tom Paine and Revolutionary America* (New York: Oxford University Press, 1976); Jack Fruchtman, *Thomas Paine and the Religion of Nature* (Baltimore: Johns Hopkins University Press, 1993); and Jack P. Greene, "Paine, America, and the 'Modernization' of Political Consciousness," *Political Science Quarterly* 93 (1978): 73–92.

73. Paine's closest attempt to define talents may have been the following: "if we examine, with attention, into the composition and constitution of man, the diversity of his wants, and the diversity of talents in different men for reciprocally accommodating the wants of each other, his propensity to society, and consequently to preserve the advantages resulting from it, we shall easily discover, that a great part of what is called government is mere imposition." Paine, *Rights of Man,* pp. 163–64.

74. See ibid., pp. 175–76, 172.

75. Samuel Johnson, *A Dictionary of the English Language . . .* (London: W. Strahan, 1755), n.p.; *Le Grand vocabulaire françois,* vol. 27 (Paris: n.p., 1773), pp. 324–25; and *Dictionnaire de l'Académie françoise, nouvelle édition,* vol. 2 (Nismes: Pierre Beaume, 1786), p. 554.

76. Although, as David Bien has insisted, the meaning of "*talent*" was changing in the latter part of the eighteenth century in France, when it came to include the possibility that talents could be acquired by hard work and not just by birth. See "Army in the French Revolution," pp. 80–81.

77. *Le Grand vocabulaire françois,* vol. 27, pp. 324–25.

78. John Adams, Letter to Thomas Jefferson, 15 November 1813, in *Adams-Jefferson Letters,* pp. 397–98.

79. On Wollstonecraft, see Miriam Brody, introduction to Wollstonecraft, *Vindication of the Rights of Woman,* pp. 4–20; Porter, *Enlightenment Britain,* chap. 14; and Virginia Sapiro, *A Vindication of Political Virtue: The Political Theory of*

Mary Wollstonecraft (Chicago: University of Chicago Press, 1992), chap. 1. On the response to the *Rights of Woman* in the United States, see Chandos Michael Brown, "Mary Wollstonecraft, or, The Female Illuminati: The Campaign against Women and 'Modern Philosophy' in the Early Republic," *Journal of the Early Republic* 15 (1995): 389–424; and Zagarri, "Rights of Man and Woman in Post-Revolutionary America."

80. One of Wollstonecraft's first rhetorical moves was to represent military officers as "idle superficial young men" because of the nature of their educations, and then to assert that women and officers were in completely similar states. See Wollstonecraft, *Rights of Woman,* pp. 97, 104–108; also chap. 4. Brody remarks on this feature of Wollstonecraft's argument in her introduction, pp. 44–45.

81. See ibid., esp. chap. 6.

82. On this general issue, see Scott, "French Feminists and the Rights of 'Man.'"

83. Wollstonecraft, *Rights of Woman,* pp. 147–48.

84. Paine, *Rights of Man,* p. 125; Wollstonecraft, *Rights of Woman,* pp. 143–44; and Rousseau, *Discourse on Inequality,* p. 16. See also Rousseau, *Emile,* pp. 157–58, although throughout *Emile* Rousseau also praised the power of sentiment as a necessary adjunct to reason; and Publius [James Madison], "The Foederalist. No. LVIII" (22 February 1878), in *The Federalist: A Collection of Essays, Written in Favor of the New Constitution, As Agreed Upon by the Foederal Convention, September 17, 1787,* ed. Henry B. Dawson (New York: Scribner, Armstrong, 1876), pp. 410–16.

85. See esp. Goodman, *Republic of Letters;* Carla Hesse, "Kant, Foucault, and *Three Women,*" in *Foucault and the Writing of History,* ed. Jan Goldstein (Cambridge: Blackwell, 1994), pp. 81–98; Lynn Hunt, *The Family Romance of the French Revolution* (Berkeley: University of California Press, 1992); Joan B. Landes, *Women and the Public Sphere in the Age of the French Revolution* (Ithaca: Cornell University Press, 1988); Scott, "French Feminists and the Rights of 'Man'"; Steinbrügge, *Moral Sex;* Sylvana Tomaselli, "The Enlightenment Debate on Women," *History Workshop Journal* 20 (1985): 101–24; and Wingrove, *Rousseau's Republican Romance.*

86. See, for example, Paine, *Rights of Man,* p. 176.

87. See Rousseau, *Emile,* pp. 192, 245, 61.

88. "If women are by nature inferior to men," Wollstonecraft argued, "their virtues must be the same in quality, if not in degree." Wollstonecraft, *Rights of Woman,* pp. 108, 162.

89. Godwin, *Enquiry,* vol. 1, pp. 42–43.

90. For an example, see ibid., p. 38.

91. Ibid., p. 44.

92. On this general issue, see Vyverberg, *Human Nature, Cultural Diversity, and the French Enlightenment.*

93. Paine, *Rights of Man,* pp. 173, 175.

94. See Jefferson to Adams, 28 October 1813, in *Adams-Jefferson Letters,* pp. 387–92.

95. Benjamin Rush, "On the Influence of Physical Causes in Promoting an Increase in the Strength and Activity of the Intellectual Faculties of Man" (1799), in *Two Essays on Mind* (New York: Brunner/Mazel, 1972), p. 119.

96. On ideas about heredity in America, see Charles E. Rosenberg, "The Bitter Fruit: Heredity, Disease, and Social Thought," in *No Other Gods: On Science and American Social Thought* (Baltimore: Johns Hopkins University Press, 1978), pp. 25–53. For France, see Carlos López-Beltrán, "'*Les maladies héréditaires*': 18th Century Disputes in France," *Revue d'histoire des sciences* 48 (1995): 307–50.

97. See López-Beltrán, "'*Les maladies héréditaires.*'"

98. See Godwin, *Enquiry,* vol. 2, p. 86.

99. Smith, *Wealth of Nations,* vol. 1, bk. 1, chap. 2, esp. pp. 16–17.

100. Diderot, "Refutation . . . *On Man,*" pp. 283–98; and "Réflexions sur le livre *De l'esprit* par M. Helvétius" (1758), in *Œuvres complètes,* vol. 9, *L'interprétation de la nature (1753–1765), Idées III,* ed. Jean Varloot (Paris: Hermann, 1981), pp. 261–312.

101. See Paine, *Rights of Man,* p. 175.

102. Steinbrügge makes this point in her very interesting analysis of Diderot's views. See *Moral Sex,* chap. 4.

103. On eighteenth-century ideas about perfectibility, see Victor Hilts, "Enlightenment Views on the Genetic Perfectibility of Man," in *Transformation and Tradition in the Sciences: Essays in Honor of I. Bernard Cohen,* ed. Everett Mendelsohn (Cambridge: Cambridge University Press, 1984), pp. 255–71. Even Adams, Wood noted, often proclaimed the essential equality of all human beings. Wood, *Radicalism of the American Revolution,* pp. 237–38.

104. On Condorcet, see esp. Keith Michael Baker, *Condorcet: From Natural Philosophy to Social Mathematics* (Chicago: University of Chicago Press, 1975); Ernst Cassirer, *The Philosophy of Enlightenment,* trans. Fritz C. A. Koelln and James P. Pettegrove (Boston: Beacon Press, 1955); Frank E. Manuel, *The Prophets of Paris* (Cambridge: Harvard University Press, 1962); and Martin S. Staum, *Minerva's Message: Stabilizing the French Revolution* (Montreal: McGill–Queen's University Press, 1996).

105. See Jean-Antoine-Nicolas de Caritat, marquis de Condorcet, *Esquisse d'un tableau historique des progrès de l'esprit humain* (1795; reprint, Paris: Dubuisson et Cie., 1864); Condorcet, *Sketch for a Historical Picture of the Progress of the Human Mind,* trans. June Barraclough, intro. Stuart Hampshire (Westport: Hyperion Press, 1955); Condorcet, *Écrits sur l'instruction publique,* 2 vols., eds. Charles Coutel and Catherine Kintzler, ([n.p.]: Les classiques de la république, 1989); and Condorcet, *Condorcet: Selected Writings,* ed. and trans. Keith Michael Baker (Indianapolis: Library of Liberal Arts, 1976).

106. See Condorcet, *Sketch,* pp. 3, 184, 177.

107. See ibid., "The Nature and Purpose of Public Instruction (1791)," in *Selected Writings,* pp. 111–13. For an insightful analysis of Condorcet's understanding of the natural inequalities and how a republic could be constructed on their basis, see Jean-Fabien Spitz, *L'amour de l'égalité: Essai sur la critique de l'égalitarianisme républicain en France 1770–1830* (Paris: EHESS and J. Vrin, 2000), chap. 3.

108. Condorcet, "Nature and Purpose of Public Instruction," pp. 117, 108.

109. "It is therefore important to establish a system of public instruction," Condorcet observed in 1791, "that allows no talent to go unnoticed, offering it all help heretofore reserved for the children of the wealthy." Condorcet, "Nature

and Purpose of Public Instruction," p. 111. Jefferson proposed a very similar scheme in 1787 in *Notes on the State of Virginia,* pp. 193–99.

110. Condorcet, "Nature and Purpose of Public Instruction," p. 116.

111. Ibid., *Sketch,* pp. 130–31.

112. Ibid., "Nature and Purpose of Public Instruction," p. 119.

113. Ibid., "On the Admission of Women to the Rights of Citizenship (1790)," in *Condorcet: Selected Writings,* p. 98.

114. See ibid., *Sketch,* p. 201; and ibid., "Nature and Purpose of Public Instruction," p. 113.

115. Ibid., "Nature and Purpose of Public Instruction," p. 120.

116. See Wollstonecraft, *Rights of Woman,* p. 92; and Sapiro, *A Vindication of Political Virtue,* chaps. 3 and 4.

117. See, for example, Wollstonecraft, *Rights of Woman,* pp. 119, 172.

Chapter 2

1. Important explorations of these concepts in the context of the new metropolis include James W. Cook, *The Arts of Deception: Playing with Fraud in the Age of Barnum* (Cambridge: Harvard University Press, 2001); Ann Fabian, *The Unvarnished Truth: Personal Narratives in Nineteenth-Century America* (Berkeley: University of California Press, 2000); Karen Halttunen, *Confidence Men and Painted Women: A Study of Middle-Class Culture in America, 1830–1870* (New Haven: Yale University Press, 1982); John F. Kasson, *Rudeness and Civility: Manners in Nineteenth-Century Urban America* (New York: Hill & Wang, 1990); T. J. Jackson Lears, *Something for Nothing: Luck in America* (New York: Viking, 2003). See also James L. Huston, "Virtue Besieged: Virtue, Equality, and the General Welfare in the Tariff Debates of the 1820s," *Journal of the Early Republic* 14 (1994): 523–47; James T. Kloppenberg, "The Virtues of Liberalism: Christianity, Republicanism, and Ethics in Early American Political Discourse," *Journal of American History* 74 (1987): 9–33; and Donald H. Meyer, *The Instructed Conscience: The Shaping of the American National Ethic* (Philadelphia: University of Pennsylvania Press, 1972). On the notion of "character" in the English context, see Stefan Collini, "The Idea of 'Character' in Victorian Political Thought," *Transactions of the Royal Historical Society,* 5th ser., 35 (1985): 29–50.

2. I am indebted to my colleague Mary Kelley for helping me clarify my ideas on this matter. See also Gordon S. Wood, *The Radicalism of the American Revolution* (New York: Vintage, 1991), esp. pp. 215–21, 356–57.

3. See Ruth H. Bloch, "The Gendered Meanings of Virtue in Revolutionary America," *Signs: Journal of Women in Culture and Society* 13 (1987): 37–58; Fabian, *Unvarnished Truth;* Halttunen, *Confidence Men and Painted Women;* Mary Kelley, *Private Woman, Public Stage: Literary Domesticity in Nineteenth-Century America* (New York: Oxford University Press, 1984); and Carroll Smith-Rosenberg, "The Cross and the Pedestal: Women, Anti-Ritualism, and the Emergence of the American Bourgeoisie," in *Disorderly Conduct: Visions of Gender in Victorian America* (New York: Knopf, 1985), pp. 129–64.

4. See particularly Collini, "Idea of 'Character' in Victorian Political Thought"; Kasson, *Rudeness and Civility;* Judy Hilkey, *Character Is Capital: Suc-*

cess Manuals and Manhood in Gilded Age America (Chapel Hill: University of North Carolina Press, 1997); Lears, *Something for Nothing;* E. A. Rotundo, *American Manhood: Transformations in Masculinity from the Revolution to the Modern Era* (New York: Basic Books, 1993); and Warren Susman, "Personality and the Making of Twentieth-Century Culture," in *Culture as History: The Transformation of American Society in the Twentieth Century* (New York: Pantheon, 1984).

5. On the power of such flexible concepts, see Cook, *Arts of Deception;* Barbara J. Fields, "Ideology and Race in American History," in *Region, Race, and Reconstruction: Essays in Honor of C. Vann Woodward,* eds. J. Morgan Kousser and James M. McPherson (New York: Oxford University Press, 1982), pp. 143–77; Daniel T. Rodgers, *Contested Truths: Keywords in American Politics since Independence* (New York: Basic Books, 1987); and Rodgers, *The Work Ethic in Industrializing America, 1850–1920* (Chicago: University of Chicago Press, 1978). I thank my colleague Jay Cook for underscoring the importance of this phenomenon to my account.

6. For an interesting example, see Ralph Waldo Emerson, "Aristocracy" (1848), in *Lectures and Biographical Sketches* (Boston: Houghton, Mifflin, 1895), pp. 35–67.

7. The information about Miller largely comes from Henry F. May, *The Enlightenment in America* (New York: Oxford University Press, 1976), pp. 338–41. Miller's importance as an Old School Presbyterian minister is underscored by Theodore Dwight Bozeman, *Protestants in an Age of Science: The Baconian Ideal and Antebellum American Religious Thought* (Chapel Hill: University of North Carolina Press, 1977), esp. chap. 3.

8. Samuel Miller, *A Brief Retrospect of the Eighteenth Century,* vol. 1 (New York: T. and J. Swords, 1803), p. vii.

9. Ibid., vol. 2, pp. 28–30, 300, 297.

10. Ibid., pp. 285, 290–91.

11. See esp. Daniel Walker Howe, *Making the American Self: Jonathan Edwards to Abraham Lincoln* (Cambridge: Harvard University Press, 1997), chap. 4; Jean V. Matthews, *Toward a New Society: American Thought and Culture, 1800–1830* (Boston: Twayne, 1991), pp. 3–46; May, *Enlightenment in America,* pp. 307–62; and Steven Mintz, *Moralists & Modernizers* (Baltimore: Johns Hopkins University Press, 1995).

12. My interpretation of the American Whigs has been strongly influenced by Daniel Walker Howe, *The Political Culture of the American Whigs* (Chicago: University of Chicago Press, 1979) and Daniel T. Rodgers, *Contested Truths,* esp. chap. 4. See also John Higham, *From Boundlessness to Consolidation: The Transformation of American Culture, 1848–1860* (Ann Arbor: William L. Clements Library, 1969); Paul Johnson, *A Shopkeeper's Millennium: Society and Revivals in Rochester, New York, 1815–1837* (New York: Hill and Wang, 1978); Lawrence F. Kohl, *The Politics of Individualism: Parties and the American Character in the Jacksonian Era* (New York: Oxford University Press, 1989); Meyer, *The Instructed Conscience;* Kathryn Kish Sklar, *Catherine Beecher: A Study in American Domesticity* (New Haven: Yale University Press, 1973); and Anthony F. C. Wallace, *Rockdale: The Growth of an American Village in the Early Industrial Revolution* (New York: W. W. Norton., 1978).

13. [James Reynolds], *Equality; A History of Lithconia* (1837; reprint, Philadelphia: Prime Press, 1947), pp. 17–18, 23–24.

14. See ibid., pp. 28, 37–40.

15. On gender inflections in Jeffersonian-Jacksonian discourse, see Mary Kelley, "'Vindicating the Equality of Female Intellect': Women and Authority in the Early Republic," *Prospects: An Annual of American Cultural Studies* 17 (1992): 1–27. Also valuable are Catherine Allgor, *Parlor Politics: In Which the Ladies of Washington Help Build a City and a Government* (Charlottesville: University Press of Virginia, 2000); and Susan Branson, *These Fiery Frenchified Dames: Women and Political Culture in Early National Philadelphia* (Philadelphia: University of Pennsylvania Press, 2001).

16. On Jacksonian thought and culture, see Lee Benson, *The Concept of Jacksonian Democracy: New York as a Test Case* (Princeton: Princeton University Press, 1973); Kohl, *Politics of Individualism;* Matthews, *Toward a New Society;* Marvin Meyers, *The Jacksonian Persuasion: Politics & Belief* (New York: Vintage, 1960); Edward Pessen, *Jacksonian America: Society, Personality, and Politics* (Homewood: Dorsey Press, 1969); Rodgers, *Contested Truths,* esp. chap. 3; Sean Wilentz, *Chants Democratic: New York City & the Rise of the American Working Class, 1788–1850* (New York: Oxford University Press, 1986); and Wilentz, "On Class and Politics in Jacksonian America," in *The Promise of American History: Progress and Prospects,* eds. Stanley I. Kutler and Stanley N. Katz (Baltimore: Johns Hopkins University Press, 1982), pp. 45–63.

17. As Marvin Meyers has noted, "Morals, habits, character, are key terms in Jackson's discussion of the people—and almost every other subject." Meyers, *Jacksonian Persuasion,* p. 22.

18. Andrew Jackson, "Veto Message," quoted in Richard Hofstadter, *The American Political Tradition* (New York: Vintage, 1948), p. 62 (emphasis added).

19. For background on Common Sense in America, see Joseph L. Blau, introduction to Francis Wayland, *The Elements of Moral Science,* ed. Joseph L. Blau (Cambridge: Harvard University Press, 1963); Bozeman, *Protestants in an Age of Science,* esp. chap. 1; Merle Curti, *Human Nature in American Thought* (Madison: University of Wisconsin Press, 1980), esp. chaps. 4–5; Rand Evans, "The Origins of American Academic Psychology," in *Explorations in the History of Psychology in America,* ed. Josef Brozek (Lewisburg: Bucknell University Press, 1984), pp. 17–60; Alfred H. Fuchs, "Contributions of American Mental Philosophers to Psychology in the United States," *History of Psychology* 3 (2000): 3–19; E. Brooks Holifield, *The Gentlemen Theologians: American Theology in Southern Culture, 1795–1860* (Durham: Duke University Press, 1978), esp. chaps. 5–6; Daniel Walker Howe, *The Unitarian Conscience: Harvard Moral Philosophy, 1805–1861* (Cambridge: Harvard University Press, 1970); Bruce Kuklick, *Churchmen and Philosophers: From Jonathan Edwards to John Dewey* (New Haven: Yale University Press, 1985); and Meyer, *Instructed Conscience.*

20. Much of the argument in this section of the chapter is based on Evans's discussion in "Origins of American Academic Psychology," esp. p. 33.

21. See ibid., p. 34; Kuklik, *Churchmen and Philosophers,* pp. 131–32; and Louis F. Snow, *The College Curriculum in the United States,* (New York: Teachers College, 1907) pp. 126–29.

22. See, for example, the lectures produced by Mark Hopkins, president of Williams College, as a text for his course. Mark Hopkins, *An Outline Study of Man; or, The Body and the Mind in One System* (New York: Scribner, Armstrong, 1874). See also Holifield, *Gentlemen Theologians,* chap. 5.

23. See Dugald Stewart, *Outlines of Moral Philosophy,* in *The Collected Works of Dugald Stewart,* vol. 2 (Edinburgh: Thomas Constable, 1854), p. 11.

24. See, for example, Thomas Reid, *Essays on the Intellectual Powers of Man* (1785), in *Philosophical Works* (Hildesheim: Georg Ohms Verlag, 1983) p. 481.

25. For an example of this dual move, see Thomas Upham, *Elements of Mental Philosophy,* vol. 1 (Boston: Hilliard Gray, 1831).

26. See Stewart, *Outlines of Moral Philosophy,* pp. 12–13; and Francis Wayland, *The Elements of Intellectual Philosophy* (New York: Sheldon, 1864), pp. 9–13.

27. See Stewart, *Outlines of Moral Philosophy,* p. 31.

28. The major works of Thomas C. Upham include: *Elements of Intellectual Philosophy: Designed as a Text–Book* (Portland: William Hyde, 1827), expanded in 1831 to *Elements of Mental Philosophy,* vols. 1–2, and in 1840 to the three-volume *Mental Philosophy; Embracing the Three Departments of the Intellect, Sensibilities, and Will,* 3 vols. (1840; reprint, New York: Harper & Brothers, 1869); and *Outlines of Imperfect and Disordered Mental Action* (New York: Harper & Bros., 1840). Francis Wayland's principal philosophical writings include: *The Elements of Moral Science* (1837; reprint, Cambridge: Harvard University Press, 1963); and *The Elements of Intellectual Philosophy* (1854; reprint, New York: Sheldon, 1864). Other important works in the American Common Sense tradition were Joseph Haven, *Mental Philosophy: Including the Intellect, Sensibilities, and Will* (Boston: Gould and Lincoln, 1857); Hopkins, *An Outline Study of Man*; Noah Porter, *The Human Intellect* (New York: Scribner, 1868); and Leicester A. Sawyer, *Mental Philosophy* (New Haven: Durrie and Peck, 1839).

29. Upham acknowledged the eclectic nature of his approach in the preface to his *Mental Philosophy,* vol. 1, p. iii. Robert Vaughn also commented on the eclecticism of American Common Sense in "American Philosophy," *British Quarterly Review* 5 (1847): 88–119.

30. Francis Wayland, *The Education Demanded by the People of the U. States . . .* (Boston: Phillips, Sampson, 1855), p. 23.

31. George Peck, "Upham's Mental Philosophy," *Methodist Quarterly Review* 23 (1841): 263–82, quote on p. 269; and Wayland, *Education Demanded,* p. 10.

32. Wayland, *Elements of Intellectual Philosophy,* p. 261. See also Sawyer, *Elements of Mental Philosophy,* p. 89.

33. See Upham, *Elements of Intellectual Philosophy,* p. 494.

34. See, for example, Upham's discussion of deaf or blind people recovering their senses in *Mental Philosophy,* vol. 1, pp. 72–80.

35. See, for example, Wayland, *Elements of Intellectual Philosophy,* pp. 138–39; Upham, *Elements of Mental Philosophy,* vol. 1, pp. 206–209; and Sawyer, *Elements of Mental Philosophy,* p. 133.

36. See Wayland, *Elements of Intellectual Philosophy,* p. 22; and Upham, *Elements of Mental Philosophy,* vol. 1, p. 159.

37. Upham, *Elements of Intellectual Philosophy,* p. 494. See also Upham, *Elements of Mental Philosophy,* vol. 2, pp. 30–32.

38. As an anonymous author in the Jacksonian party house organ, the *U.S. Magazine and Democratic Review,* noted, "The differences, therefore, [that] abound in the manifestations of intellectual power, which are the subjects of daily observation . . . are not necessarily the result of differences in *mental capacity,* but may depend upon secondary and adventitious circumstances." "The Absolute Equality of Mind," *U.S. Magazine and Democratic Review* 24 (1849): 28.

39. Peck, "Upham's Mental Philosophy," p. 281 (emphasis in original).

40. Upham, *Elements of Intellectual Philosophy,* p. 19.

41. For general background, see Louise Stevenson, *Scholarly Means to Evangelical Ends: The New Haven Scholars and the Transformation of Higher Learning in America, 1830–1890* (Baltimore: Johns Hopkins University Press, 1986).

42. [Noah Porter], "Recent Works on Psychology," *New Englander* 13 (1855): 129–44, quote on p. 138.

43. See Benson, *Concept of Jacksonian Democracy,* p. 86.

44. See, for example, Lyman Beecher, *Six Sermons on the Nature, Occasions, Signs, Evils, and Remedy of Intemperance* (1828), in *Lyman Beecher and the Reform of Society: Four Sermons, 1804–1828* (New York: Arno Press, 1972), pp. 49–51; and for the concern in Common Sense with temperance, Sawyer, *Elements of Mental Philosophy,* pp. 405–408.

45. Catherine E. Beecher, *The Elements of Mental and Moral Philosophy, Founded upon Experience, Reason, and the Bible* (Hartford: n.p., 1831).

46. See Henry C. Carey, *The Harmony of Interests: Agricultural, Manufacturing & Commercial,* 2nd ed. (New York: M. Finch, 1852).

47. Calvin Colton, *The Junius Tracts, No. VI: Labor and Capital* (1844), in *Antebellum American Culture: An Interpretive Anthology,* ed. David Brion Davis (Lexington: D.C. Heath, 1979), p. 197.

48. Henry C. Carey, "Of Wealth" (1858), in *Principles of Social Science,* 3 vols. (1871), anthologized in *The American Intellectual Tradition,* vol. 1, eds. David A. Hollinger and Charles Capper (New York: Oxford University Press, 1993), p. 264 (emphasis in original).

49. Alexis de Tocqueville, *Democracy in America,* ed. Phillips Bradley, intro. Alan Ryan (1840; reprint, New York: Knopf, 1994), vol. 2, p. 159. Note also the close parallel here to the views of Adam Smith, as explained in chapter 1, and developed by Smith in *An Inquiry into the Nature and Causes of the Wealth of Nations* (1776; reprint, New York: Modern Library, 1994), vol. 1, bk. 1, chap. 2, esp. pp. 16–17.

50. For specific condemnations of the Whigs as aristocrats, see, for example, some of William Leggett's editorials in the *Plaindealer* and *Evening Post* during the 1830s, as collected in *Democratick Editorials: Essays in Jacksonian Political Economy,* ed. Lawrence H. White (Indianapolis: Liberty Press, 1984), esp. pp. 260–65, and 268–71.

51. "Introduction," *United States Magazine and Democratic Review* 1 (1837): 1–15, quote on p. 2.

52. George Bancroft, "An Oration Delivered before the Adelphi Society of Williamstown College in August, 1835," in *American Intellectual Tradition,* vol. 1, p. 237.

53. William Leggett, "True Functions of Government," *Evening Post,* November 21, 1834, in *Democratick Editorials,* p. 5.

54. "In as far as inequality of human condition is the result of natural causes it affords no just topic of complaint," William Leggett proclaimed. Leggett, "The Inequality of Human Condition," *Plaindealer,* December 31, 1836, in *Democratick Editorials,* p. 257.

55. "Introduction," *United States Magazine and Democratic Review,* pp. 4–5.

56. "Absolute Equality of Mind," pp. 24–32.

57. J. C. Hope, "Genius and Industry in Their Results," *DeBow's Review* 29 (1860): 269–80, quote on p. 270.

58. "Genius," *Southern Literary Messenger* 2 (1835): 297–300, quote on p. 300.

59. As Sean Wilentz points out, even when calls from radical sectors of the working class came for a large-scale redistribution of property in the late 1820s, their leaders continued to acknowledge the importance of natural differences in talent. See Wilentz, *Chants Democratic,* pp. 93, 187, 190–201.

60. Winthrop D. Jordan, introduction to *An Essay on the Causes of the Variety of Complexion and Figure in the Human Species,* by Samuel Stanhope Smith (Cambridge: Harvard University Press, 1965), p. xxiii. See also Frederick Rudolph, *The American College and University: A History* (New York: Vintage, 1962), p. 38; and Douglas Sloan, *The Scottish Enlightenment and the American College Ideal* (New York: Teachers College Press, 1971), p. 180.

61. For an assessment of the tumult of collegiate life at the turn of the century, see Rudolph, *American College and University,* esp. pp. 38–40; Leon Jackson, "The Rights of Man and the Rites of Youth: Fraternity and Riot at Eighteenth-Century Harvard," in *The American College in the Nineteenth Century,* ed. Roger Geiger (Nashville: Vanderbilt University Press, 2000), pp. 46–79; and Steven J. Novak, *The Rights of Youth: American Colleges and Student Revolt, 1798–1815* (Cambridge: Harvard University Press, 1977).

62. Rudolph, *American College and University,* p. 38.

63. See Jedidiah Morse's comments to Charles Nisbet on 4 April 1800, about Jefferson and his fellow American "Jacobins," as quoted in May, *Enlightenment in America,* pp. 303–304.

64. My discussion of mental philosophy in the American collegiate curriculum owes much to Rand Evans's astute analysis in "Origins of American Academic Psychology." For general background, see Richard Hofstadter and Wilson Smith, eds., *American Higher Education: A Documentary History,* vol. 1 (Chicago: University of Chicago Press, 1961); Howard Miller, *The Revolutionary College: American Presbyterian Higher Education, 1707–1837* (New York: New York University Press, 1976); Mark Noll, *Princeton and the Republic: The Search for a Christian Enlightenment in the Era of Samuel Stanhope Smith* (Princeton: Princeton University Press, 1989); David B. Potts, "Curriculum and Enrollment: Assessing the Popularity of Antebellum Colleges," in *American College in the Nineteenth Century,* pp. 37–45; Frederick Rudolph, *Curriculum: A History of the American Undergraduate Course of Study since 1636* (San Francisco: Jossey-Bass, 1977); Sloan, *Scottish Enlightenment and the American College Ideal;* and Laurence R. Veysey, *The Emergence of the American University* (Chicago: University of Chicago Press, 1965), esp. chap. 1.

65. "The object of such institutions [colleges]," Francis Wayland observed, "is to cultivate and develop to the highest perfection the best minds of the country."

Francis Wayland, *Thoughts on the Present Collegiate System in the United States* (Boston: Gould, Kendall & Lincoln, 1842), p. 79.

66. For general background on American public education during the antebellum period, see Lawrence A. Cremin, *American Education: The National Experience, 1783–1876* (New York: Harper & Row, 1980); Carl F. Kaestle, *Pillars of the Republic: Common Schools and American Society, 1780–1860* (New York: Hill & Wang, 1983); David Nasaw, *Schooled to Order: A Social History of Public Schooling in the United States* (New York: Oxford University Press, 1979), pp. 7–84; William J. Reese, *The Origins of the American High School* (New Haven: Yale University Press, 1995), pp. 1–58; Stanley K. Schultz, *The Culture Factory: Boston Public Schools, 1789–1860* (New York: Oxford University Press, 1973); and Caroline Winterer, "Avoiding a 'Hothouse System of Education': Nineteenth-Century Early Childhood Education from the Infant School to the Kindergartens," *History of Education Quarterly* 32 (1992): 288–314.

67. Joseph Perkins, *An Oration upon Genius* (Boston: Joseph Nancrede, 1797), pp. 11, 15.

68. Benjamin Rush, "Of the Mode of Education Proper in a Republic," in *Essays Literary, Moral, and Philosophical* (1798; reprint, Schenectady: Union College Press, 1988), p. 9 (emphasis added). See also Rush, *Thoughts upon Female Education* (Boston: Samuel Hall, 1787).

69. Perkins, *Oration upon Genius,* pp. 15–16. See also Thomas Jefferson's elaborate plan in *Notes on the State of Virginia* for a system of education.

70. See, for example, Wayland, *Education Demanded by the People of the U. States,* p. 25.

71. E. H. Chapin, "Anniversary Address," *Southern Literary Messenger* 5 (1839): 725–33, quote on p. 730.

72. James G. Carter, *Essays upon Popular Education, Containing a Particular Examination of the Schools of Massachusetts and an Outline of an Institution for the Education of Teachers* (1826; reprint, New York: Arno Press, 1969), p. 20.

73. Upham, *Elements of Intellectual Philosophy,* p. 490.

74. Joseph Henry, *The Papers of Joseph Henry, 1844–1846,* ed. Marc Rothenberg (Washington: Smithsonian Institution Press, 1992), p. 429. See also James G. Carter, *Letters to the Hon. William Prescott, LL.D., on the Free Schools of New England, with Remarks upon the Principles of Instruction* (1824), reprinted in *Enlightenment and Social Progress: Education in the Nineteenth Century,* ed. J. J. Chambliss (Minneapolis: Burgess, 1971), p. 156.

75. Upham, *Elements of Intellectual Philosophy,* pp. 494–96.

76. In *Outlines of Imperfect and Disordered Mental Action* (1840), for example, Upham explained in great detail the various ways in which an over- or underdeveloped faculty could have deleterious consequences, most often insanity, for the individual so afflicted.

77. [Jeremiah Day and James L. Kingsley], "Original Papers in Relation to a Course of Liberal Education" ("Yale Report," 1828), *American Journal of Science and Arts* 15 (1829): 297–351. On the "Yale Report," see Rudolph, *American College and University,* pp. 130–35; and Rudolph, *Curriculum,* pp. 65–83. For a vigorous criticism of the assumption that the New England college model

was widely followed, see Colin B. Burke, *American Collegiate Populations: A Test of the Traditional View* (New York: New York University Press, 1982).

78. "Yale Report," p. 323.

79. For more on the importance of maintaining balance among the faculties, especially within Whig ideology, see Howe, *Political Culture of the American Whigs,* esp. chap. 2.

80. "Yale Report," pp. 308–309.

81. Wayland, *Thoughts on the Present Collegiate System in the United States,* p. 7.

82. My thanks to Mary Kelley for helping me to rethink aspects of this section. There is an enormous and rich secondary literature on domesticity and the so-called separate-spheres ideology in nineteenth-century America. Some of the most significant contributions to my understanding include Bloch, "Gendered Meanings of Virtue"; Nancy F. Cott, *The Bonds of True Womanhood: "Woman's Sphere" in New England, 1780–1835* (New Haven: Yale University Press, 1977); Sara Deutsch, *Women and the City: Gender, Space, and Power in Boston, 1870–1940* (New York: Oxford University Press, 2000); Ann Douglas, *The Feminization of American Culture* (New York: Avon Books, 1977); Lori D. Ginzberg, *Women and the Work of Benevolence: Morality, Politics, and Class in the Nineteenth-Century United States* (New Haven: Yale University Press, 1990); Kelley, *Private Woman, Public Stage;* Mary P. Ryan, *Cradle of the Middle Class: The Family in Oneida County, New York, 1790–1865* (Cambridge: Cambridge University Press, 1983); Ryan, *Women in Public: Between Banners and Ballots, 1825–1880* (Baltimore: Johns Hopkins University Press, 1990); Sklar, *Catherine Beecher;* Smith-Rosenberg, "Cross and the Pedestal"; and Barbara Welter, "The Cult of True Womanhood, 1820–1860," *American Quarterly* 18 (1966): 151–74.

83. Catherine E. Beecher, "Letter XVII," in *The True Remedy for the Wrongs of Woman* (1851), in *The Limits of Sisterhood: The Beecher Sisters on Women's Rights and Woman's Sphere,* eds. Jeanne Boydston, Mary Kelley, and Anne Margolis (Chapel Hill: University of North Carolina Press, 1988), pp. 141–42 (emphasis in original).

84. A. B. Muzzey, *The Young Maiden* (Boston: William Crosby, 1840), p. 4.

85. Thomas Branagan, *The Excellency of the Female Character Vindicated; Being an Investigation Relative to the Cause and Effects of the Encroachments of Men upon The Rights of Women, and the Too Frequent Degradation and Consequent Misfortunes of the Fair Sex* (1808; reprint, New York: Arno Press, 1972), pp. 74–75.

86. Sarah Grimké, "The Education of Women" (n.d.), in *Letters on the Equality of the Sexes and Other Essays,* ed. Elizabeth Ann Bartlett (New Haven: Yale University Press, 1988), p. 111; see also "Letter III [1837]," in *Letters on the Equality of the Sexes and Other Essays,* ed. Bartlett, p. 39. Judith Sargent Murray argued much the same point in "Observations on Female Abilities" (1798), in *The Gleaner,* introduction by Nina Baym (Schenectady: Union College Press, 1992), p. 710, as did Sarah C. Edgerton, "Female Culture," *Mother's Assistant* 3 (1843): 94–95.

87. Catherine E. Beecher, *A Treatise on Domestic Economy* (1841), excerpted in *The Limits of Sisterhood,* eds. Boydston, Kelley, and Margolis, p. 133. Even Sarah Grimké employed similar arguments; see "Letter VIII [1837]," in *Letters on the Equality of the Sexes and Other Essays,* ed. Bartlett, p. 58.

88. On the question of higher education for women, see Mary Kelley, *Empire of Reason: The Making of Learned Women in America's Republic* (Chapel Hill: University of North Carolina Press, 2006).

89. Tocqueville, *Democracy in America,* vol. 2, pp. 211–12.

90. See, for example, Tocqueville, *Democracy in America,* vol. 2, pp. 211–14.

91. Hannah Mather Crocker, *Observations on the Real Rights of Women, With Their Appropriate Duties, Agreeable to Scripture, Reason and Common Sense* (1818), reprinted in *Sex and Equality* (New York: Arno Press, 1974), p. 5.

92. Judith Sargent Murray and, indeed, almost all of the advocates of women's equality, also used Christianity to justify their positions. "Our souls are by nature *equal* to yours," Murray declared, "the same breath of God animates, enlivens, and invigorates us." Murray, "On the Equality of the Sexes" (1790), in *Selected Writings of Judith Sargent Murray,* ed. Sharon M. Harris (New York: Oxford University Press, 1995), p. 8. See also Margaret Fuller, *Woman in the Nineteenth Century* (1845; reprint, with introduction by Bernard Rosenthal, New York: Norton, 1971).

93. Catherine E. Beecher, *Suggestions Respecting Improvements in Education* (1829), excerpted in *Limits of Sisterhood,* eds. Boydston, Kelley, and Margolis, p. 43. See also Beecher, *Treatise on Domestic Economy;* and Beecher, *Elements of Mental and Moral Philosophy.*

94. Crocker, *Observations on the Real Rights of Women,* pp. 15–16.

95. Sarah Grimké, for one, exploited this possibility vigorously in "Education of Women."

96. As Judith Sargent Murray observed in 1790: "What can she [the uncultivated woman] do? to books she may not apply; or if she doth, *to those only of the novel kind,* lest she merit the appellation of *learned lady;* and what ideas have been affixed to this term, the observation of many can testify." Murray, "On the Equality of the Sexes," p. 6. See as well Catherine Beecher's very unflattering description of Fanny Wright as masculine because she deviated from the "ap[p]ropriate character of a woman" in her *Letters on the Difficulties of Religion* (1836), excerpted in *The Limits of Sisterhood,* eds. Boydston, Kelley, and Margolis, pp. 236–37.

97. For an important analysis of how Napoleon used the system of merit while simultaneously seeking to undercut it, see Rafe Blaufarb, "The Social Contours of Meritocracy in the Napoleonic Officer Corps," in *Taking Liberties: Problems of a New Order from the French Revolution to Napoleon,* eds. Howard G. Brown and Judith A. Miller (Manchester: Manchester University Press, 2002), pp. 126–46.

98. C. L. Masuyer, *Discours sur l'organisation de l'instruction publique et de l'éducation nationale en France* (Paris: Imprimerie nationale, 1793), pp. 5–6 (emphasis in original).

99. The most comprehensive secondary account of the debates over the French educational system during the Revolution and Directory is Robert R. Palmer, *The Improvement of Humanity: Education and the French Revolution* (Princeton:

Princeton University Press, 1985). It can be supplemented with Jean-François Chassaing, "Les manuels de l'enseignement primaire de la Révolution et les idées révolutionnaires," in *Le mouvement de réforme de l'enseignement en France, 1760–1798* (Paris: Presses universitaires de France, 1974), pp. 97–165; Jean-Pierre Gross, *Fair Shares for All: Jacobin Egalitarianism in Practice* (Cambridge: Cambridge University Press, 1997), chap. 7; T. E. Kaiser, "Enlightenment and Public Education during the French Revolution: The Views of the Idéologues," *Studies in 18th-Century Culture* 10 (1980): 95–111; Françoise Mayeur, *Histoire générale de l'enseignement et de l'éducation en France,* tome III: *De la Révolution à l'école républicaine* (Paris: Nouvelle librairie de France, 1981); and Isser Woloch, *The New Regime: Transformations of the French Civic Order, 1789–1820s* (New York: Norton, 1994), esp. chaps. 6–7. The most important collection of primary documents is James Guillame, *Procès-verbaux du Comité d'instruction publique de la Convention Nationale,* 7 vols. (Paris: Imprimerie nationale, 1891–1955). Also valuable are Alfred de Beauchamp, *Recueil des lois et règlements sur l'enseignement supérieur: Comprenant les décisions de la jurisprudence et les avis des conseils de l'instruction publique et du conseil d'état,* 2 vols. (Paris: Typographie de Delalain Frères, 1880); Bronislaw Baczko, ed., *Une éducation pour la démocratie: Textes et projets de l'époque révolutionnaire* (Paris: Editions Garnier Frères, 1982); and Octave Gréard, ed., *La législation de l'instruction primaire en France depuis 1789 jusqu'à nos jours,* vol. 1 (Paris: Charles de Morgues Frères, 1874).

100. Charles-Maurice de Talleyrand-Périgord, *Rapport sur l'instruction publique, fait au nom du Comité de Constitution à l'Assemblée Nationale, les 10, 11 et 19 septembre 1791* (Paris: Baudoin, 1791).

101. Jean-Antoine-Nicolas de Caritat, marquis de Condorcet, *Report on the General Organization of Public Instruction* (1792), in *French Liberalism and Education in the Eighteenth Century: The Writings of La Chalotais, Turgot, Diderot, and Condorcet on National Education,* ed. and trans. F. de la Fontainerie (New York: McGraw-Hill, 1932), p. 323.

102. Notice how different the situation was in the United States, where the origin of talents was largely deemed to be a mixture of endowment and training.

103. Masuyer, *Discours sur l'organisation de l'instruction publique,* p. 76 below. Or, as Jean-Bon Saint André observed during the same debate, "Le génie est un don de la nature." Jean-Bon Saint-André, *Sur l'éducation nationale* (Paris: Imprimerie nationale, 1793), p. 8.

104. See Palmer, *Improvement of Humanity,* esp. chaps. 3 and 4. See also François Lanthenas, *Rapport et projet de décret sur l'organisation des écoles primaires présentés à la Convention nationale, au nom du comité d'instruction publique par F. Lanthenas député à la Convention nationale, le 18 décembre 1792,* in *L'instruction publique en France pendant la révolution,* p. 153.

105. See Condorcet, *Report on the General Organization of Public Instruction,* p. 355.

106. R. R. Palmer argues at length for the antidemocratic character of the proposals of Talleyrand and Condorcet in *Improvement of Humanity,* pp. 94–100, 124–29.

107. Pierre-Louis Lacretelle, *De l'établissement des connoissances humaines, et de l'instruction publique, dans la constitution française* (Paris: Chez Desenne, 1791), p. 117. See also Talleyrand, *Rapport sur l'instruction publique,* p. 9. For an important analysis of the place of *émulation* in nineteenth-century French culture, see Carol E. Harrison, *The Bourgeois Citizen in Nineteenth-Century France: Gender, Sociability, and the Uses of Emulation* (Oxford: Oxford University Press, 1999). Harrison argues that émulation served to unite egalitarianism with hierarchy around the image of a meritocratic society whose competitions, in theory, were open to all.

108. See Talleyrand, *Rapport sur l'instruction publique,* pp. 17–18, 82–84. Condorcet, however, was rather ambivalent about émulation per se. See Condorcet, *Report on the General Organization of Public Instruction,* pp. 350, 354–55.

109. Louis-Pierre Dufourny, as quoted in Palmer, *Improvement of Humanity,* p. 79. On émulation, see also Anne Louise Germaine (Necker), baronne de Staël-Holstein, "Chapitre III: De l'Emulation," in *De la littérature considérée dans ses rapports avec les institutions sociales* (1800; reprint, Geneva: Librairie Droz, 1959), pp. 317–31.

110. *Plan d'éducation nationale de Michel Lepelletier*[sic], *présenté à la Convention par Maximilien Robespierre, au nom de la Commission d'instruction publique* (Paris: Imprimerie nationale, 1793), p. 41.

111. Condorcet, *Report on the General Organization of Public Instruction,* pp. 334–35.

112. Babeuf, as quoted in Palmer, *Improvement of Humanity,* p. 145.

113. Masuyer, *Discours sur l'organisation de l'instruction publique,* p. 23 (emphasis in original). Jean-Bon Saint André concurred in *Sur l'éducation nationale,* p. 7.

114. See Pierre-Claude-François Daunou, *Essai sur l'instruction publique (juillet 1793),* in *Une éducation pour la démocratie,* pp. 334–35.

115. Masuyer, *Discours sur l'organisation de l'instruction publique,* pp. 8–9.

116. See Talleyrand, *Rapport sur l'instruction publique,* pp. 9–10; and Condorcet, *Report on the General Organization of Public Instruction,* p. 326. On Condorcet's balancing of talent and equality, see Jean-Fabien Spitz, *L'amour de l'égalité: Essai sur la critique de l'égalitarianisme républicain en France, 1770–1830* (Paris: EHESS and J. Vrin, 2000), chap. 3.

117. *Plan d'éducation nationale de Michel Lepelletier,* p. 7. Michel Le Peletier had been assassinated on January 20, 1793, and was revered as a martyr to the republic. It was thus his brother Félix who had conveyed his educational plan to Robespierre, who then reported it to the National Convention.

118. *Plan d'éducation nationale de Michel Lepelletier,* pp. 18, 22, 35. On Le Pelletier's plan, see Palmer, *Improvement of Humanity,* pp. 137–46.

119. See Antoine Fourcroy, *Opinion de Fourcroy, député du département de Paris, sur le projet d'éducation nationale de Michel Le Pelletier*[sic], *Prononcée dans la séance du 30 juillet 1793* (Paris: Imprimerie nationale, 1793).

120. See *Décret sur l'organisation de l'instruction publique du 3 brumaire an IV [24 octobre 1795],* in *Une éducation pour la démocratie,* ed. Baczko, pp. 514–15. Isser Woloch, however, argues that the Lakanal Law of 27 brumaire An III (17

November 1794) was the most significant, even though it was superseded less than one year later by the Daunou Law. See Woloch, *New Regime,* pp. 181–94. Preceding both was the Bouquier Law, of 29 frimaire An II (December 19, 1793), which promised to pay the tuition of students for elementary education, but did not establish a system of public schools. See Gabriel Bouquier, *Rapport sur le plan général d'instruction publique (22 frimaire an II),* in *Une éducation pour la démocratie,* ed. Baczko, pp. 418–20.

121. On this point, see Palmer, *Improvement of Humanity,* pp. 225–36; and Woloch, *New Regime,* pp. 191–207.

122. See *Loi générale sur l'Instruction publique. 11 Floréal An X (1er Mai 1802),* in Beauchamp, *Recueil des lois et règlements sur l'enseignement supérieur,* vol. 1, pp. 81–87. For accessibility to secondary education, see Robert Gildea, "Education and the Classes Moyennes in the Nineteenth Century," in *The Making of Frenchmen: Current Directions in the History of Education in France, 1679–1979,* eds. Donald N. Baker and Patrick J. Harrigan (Waterloo: Historical Reflections Press, 1980), pp. 275–99.

123. *Décret portant organisation de l'Université. 17 Mars 1808,* in Beauchamp, *Recueil des lois et règlements sur l'enseignement supérieur,* vol. 1, pp. 171–86. See also Jean Tulard, "L'Université Napoléonienne," in *Histoire de l'administration de l'enseignement en France, 1789–1981,* eds. Pierre Bousquet, et al. (Geneva: Librairie Droz, 1983), pp. 11–36.

124. This was true even though, as Woloch points out, the Napoleonic regime actually accomplished few of its educational goals. See Woloch, *New Regime,* pp. 208–16; and Joseph N. Moody, *French Education since Napoleon* (Syracuse: Syracuse University Press, 1978), esp. p. 15. Other important sources include Robert D. Anderson, *Education in France, 1848–1870* (Oxford: Clarendon Press, 1975); Robert Fox, "Science, the University, and the State in Nineteenth-Century France," in *Professions and the French State, 1700–1900,* ed. Gerald L. Geison (Philadelphia: University of Pennsylvania Press, 1984), pp. 66–145; Fritz Ringer, *Fields of Knowledge: French Academic Culture in Comparative Perspective, 1890–1920* (Cambridge: Cambridge University Press, 1992), chap. 1; Alan B. Spitzer, *The French Generation of 1820* (Princeton: Princeton University Press, 1987), esp. chap. 8; and George Weisz, *The Emergence of the Modern University in France, 1863–1914* (Princeton: Princeton University Press, 1983).

125. François Guizot, *Essai sur l'histoire et sur l'état actuel de l'instruction publique en France* (Paris: Maradan, 1816), p. 4. For the Guizot Law, see *Loi portant organisation de l'Instruction primaire. 28 juin 1833,* in Gréard, *La législation de l'instruction primaire en France,* vol. 1, p. 244; for the Falloux Law, see *Loi relative à l'enseignement. 15 Mars 1850,* in Beauchamp, *Recueil des lois et règlements sur l'enseignement supérieur,* vol. 2, p. 91.

126. On the Ecole Polytechnique, see Terry Shinn, *L'Ecole Polytechnique* (Paris: Presses de la fondation nationale des sciences politiques, 1980); and John H. Weiss, *The Making of Technological Man: The Social Origins of French Engineering Education* (Cambridge: MIT Press, 1982). On the Ecole Normale, see *Le Centenaire de L'Ecole Normale, 1795–1895* (Paris: Hachette, 1895); Victor Cousin, ed., *Ecole Normale: Règlements, Programmes et Rapports* (Paris: Ha-

chette, 1837); Robert J. Smith, *The Ecole Normale Supérieure and the Third Republic* (Albany: SUNY Press, 1982); and Craig Zwerling, *The Emergence of the Ecole Normale Supérieure as a Centre of Scientific Education in Nineteenth-Century France* (New York: Garland, 1990).

127. Ambroise Rendu, *Système de l'Université de France, ou Plan d'une éducation nationale, essentiellement monarchique et religieuse* (Paris: H. Nicolle, 1816), pp. vi–vii.

128. As Alan Spitzer has noted, "The competitive element was not merely an ornament of this system but mortared into its very structure." Spitzer, *French Generation of 1820,* p. 211, and, more generally, chap 8.

129. See Terry Shinn's description in *L'Ecole Polytechnique,* p. 50.

130. Guizot, *Essai sur . . . l'instruction publique en France,* pp. 89–90.

131. *Règlement pour l'admission à l'Ecole normale supérieure. 7 Décembre 1850,* in Beauchamp, *Recueil des lois et règlements sur l'enseignement supérieur,* vol. 2, pp. 186, 188.

132. See Anderson, *Education in France,* pp. 8–15; Blaufarb, "The Social Contours of Meritocracy"; Shinn, *L'Ecole Polytechnique,* p. 49; and Smith, *Ecole Normale Supérieure,* p. 25.

133. See especially Theodore Porter, *Trust in Numbers: The Pursuit of Objectivity in Science and Public Life* (Princeton: Princeton University Press, 1995), chap. 6; and Spitzer, *French Generation of 1820,* p. 213.

134. See Shinn, *L'Ecole Polytechnique,* pp. 80–99.

135. See Spitzer, *French Generation of 1820,* p. 215.

136. *Arrêté qui établit un concours entre les élèves de l'Ecole normale qui terminent leurs cours. 12 Juillet 1820,* in Beauchamp, *Recueil des lois et règlements sur l'enseignement supérieur,* vol. 1, p. 444.

137. See, for example, Victor Cousin's un-self-conscious description of secondary education producing the nation's "legitimate aristocracy" in *Défense de l'Université et de la philosophie* (Paris: Joubert, 1844), p. 51. Also Blaufarb, "The Social Contours of Meritocracy"; Jan Goldstein, *Console and Classify: The French Psychiatric Profession in the Nineteenth Century* (Cambridge: Cambridge University Press, 1990), p. 264; and Porter, *Trust in Numbers,* pp. 137–47.

138. Compare this system of minute surveillance with that described by Michel Foucault in *Discipline and Punish: The Birth of the Prison* (New York: Vintage, 1979), pp. 170–228.

139. Although, as Theodore Porter has demonstrated, the single standard at all levels of the French administrative bureaucracy was one defined by expert judgment, not by simple numerical scores.

140. See, for example, Shinn's description of the sense of entitlement of the *polytechnicien.* Shinn, *L'Ecole Polytechnique,* pp. 52–60.

141. Staël-Holstein, *De la littérature considérée dans ses rapports avec les institutions sociales,* pp. 328–29.

142. For biographical information on Cousin, see Emile Bréhier, *The History of Philosophy,* vol. 6, *The Nineteenth Century: Period of Systems, 1800–1850* (Chicago: University of Chicago Press, 1968); George Boas, *French Philosophies of the Romantic Period* (Baltimore: Johns Hopkins University Press, 1925), esp. chap. 5; John H. Randall, *The Career of Philosophy,* vol. 2 (New York: Colum-

bia University Press, 1965), esp. bk. 6; and Jules Simon, *Victor Cousin* (Paris: Hachette, 1891).

143. Spitzer, *French Generation of 1820,* p. 76.

144. On Royer-Collard, see Bréhier, *Nineteenth Century,* pp. 75–78.

145. For a thoughtful discussion of Cousin as cult figure, see Spitzer, *French Generation of 1820,* esp. chap. 3. Spitzer also analyzes the repudiation of Cousin during the Restoration (*French Generation of 1820,* pp. 92–96).

146. As John Randall has noted: "From 1830 to 1848 Cousin ran philosophy in France, with a soupçon of the tyrant. Eclecticism or Spiritualism was the official philosophy of the bourgeois establishment." Randall, *Career of Philosophy,* vol. 2, p. 448. The definitive interpretation of Cousin's philosophy within its cultural setting is Jan Goldstein's *The Post-Revolutionary Self: Politics and Psyche in France, 1750–1850* (Cambridge: Harvard University Press, 2005). Her basic position is staked out in *Console and Classify,* esp. chap. 7; "The Advent of Psychological Modernism in France: An Alternative Narrative," in *Modernist Impulses in the Human Sciences, 1870–1930,* ed. Dorothy Ross (Baltimore: Johns Hopkins University Press, 1994), pp. 190–209, 342–46; "Saying 'I': Victor Cousin, Caroline Angebert, and the Politics of Selfhood in Nineteenth-Century France," in *Rediscovering History: Culture, Politics, and the Psyche,* ed. Michael S. Roth (Stanford: Stanford University Press, 1994), pp. 321–35, 496–99; and esp. "Mutations of the Self in Old Regime and Postrevolutionary France: From *Ame* to *Moi* to *Le Moi,*" in *Biographies of Scientific Objects,* ed. Lorraine Daston (Chicago: University of Chicago Press, 2000), pp. 86–116. Goldstein can usefully be supplemented with R. R. Bolgar, "Victor Cousin and Nineteenth-Century Education," *Cambridge Journal* 2 (1949): 357–68; Doris S. Goldstein, "'Official Philosophies' in Modern France: The Example of Victor Cousin," *Journal of Social History* 1 (1968): 259–79; W. M. Simon, "The 'Two Cultures' in Nineteenth-Century France: Victor Cousin and Auguste Comte," *Journal of the History of Ideas* 26 (1965): 45–58; and Spitzer, *French Generation of 1820.*

147. For more on psychology in France during the nineteenth century, see in addition to the material cited on Cousin, François Azouvi, "Psychologie et physiologie en France, 1800–1830," *History and Philosophy of the Life Sciences* 6 (1984): 151–70; L. S. Jacyna, "The Cultural Context of Localization Debates in Early Nineteenth-Century France," in Edwin Clarke and L. S. Jacyna, *Nineteenth-Century Origins of Neuroscientific Concepts,* (Berkeley: University of California Press, 1987), pp. 267–85; Jacyna, "Medical Science and Moral Science: The Cultural Relations of Physiology in Restoration France," *History of Science* 25 (1987): 111–46; William Logue, *From Philosophy to Sociology: The Evolution of French Liberalism, 1870–1914* (DeKalb: Northern Illinois University Press, 1983), esp. chaps. 2–4; and W. Jay Reedy, "Language, Counter-Revolution and the 'Two Cultures': Bonald's Traditionalist Scientism," *Journal of the History of Ideas* 44 (1983): 579–97.

148. Stephen Jacyna and Jan Goldstein both see a tripartite division in French philosophy during the nineteenth century, and each argues that idéologie and eclecticism formed two of the pieces. Where Goldstein (and François Azouvi as well) argues that the other major contender was phrenology, however, Jacyna contends that it was ultra-Catholic spiritualism. See Azouvi, "Psychologie et physiolo-

gie en France"; Goldstein, "Advent of Psychological Modernism in France"; Goldstein, "Foucault and the Post-Revolutionary Self"; and Jacyna, "Medical Science and Moral Science."

149. On de Maistre and Bonald, see Bréhier, *Nineteenth Century,* chap. 2; Boas, *French Philosophies of the Romantic Period,* esp. chap. 3; Jacyna, "Medical Science and Moral Science," esp. pp. 123–34; Randall, *Career of Philosophy,* vol. 2, bk. 6, chap. 1; and Reedy, "Language, Counter-Revolution, and the 'Two Cultures.'"

150. See Goldstein, "'Official Philosophies' in Modern France," p. 260. For more on Cousin's desire to avoid the problems of idéologie, see Spitzer, *French Generation of 1820,* pp. 84–87.

151. On idéologie and the nature of the mind, see Goldstein, *Console and Classify;* Jacyna, "Medical Science and Moral Science"; Boas, *French Philosophies of the Romantic Period,* esp. chap. 2; Georges Gusdorf, *Les sciences humaines et la pensée occidentale,* vol. 8, *La conscience révolutionnaire: Les idéologues* (Paris: Payot, 1978); Claude Nicolet, *Histoire, nation, république* (Paris: Editions Odile Jacob, 2000), chap. 9; Martin S. Staum, *Cabanis: Enlightenment and Medical Philosophy in the French Revolution* (Princeton: Princeton University Press, 1980); and Staum, *Minerva's Message: Stabilizing the French Revolution* (Montreal: McGill-Queen's University Press, 1996).

152. See Jacyna, "Medical Science and Moral Science," pp. 119–27; Simon, "The 'Two Cultures' in Nineteenth-Century France," pp. 49–50; Goldstein, *Console and Classify,* pp. 257–63; and Goldstein, "Mutations of the Self in Old Regime and Postrevolutionary France," pp. 101–109.

153. Emile Bréhier points out, following Cousin's own analogy, that "eclecticism is like a representative government which reconciles all the diverse elements in society." Bréhier, *Nineteenth Century,* p. 83. See also Victor Cousin, "Treizième Leçon," in *Introduction á l'histoire de la philosophie* (1828; reprint, Paris: Didier et Cie, 1861), pp. 279–309.

154. Cousin, *Introduction á l'histoire de la philosophie,* p. 290.

155. Ibid., *Lectures on the True, the Beautiful, and the Good* (1818; reprint, New York: D. Appleton, 1875), pp. 9–10.

156. On this point, see Spitzer, *French Generation of 1820,* p. 88. On the split between "psychology" and "ideology," see Azouvi, "Psychologie et physiologie en France"; and Goldstein, *Console and Classify,* p. 257. On Cousin's attempt to merge empirical observation with a spiritualist metaphysics, see Bréhier, *Nineteenth Century,* pp. 85–87; Spitzer, *French Generation of 1820,* pp. 81–84.

157. Cousin, *Lecture,* p. 245.

158. Ibid., *Elements of Psychology: Included in A Critical Examination of Locke's Essay on the Human Understanding,* trans. C. S. Henry (New York: Dayton & Saxton, 1842), p. 160.

159. Ibid., *Introduction á l'histoire de la philosophie,* p. 111.

160. See esp. Jan Goldstein, "Saying 'I': Victor Cousin, Caroline Angebert, and the Politics of Selfhood in Nineteenth-Century France," in *Rediscovering History: Culture, Politics, and the Psyche,* ed. Michael S. Roth (Stanford: Stanford University Press, 1994), pp. 321–35.

161. See Cousin, *Introduction á l'histoire de la philosophie,* pp. 203, 93.

162. Doris Goldstein has described in detail Cousin's success in gathering power within the French educational system in "'Official Philosophies' in Modern France." Jan Goldstein has argued, along similar lines, that Cousin's imposing presence in mainstream French philosophy can be detected at least until the end of the century. See Jan Goldstein, "Advent of Psychological Modernism," pp. 198–206.

163. See, for example, Théodore Jouffroy, "On the Faculties of the Human Soul" (1828), in *Philosophical Miscellanies, Translated from the French of Cousin, Jouffroy, and B. Constant,* ed. George Ripley, vol. 1 (Boston: Hilliard, Gray, 1838).

164. L. E. Bautain, *Psychologie expérimentale,* 2 vols. (Paris: Lagny Frères, 1839), vol. 2, pp. 382–83.

165. Cousin, *Introduction á l'histoire de la philosophie,* pp. 9, 307.

166. See Reedy, "Language, Counter-Revolution and the Two Cultures," esp. p. 592, and for general background, Harry W. Paul, *The Edge of Contingency: French Catholic Reaction to Scientific Change from Darwin to Duhem* (Gainesville: University of Florida Press, 1979).

167. For a comprehensive account of the Class of Moral and Political Sciences, see Staum, *Minerva's Message.*

168. On Cabanis, see Gusdorf, *La conscience révolutionnaire;* François Picavet, *Les idéologues: Essai sur l'histoire des idées et des théories scientifiques, philosophiques, religieuse etc. en France depuis 1789* (Paris: Alcan, 1894); Robert J. Richards, *Darwin and the Emergence of Evolutionary Theories of Mind and Behavior* (Chicago: University of Chicago Press, 1987), pp. 40–45; and esp. Staum, *Cabanis.*

169. Pierre-Jean-Georges Cabanis, *On the Relations between the Physical and Moral Aspects of Man,* 2 vols.; 2nd ed. (An. XIII [1805]; reprint, Baltimore: Johns Hopkins University Press, 1981), vol. 1, p. 10 (capitals in original); vol. 1, p. 50; vol. 2, pp. 568–76; vol. 1, pp. 60, 116, 141, 280–81; vol. 2., p. 671.

170. See Staum, *Minerva's Message,* chap. 12.

171. On physiology in France during the nineteenth century, see John E. Lesch, *Science and Medicine in France: The Emergence of Experimental Physiology, 1790–1855* (Cambridge: Harvard University Press, 1984); and Russell C. Maulitz, *Morbid Appearances: The Anatomy of Pathology in the Early Nineteenth Century* (Cambridge: Cambridge University Press, 1987), chaps. 1–4. On physiological approaches to the brain, the most important sources are Clarke and Jacyna, *Nineteenth-Century Origins of Neuroscientific Concepts;* and Robert M. Young, *Mind, Brain and Adaptation in the Nineteenth Century: Cerebral Localization and Its Biological Context from Gall to Ferrier* (Oxford: Clarendon Press, 1970). For a good introduction to the rich literature on nineteenth-century French medical psychiatry/psychology, see Ian R. Dowbiggin, *Inheriting Madness: Professionalization and Psychiatric Knowledge in Nineteenth-Century France* (Berkeley: University of California Press, 1991); Goldstein, *Console and Classify;* and Dora B. Weiner, *The Citizen-Patient in Revolutionary and Imperial Paris* (Baltimore: Johns Hopkins University Press, 1993).

172. See esp. Jacyna's account in "Cultural Context of the Localization Debates in Early Nineteenth-Century France." It can be supplemented with Goldstein, "Advent of Psychological Modernism in France"; and Georges Lanteri-Laura,

Histoire de la phrénologie: L'homme et son cerveau selon F. J. Gall (Paris: Presses universitaires de France, 1970).

Chapter 3

1. Benjamin Rush, "Account of a Wonderful Talent for Arithmetical Calculation, in an African Slave, Living in Virginia," *American Museum* 5 (1789): 62; reprinted in John Flauvel and Paulus Gerdes, "African Slave and Calculating Prodigy: Bicentenary of the Death of Thomas Fuller," *Historia Mathematica* 17 (1990): 141–51, quote on p. 144.

2. For an early abolitionist argument promoting human equality, see Benjamin Franklin, "Petition from Pennsylvania Abolition Society to Congress" (1790), in *Race and Revolution,* ed. Gary B. Nash (Madison: Madison House, 1990), pp. 144–45. On transatlantic calls for emancipation, see David Brion Davis, *The Problem of Slavery in the Age of Revolution, 1770–1823* (Ithaca: Cornell University Press, 1975).

3. Barbara J. Fields, "Ideology and Race in American History," in *Region, Race, and Reconstruction: Essays in Honor of C. Vann Woodward,* eds. J. Morgan Kousser and James M. McPherson (New York: Oxford University Press, 1982), pp. 143–77, quotes on p. 152.

4. See Eugen Weber, *Peasants into Frenchmen: The Modernization of Rural France, 1870–1914* (Stanford: Stanford University Press, 1976).

5. On this rival discourse in arguments about American slavery, see Elizabeth B. Clark, "'The Sacred Rights of the Weak': Pain, Sympathy, and the Culture of Individual Rights in Antebellum America," *Journal of American History* 82 (1995): 463–93.

6. See *Trésor de la Langue Française,* vol. 10 (Paris: Editions du CNRS, 1983), pp. 385–86; and *Le Dictionnaire de l'Académie françoise, dédié au roi,* vol. 1 (Paris: Jean Baptiste Coignard, 1694), p. 601.

7. *Dictionnaire de l'Académie françoise, cinquième édition,* vol. 1 (Paris: J. J. Smits et Cie, An VII [1798]), p. 738.

8. See the *Dictionnaire de l'Académie française, sixième édition,* vol. 2 (Paris: Didot Frères, 1835).

9. In addition to Johnson's *Dictionary,* I have consulted Noah Webster, *An American Dictionary of the English Language* (1828); Webster, *An American Dictionary of the English Language,* revised 2nd ed. (1841); *Johnson's English Dictionary, as Improved by Todd and Abridged by Chalmers . . .* (1844); Chauncey A. Goodrich, *An American Dictionary of the English Language . . . by Noah Webster, LL.D. . . . Revised and Enlarged,* 3rd edition of Webster's (1856); Alexander Reid, *A Dictionary of the English Language . . .* (1857); Joseph E. Worcester, *A Dictionary of the English Language* (1860); Noah Porter, ed., *An American Dictionary of the English Language, Unabridged,* 5th ed. of Webster's (1864); William A. Wheeler, *A Dictionary of the English Language . . . of Noah Webster, LL.D., as Revised by Chauncey A. Goodrich, D.D., and Noah Porter, D.D.* (1872); and Robert Gordon Latham, *A Dictionary of the English Language. Founded on that of Dr. Samuel Johnson as Edited by the Rev. H. J. Todd, M.A. with Numerous Emendations and Additions* (1876).

10. Samuel Johnson, *A Dictionary of the English Language* (1755), ed. Anne McDermott (http://ets.umdl.umich.edu/j/johnson/: Cambridge University Press, 1966).

11. *Oxford English Dictionary,* 2nd ed. (Oxford: Clarendon Press, 1989), p. 1069; and Webster, *American Dictionary of the English Language,* revised 2nd ed. (New York: White & Sheffield, 1841)

12. *Dictionnaire de l'Académie française, sixième édition,* vol. 2.

13. On Linnaeus's work, see Lisbet Koerner, *Linnaeus: Nature and Nation* (Cambridge: Harvard University Press, 1999).

14. Nicholas Hudson makes a similar point about the evolution in the meaning of "race" in "From 'Nation' to 'Race': The Origin of Racial Classification in Eighteenth–Century Thought," *Eighteenth-Century Studies* 29 (1996): 247–64. See also Philip Sloan, "The Gaze of Natural History," in *Inventing Human Science: Eighteenth-Century Domains,* eds. Christopher Fox, Roy Porter, and Robert Wokler (Berkeley: University of California Press, 1995), pp. 112–51; and Jacques Roger, *The Life Sciences in Eighteenth-Century French Thought* (Stanford: Stanford University Press, 1997).

15. For an extraordinary account of eighteenth- and nineteenth-century theories of mind, see Robert J. Richards, *Darwin and the Emergence of Evolutionary Theories of Mind and Behavior* (Chicago: University of Chicago Press, 1987), esp. chap. 1. This section is highly indebted to his work. See also Richards, "Influence of Sensationalist Tradition on Early Theories of the Evolution of Behavior," *Journal of the History of Ideas* 40 (1979): 85–105; and Elizabeth A. Williams, *The Physical and the Moral: Anthropology, Physiology, and Philosophical Medicine in France, 1750–1850* (New York: Cambridge University Press, 1994), pp. 188–213.

16. See Etienne Bonnot de Condillac, *Traité des animaux* (1755; reprint, Paris: Vrin, 1981); and also Richards, "Influence of Sensationalist Tradition."

17. Charles-Georges Le Roi, "Instinct des animaux," in *Encyclopédie méthodique. Philosophie ancienne et moderne,* vol 3 (Paris: Agasse, An 2 [1795]), p.16. See also Le Roi, "Instinct," in *Encyclopédie ou dictionnaire raisonné des sciences, des arts et des métiers,* ed. Denis Diderot (Paris: Faulch, 1751–65), pp. 795–99; and *Lettres philosophiques sur l'intelligence et la perfectibilité des animaux, avec quelques lettres sur l'homme* (1768, rev. 1802), in *Studies on Voltaire and the Eighteenth Century* 316 (1994): 75–210.

18. John Bascom, "Instinct," *The Bibliotheca Sacra and Theological Eclectic* 28 (1871): 654–85, quote on pp. 662–63. On Bascom and studies of animal intellect in America, see Timothy D. Johnston, "Three Pioneers of Comparative Psychology in America, 1843–1890: Lewis H. Morgan, John Bascom, and Joseph Le Conte," *History of Psychology* 6 (2003): 14–51.

19. Richards, *Darwin and the Emergence of Evolutionary Theories,* p. 67.

20. Jean-Baptiste Lamarck, *Recherches sur l'organisation des corps vivans* [1802], as quoted and translated in Richards, *Darwin and the Emergence of Evolutionary Theories,* p. 53.

21. Lewis Henry Morgan, "Mind or Instinct: An Inquiry Concerning the Manifestations of Mind by the Lower Animals," *Knickerbocker* 22 (1843): 414–20, 507–15, quote on p. 515.

22. Charles Darwin, *The Descent of Man and Selection in Relation to Sex* (1871; reprint, Princeton: Princeton University Press, 1981), p. 35. See also Richards, *Darwin and the Emergence of Evolutionary Theories,* esp. chaps. 2–5;

and Johnston, "Three Pioneers of Comparative Psychology in America." Robert Boakes and other scholars have seen Darwin as much more revolutionary in regard to animal mentality. See Boakes, *From Darwin to Behaviorism: The Psychology of Animals* (Cambridge: Cambridge University Press, 1984), esp. chap. 1.

23. Although Lewis Henry Morgan argued against the "instinct hypothesis" well into the latter half of the nineteenth century. See, for example, [Morgan], "Chadbourne on Instinct," *Nation* 14 (1872): 291–92. On the sensationist position on instinct, see Le Roi, "Instinct," p. 796. On the later nineteenth-century debate, see Boakes, *From Darwin to Behaviorism,* chap. 2. On Morgan, see Thomas R. Trautmann, *Lewis Henry Morgan and the Invention of Kinship* (Berkeley: University of California Press, 1987).

24. Joseph Le Conte, "Instinct and Intelligence," *Popular Science Monthly* 7 (1875): 653–64, quote on p. 659.

25. Darwin, *Descent of Man,* p. 37. For an analysis of Darwin's account of instincts, see Richards, *Darwin and the Emergence of Evolutionary Theories,* chap. 2.

26. Darwin, *Descent of Man,* p. 105.

27. Samuel G. Morton, *Crania Americana, or a Comparative View of the Skulls of Various Aboriginal Nations of North and South America, to Which is Prefixed an Essay on the Varieties of the Human Species* (Philadelphia: Pennington, 1839).

28. On the American school of anthropology, see Mia Bay, *The White Image in the Black Mind: African-American Ideas about White People, 1830–1925* (New York: Oxford University Press, 2000), chaps. 1–2; Bruce Dain, *A Hideous Monster of the Mind: American Race Theory in the Early Republic* (Cambridge: Harvard University Press, 2002); George M. Fredrickson, *The Black Image in the White Mind: The Debate on Afro-American Destiny, 1817–1914* (New York: Harper & Row, 1971), chaps. 1–5; Thomas F. Gossett, *Race: The History of an Idea in America* (Dallas: Southern Methodist University Press, 1963), chap. 4; Stephen Jay Gould, *The Mismeasure of Man* (New York: W. W. Norton, 1981), chap. 2; Winthrop D. Jordan, *White Over Black: American Attitudes toward the Negro, 1550–1812* (New York: W. W. Norton, 1977), esp. chaps. 1, 6, 12–16; and William Stanton, *The Leopard's Spots: Scientific Attitudes Toward Race in America, 1815–1859* (Chicago: University of Chicago Press, 1960). For details about Nott's life and career, see Reginald Horsman, *Josiah Nott of Mobile: Southerner, Physician, and Racial Theorist* (Baton Rouge: Louisiana State University Press, 1987). On Agassiz and race, see Edward Lurie, "Louis Agassiz and the Races of Man," *Isis* 45 (1954): 227–42.

29. On the mixed success of the American School in dominating anthropology in the United States, see Curtis M. Hinsley, *The Smithsonian and the American Indian: Making a Moral Anthropology in Victorian America* (Washington, D.C.: Smithsonian Institution Press, 1981), esp. chap. 1. See also Stanton, *Leopard's Spots;* and Fredrickson, *Black Image in the White Mind,* esp. chap. 3.

30. On the African American response, see Bay, *White Image in the Black Mind,* chap. 2; and Dain, *Hideous Monster of the Mind,* chap. 7.

31. George Fredrickson, however, contends that polygenist arguments were more widely adopted by southern defenders of slavery than Stanton suggests. See Fredrickson, *Black Image in the White Mind.* And for a fascinating attempt to reconcile orthodoxy and polygenism, see William Frederick Van Amringe, *An In-*

vestigation of the Theories of the Natural History of Man, By Lawrence, Prichard, and Others . . . (New York: Baker & Scribner, 1848).

32. G., "The Unity of the Race Question," *National Era* 8 (August 17, 1854).

33. Frederick Douglass, "The Claims of the Negro Ethnologically Considered" (1854), in *Life and Writings of Frederick Douglass,* ed. Philip S. Foner, vol. 2 (New York: International Publishers, 1950), pp. 289–309, 295. See also Bay, *White Image in the Black Mind,* pp. 66–71.

34. See Jordan, *White Over Black,* esp. chaps. 1, 6, and 12–16. Also Richard H. Popkin, "Medicine, Racism, Anti-Semitism: A Dimension of Enlightenment Culture," in *The Languages of Psyche: Mind and Body in Enlightenment Thought,* ed. George S. Rousseau (Berkeley: University of California Press, 1991), pp. 405–42; Londa Schiebinger, *Nature's Body: Gender in the Making of Modern Science* (Boston: Beacon Press, 1993), chap. 4; and Robert Wokler, "Anthropology and Conjectural History in the Enlightenment," in *Inventing Human Science,* pp. 31–52.

35. On this general project, see Thomas L. Hankins, *Science and the Enlightenment* (Cambridge: Cambridge University Press, 1985); Koerner, *Linneaus;* and Sloan, "Gaze of Natural History."

36. Linneaus's 1758 classification of *Homo sapiens* listed seven varieties, the four mentioned plus Wild Man, Monstrous, and Troglodytes. Blumenbach proposed four varieties in 1775, and five in 1781. By 1795 he denominated them Caucasian, American, Mongolian, Malay, and Ethiopian. See Jordan, *White Over Black,* pp. 218–23.

37. In addition to Jordan, important studies of early nineteenth-century racial science include Anthony J. Barker, *The African Link: British Attitudes to the Negro in the Era of the Atlantic Slave Trade, 1550–1807* (London: Cass, 1978); Bay, *White Image in the Black Mind,* chap. 1; Fredrickson, *Black Image in the White Mind,* chaps. 1–3; Gossett, *Race,* esp. chaps. 3–4; Winthrop D. Jordan, introduction to Samuel Stanhope Smith, *An Essay on the Causes of the Variety of Complexion and Figure in the Human Species,* ed. Winthrop D. Jordan (Cambridge: Harvard University Press, 1965), pp. vii–liii; Nancy Leys Stepan, *The Idea of Race in Science: Great Britain, 1800–1960* (Hamden: Archon Books, 1982); and William H. Tucker, *The Science and Politics of Racial Research* (Urbana: University of Illinois Press, 1994), chap. 1.

38. Environmentalists argued that physical characteristics demarcating race resulted from environmental conditions, specifically climate, and could alter; essentialists disputed this, contending that racial characteristics were permanent and unvarying. See John Bachman, *The Doctrine of the Unity of the Human Race Examined on the Principles of Science* (Charleston: C. Canning, 1850); Smith, *Essay on the Causes;* and Charles White, *An Account of the Regular Gradation in Man, and in Different Animals and Vegetables; and from the Former to the Latter* (London: C. Dilly, 1799). For a reminder of the continuing importance of skin color, esp. in popular culture, see Charles D. Martin, *The White African American Body: A Cultural and Literary Exploration* (New Brunswick: Rutgers University Press, 2002), esp. chap. 2.

39. On craniometry in general, see Claude Blanckaert, "'Les vicissitudes de l'angle facial,' et les débuts de la craniométrie (1765–1875)," *Revue de synthèse* 108 (1987): 417–53; Gould, *Mismeasure of Man,* chap. 3; Stanton, *Leopard's Spots;* and George W. Stocking, "The Persistence of Polygenist Thought in Post-

Darwinian Anthropology," in *Race, Culture, and Evolution* (Chicago: University of Chicago Press, 1968), pp. 42–68. For important analyses of the gendered nature of this program, see Schiebinger, *Nature's Body,* chap. 5; and Elizabeth Fee, "Nineteenth-Century Craniology: The Study of the Female Skull," *Bulletin of the History of Medicine* 53 (1978): 415–33.

40. For more on the facial angle and other early measures, see Blanckaert, "'Les vicissitudes de l'angle facial.'" For Camper's reception, or rather nonreception in England, see Barker, *African Link.*

41. See Peter Camper, *The Works of the Late Professor Camper, on the Connexion Between the Science of Anatomy and the Arts of Drawing, Painting, Statuary, etc.,* trans. T. Cogan (London: C. Dilly, 1794).

42. On Camper, see Miriam C. Meijer, *Race and Aesthetics in the Anthropology of Petrus Camper (1722–1789)* (Amsterdam: Rodopi, 1999).

43. See Johann Friedrich Blumenbach, "On the Natural Variety of Mankind" (1795), in *The Anthropological Treatises of Johann Friedrich Blumenbach,* ed. and trans. Thomas Bendyshe (London: Longman, Green, Longman, Roberts, & Green, 1865), p. 193. Blumenbach specifically critiques Camper on pp. 235–36 and describes the five principal varieties of mankind on pp. 265–76.

44. Georges Cuvier and Etienne Geoffroy Saint-Hilaire, "Histoire naturelle des Orangs-Outangs," *Magasin encyclopédique* 3 (1795): 451–63. See also Blanckaert, "'Les vicissitudes de l'angle facial'"; William Coleman, *Georges Cuvier, Zoologist: A Study in the History of Evolution Theory* (Cambridge: Harvard University Press, 1964); and George W. Stocking, "French Anthropology in 1800," in *Race, Culture, and Evolution,* pp. 13–41, 313–19.

45. Cuvier and Geoffroy, "Histoire naturelle des Orangs-Outangs," p. 457.

46. For the relations of these theories to the issue of race, see Gossett, *Race,* esp. chap. 4.

47. Although, as Curtis Hinsley shows, by no means the only important research area in the field. See Hinsley, *Smithsonian and the American Indian.*

48. The merits of the facial angle continued to be debated throughout the nineteenth century. See, for example, Ransom Dexter, "The Facial Angle," *Popular Science Monthly* 4 (1874): 587–92.

49. For more details on Morton's methods, see John Carson, "Minding Matter/Mattering Mind: Knowledge and the Subject in Nineteenth–Century Psychology," *Studies in the History and Philosophy of the Biological and Biomedical Sciences* 30 (1999): 345–76.

50. Morton, *Crania Americana,* p. 260.

51. See Gould, *Mismeasure of Man,* chap. 2; and, for Morton's new table, Josiah C. Nott and George R. Gliddon, *Types of Mankind or, Ethnological Researches Based Upon the Ancient Monuments, Paintings, Sculptures, and Crania of Races, and Upon Their Natural, Geographical, Philological, and Biblical History* (Philadelphia: Lippincott, 1854), p. 450.

52. Josiah Nott, "Comparative Anatomy of Races," in Nott and Gliddon, *Types of Mankind,* p. 457.

53. George Combe, "Phrenological Remarks on the Relation Between the Natural Talents and Dispositions of Nations, and the Developments of Their Brains," in Morton, *Crania Americana,* p. 277.

54. James Cowles Prichard, "Review of *Crania Americana* . . ." *Journal of the Royal Geographical Society* 10 (1840): 552–61, quote on p. 561. See also Jeffries Wyman, "Review of *Crania Americana* . . ." *North American Review* 51 (1840): 173–86, quote on p. 173; and Ephraim George Squier, "American Ethnology: Being a Summary of Some of the Results Which Have Followed the Investigation of this Subject," *American Review,* n.s., 3 (1849): 385–98, quote on p. 387.

55. The primary issue debated with Morton was not the quality of his data, but its meaning. See, for example, Bachman, *Doctrine of the Unity of the Human Race;* and Daniel Wilson, "Physical Ethnology," *Annual Report of the Board of Regents of the Smithsonian Institution for 1862* (1863): 240–302.

56. On cerebral localization, see Robert M. Young, *Mind, Brain, and Adaptation in the Nineteenth Century: Cerebral Localization and Its Biological Context from Gall to Ferrier* (New York: Oxford University Press, 1990), esp. chaps. 1–2.

57. Nott, *Types of Mankind,* p. 463.

58. Morton, *Crania Americana,* p. 99.

59. The classic history of the concept of the chain of being is Arthur O. Lovejoy's *The Great Chain of Being: A Study of the History of an Idea* (1936; reprint Cambridge: Harvard University Press, 1973). The best study of its application to issues of race in America is Jordan, *White Over Black,* esp. chap. 13. See also Barker, *African Link,* William F. Bynum, "The Great Chain of Being after Forty Years: An Appraisal," *History of Science* 13 (1975): 1–28; and Stepan, *Idea of Race in Science,* esp. chap. 1.

60. See Stepan, *Idea of Race in Science,* p. 18; and Jordan, *White Over Black.*

61. On this point, see Lovejoy, *Great Chain of Being,* p. 234. Stepan emphasizes the importance of Cuvier in this change in *Idea of Race in Science,* pp. 13–19. Stepan also points out Blumenbach's and Tiedemann's resistance to reducing the gulf between humans and other animals on pp. 9–10, 17.

62. Julien Joseph Virey, *Histoire naturelle du genre humain* (An XI, 1801), vol. 1 (Paris: Crochard, 1824), pp. 45–46. Virey is praised extensively in an 1850 article, "Unity of the Human Race," *American Whig Review* 12 (1850): 567–86.

63. See, for example, Frank E. Manuel, "From Equality to Organicism," *Journal of the History of Ideas* 17 (1956): 54–69; Stocking, "French Anthropology in 1800"; and George W. Stocking, *Victorian Anthropology* (New York: Free Press, 1987), esp. pp. 224–37.

64. On this issue, see Barker, *African Link;* Popkin, "Medicine, Racism, Anti-Semitism"; and Stocking, "The Idea of Civilization before the Crystal Palace (1750–1850)," in *Victorian Anthropology,* pp. 8–45.

65. For an excellent example of the use of these dual criteria, see the ethnological sections of Morton, *Crania Americana.*

66. Smith, *Essay on the Causes,* p. 119.

67. Morton, *Crania Americana,* p. 89.

68. Ibid., p. 81.

69. Louis Agassiz, "The Diversity of Origin of the Human Races," *Christian Examiner and Religious Miscellany* 49 (1850): 110–45, quote on p. 142.

70. Morton, *Crania Americana,* pp. 149, 151.

71. Although Van Amringe's depiction of multiple differences between the races while still speaking of hierarchies is something of an exception. See Van Amringe,

Investigation of the Theories of the Natural History of Man, esp. chap. 9. See also Hinsley, *Smithsonian and the American Indian.*

72. Georges-Louis Leclerc, Comte de Buffon, "Excerpt from *Natural History, General and Particular*" (1749), in *Was America a Mistake? An Eighteenth-Century Controversy,* eds. Henry Steele Commager and Elmo Giordanetti (Columbia: University of South Carolina Press, 1967), pp. 51–74.

73. Abbé Corneille De Pauw, "From the Philosophical Investigations of the Americans" (1768), in *Was America a Mistake?,* p. 94.

74. Thomas Jefferson, *Notes on the State of Virginia* (1781–85), in *The Portable Thomas Jefferson,* ed. Merrill D. Peterson (New York: Penguin Books, 1977), pp. 94–96.

75. Ibid., pp. 98–99, 188–89. For more of Jefferson's views on African Americans, see pp. 187–93.

76. Londa Schiebinger argues that similar shifts in terms of both race and gender were occurring in Europe at the same time. See Schiebinger, "The Anatomy of Difference: Race and Sex in Eighteenth-Century Science," *Eighteenth-Century Studies* 23 (1990): 387–405. For an African American response to Jefferson, see James M'Cune Smith, "On the Fourteenth Query of Thomas Jefferson's Notes on Virginia," *Anglo-African Magazine* 1 (1859): 225–38.

77. See, for example, Benjamin Banneker's letter to Jefferson, "A Letter from Benjamin Banneker to the Secretary of State" (1792), in *Race and Revolution,* ed. Gary B. Nash (Madison: Madison House, 1990), pp. 177–81.

78. Clement C. Moore, *Observations upon Certain Passages in Mr. Jefferson's Notes on Virginia* (New York: n.p., 1804).

79. Smith, *Essay on the Causes.*

80. See Bruce Dain, "Haiti and Egypt in Early Black Racial Discourse in the United States," *Slavery and Abolition* 13 (1993): 139–61.

81. John C. Calhoun, "Speech in the U. S. Senate," in *Defending Slavery: Proslavery Thought in the Old South,* ed. Paul Finkelman (Boston: Bedford/St. Martin's, 2003), p. 58.

82. [John B. Russwurm], "On the Mutability of Human Affairs," *Freedom's Journal,* April 6, 1827, p. 3.

83. David Walker, *An Appeal . . . to the Colored Citizens of the World* (1829), in "*One Continual Cry*"*: David Walker's Appeal to the Colored Citizens of the World (1829–1830),* ed. Herbert Aptheker (New York: Humanities Press, 1965), esp. pp. 82–92; and Hosea Easton, *Treatise on the Intellectual Character, and Civil and Political Condition of the Colored People of the U. States and the Prejudice Exercised Towards Them, with a Sermon on the Duty of the Church to Them* (Boston: I. Knapp, 1837). On Easton, see James Brewer Stewart and George R. Price, "Introduction: Hosea Easton and the Agony of Race," in *To Heal the Scourge of Prejudice: The Life and Writings of Hosea Easton* (Amherst: University of Massachusetts Press, 1999), pp. 1–47; and Dain, *Hideous Monster of the Mind,* chap. 6. See also Martin R. Delany, *Principia of Ethnology: The Origin of the Races and Color, with an Archaeological Compendium of Ethiopian and Egyptian Civilization from Years of Careful Examination and Enquiry* (Philadelphia: Harper & Brother, 1879).

84. Easton, *Treatise on the Intellectual Character,* p. 10.

85. George Fredrickson has argued that there was also a group of northern "romantic racialists," such as Alexander Kinmont, who celebrated the differences between the different racial groups, but without concluding that African Americans were inferior. See Fredrickson, *Black Image in the White Mind,* chap. 4; and Alexander Kinmont, *Twelve Lectures on the Natural History of Man and the Rise and Progress of Philosophy* (Cincinnati: U. P. James, 1839).

86. Bachman, *Doctrine of the Unity of the Human Race,* p. 9. James M'Cune Smith also argued vigorously for environmentalism and the unity of the human race in "Civilization: Its Dependence on Physical Circumstances," *Anglo-African Magazine* 1 (1859): 5–17.

87. John B. Russwurm, "On the Varieties of the Human Race," *Freedom's Journal,* April, May, September 1828.

88. Easton, *Treatise on the Intellectual Character,* pp. 6, 8.

89. Agassiz, "Diversity of Origin of the Human Races," pp. 113, 143.

90. Samuel A. Cartwright, "Report on the Diseases and Physical Peculiarities of the Negro Race," *New Orleans Medical and Surgical Journal* 7 (1851): 691–715, quotes on pp. 692 and 693.

91. Thomas R. R. Cobb, *An Inquiry into the Law of Negro Slavery in the United States of America, to Which is Prefixed an Historical Sketch of Slavery* (1858), in *Defending Slavery,* p. 151.

92. James W. C. Pennington, *A Text Book of the Origin and History, &c. &c. of the Colored People* (Hartford: L. Skinner, 1841), pp. 45, 54 (emphasis in original).

93. Friedrich Tiedemann, "On the Brain of the Negro, Compared with that of the European and the Ourang Outang," *Philosophical Transactions* 126 (1836): 497–527.

94. "Intellectual Faculties of the Negro," *Colored American,* August 22, 1840.

95. See, for example, "Brain Weight and Mental Power," *Popular Science Monthly* 9 (1876): 254.

96. Agassiz, "Diversity of Origin of the Human Races," p. 144.

97. See Bay, *White Image in the Black Mind,* pp. 42–44; Dain, *Hideous Monster of the Mind,* chap. 8; Davis, *Problem of Slavery in the Age of Revolution;* Fields, "Ideology and Race in American History"; and Fredrickson, *Black Image in the White Mind.* On the claims made by white abolitionists, see Paul Goodman, *Of One Blood: Abolitionism and the Origins of Racial Equality* (Berkeley: University of California Press, 1998), esp. chaps. 11, 12, and 15; and for African Americans' use of the language of natural rights, see Patrick Rael, *Black Identity & Black Protest in the Antebellum North* (Chapel Hill: University of North Carolina Press, 2002), pp. 255–61.

98. See Cobb, *Inquiry into the Law of Negro Slavery,* pp. 143–44.

99. As quoted in "The True Foundation of Slavery," *National Era* 7 (January 27, 1853).

100. Cartwright, "Report on the Diseases," p. 715. However, for a much different understanding of the "permanence" of race, see Ralph Waldo Emerson, *English Traits* (1856) (Cambridge: Harvard University Press, 1966), esp. chap. 4.

101. John H. Van Evrie, *Negroes and Negro "Slavery": The First an Inferior Race; The Latter Its Normal Condition* (New York: Van Evrie and Horton, 1861). For a favorable review of Van Evrie, see George Fitzhugh, "The Black and White Races of Men," *DeBow's Review* 30 (1861): 446–56.

102. See Douglass, "Claims of the Negro Ethnologically Considered," p. 296.

103. See James W. Cook, *The Arts of Deception: Playing with Fraud in the Age of Barnum* (Cambridge: Harvard University Press, 2001); Eric Lott, *Love and Theft: Blackface Minstrelsy and the American Working Class* (New York: Oxford University Press, 1995); Martin, *White African American Body,* chap. 2; and David R. Roediger, *The Wages of Whiteness: Race and the Making of the American Working Class* (New York: Verso, 1991).

104. Patrick Rael elaborates on the employment of elements of racial essentialism in the arguments of African American intellectuals in *Black Identity & Black Protest,* pp. 249–54.

105. For example, during the 1866 congressional debates about suffrage, one senator asserted that "Camper, Soemmering, Lawrence, Virey, Ebel, and Blumenbach agree that the brain of the negro is smaller; and Gall, Spurzheim, and Combe, that it is so distributed as to denote less capacity for reasoning and judging than the Caucasian." As quoted in Jacob Katz Cogan, "The Look Within: Property, Capacity, and Suffrage in Nineteenth-Century America," *Yale Law Journal* 107 (1997): 473.

106. See Terry N. Clark, *Prophets and Patrons: The French University and the Emergence of the Social Sciences* (Cambridge: Harvard University Press, 1973), pp. 117–18.

107. Jennifer Hecht has demonstrated that Manouvrier, in particular, consistently opposed using anthropology to establish the inequality of the races. See Jennifer Michael Hecht, "A Vigilant Anthropologist: Léonce Manouvrier and the Disappearing Numbers," *Journal of the History of the Behavioral Sciences* 33 (1997): 221–40. More generally on the left politics of the members of the Société d'Anthropologie, see Linda L. Clark, *Social Darwinism in France* (Tuscaloosa: University of Alabama Press, 1984), chap. 9.

108. See, for example, Paul de Jouvencel, "Reprise de la discussion sur le volume et la forme du cerveau," *Bulletins de la Société d'anthropologie de Paris* 2 (1861): 283.

109. "L'anthropologie pourrait, à la rigueur, être définie l'*histoire naturelle de l'homme.*" Paul Broca, "Anthropologie," *Dictionnaire encyclopédique des Sciences médicales* (1866), in *Mémoires d'anthropologie,* vol. 1 (Paris: C. Reinwald, 1871), p. 3 (emphasis in original). See also Herbert H. Odom, "Generalizations on Race in Nineteenth-Century Physical Anthropology," *Isis* 58 (1967): 5–18.

110. On Broca, see Clark, *Prophets and Patrons,* esp. pp. 117–20; Yvette Conroy, *L'Introduction du darwinisme en France au XIXe siècle* (Paris: J. Vrin, 1974), pp. 51–89; Gould, *Mismeasure of Man,* esp. chap. 3; and Francis Schiller, *Paul Broca: Founder of French Anthropology, Explorer of the Brain* (Berkeley: University of California Press, 1979).

111. For a fuller description, see Clark, *Social Darwinism in France,* chap. 9; Patricia M. E. Lorcin, *Imperial Identities: Stereotyping, Prejudice and Race in Colonial Algeria* (London: I. B. Tauris, 1995), esp. chap. 7; Schiller, *Paul Broca,* pp. 129–35; and Williams, *Physical and the Moral,* pp. 256–72.

112. For additional general information on the history of nineteenth-century French anthropology, see Donald Bender, "The Development of French Anthropology," *Journal of the History of the Behavioral Sciences* 1 (1965): 139–51; Claude Blanckaert, "On the Origins of French Ethnology: William Edwards and the Doctrine of Race," in *Bones, Bodies, Behavior: Essays on Biological Anthro-*

pology, ed. George W. Stocking (Madison: University of Wisconsin Press, 1988), pp. 18–55; William B. Cohen, *The French Encounter with Africans* (Bloomington: Indiana University Press, 1980), pp. 210–62; Michael Hammond, "Anthropology as a Weapon of Social Combat in Late-Nineteenth-Century France," *Journal of the History of the Behavioral Sciences* 16 (1980): 118–32; Joy Harvey, "Evolutionism Transformed: Positivists and Materialists in the *Société D'Anthropologie de Paris* from Second Empire to Third Republic," in *The Wider Domain of Evolutionary Thought,* eds. David Oldroyd and Ian Langham (Dordrecht: D. Reidel, 1983), pp. 289–310; Angèle Kremer-Marietti, "L'Anthropologie physique et morale en France et ses implications idéologiques," in *Histoires de l'anthropologie: XVI–XIX siècles,* ed. Britta Rupp-Eisenreich (Paris: Klincksieck, 1984), pp. 319–52; Robert A. Nye, *The Origins of Crowd Psychology: Gustave LeBon and the Crisis of Mass Democracy in the Third Republic* (London: Sage, 1975), esp. chaps. 1–3; Stocking, "French Anthropology in 1800"; Elizabeth A. Williams, "Anthropological Institutions in Nineteenth-Century France," *Isis* 76 (1985): 331–48; and Williams, *Physical and the Moral.*

113. Harvey, "Evolutionism Transformed," p. 291. See also Michael A. Osborne, *Nature, the Exotic, and the Science of French Colonialism* (Bloomington: Indiana University Press, 1994), pp. 83–90.

114. See particularly Claude Blanckaert, "L'esclavage des Noirs et l'ethnographie américaine: Le point de vue de Paul Broca en 1858," *Nature, histoire, société: Essais en hommage à Jacques Roger,* eds. Claude Blanckaert, Jean-Louis Fischer, Roselyne Rey (Paris: Klincksieck, 1995), pp. 391–417.

115. See Claude Blanckaert, "Méthode des moyennes et notion de 'série suffisante' en anthropologie physique (1830–1880)," in *Moyenne, Milieu, Centre: Histoires et usages,* ed. J. Feldman, G. Lagneau, B. Matalon (Paris: Editions de l'Ecole des Hautes Etudes en Sciences Sociales, 1991), pp. 213–43; and Williams, *Physical and the Moral,* pp. 258–60.

116. Paul Topinard, "Mensuration des crânes des grottes de Baye (époque néolithique) d'après les registres de Broca," *Revue d'anthropologie,* 3rd ser., 1 (1886): 1–9.

117. Broca, "Anthropologie," in *Mémoires d'anthropologie,* vol. 1, p. 7.

118. Ibid., p. 16 (emphasis in the original). On Broca's statistical method, see Blanckaert, "Méthode des moyennes."

119. Broca, "Mémoire sur le craniographe et sur quelques-unes de ses applications" (1861 and 1862), in *Mémoires d'anthropologie,* vol. 1, pp. 42–43.

120. Lorraine Daston and Peter Galison, "The Image of Objectivity," *Representations* 40 (1992): 81–128. See also Ian Hacking, "Biopower and the Avalanche of Printed Numbers," *Humanities in Society* 5 (1982): 279–95.

121. See, for example, Paul Broca, "Sur le stéréographe, nouvel instrument crâniographique destiné a dessiner tous les détails du relief des corps solides," *Société d'Anthropologie de Paris: Mémoires,* vol. 3, in *Mémoires d'anthropologie,* vol. 1, pp. 118–44; Edouard Goldstein, "Des applications du calcul des probabilités a l'anthropologie," *Revue d'anthropologie,* 2nd ser., 6 (1883): 704–28; and Paul Topinard, *Anthropology,* trans. Robert T. H. Bartley (London: Chapman and Hall, 1878), esp. 218–63.

122. Elizabeth Williams discusses Broca's institutional dominance of French

anthropology in "Anthropological Institutions in Nineteenth-Century France." See also Nélia Dias, "Séries de crânes et armée de squelettes: Les collections anthropologiques en France dans la seconde moitié du XIXe siècle," *Société d'Anthropologie de Paris: Bulletins et Mémoires,* n.s., 1 (1989): 203–30.

123. For Topinard's discussion of the differences in measurement practices among various practitioners, see *Eléments d'anthropologie générale,* esp. chaps. 12–17. The goal of all this labor was to make the Société d'Anthropologie the center of an empire of measurement, in which the outflow of instructions and instruments would be matched by the frictionless inflow of observations and artifacts and calculations.

124. For an example of equivalent worries among U.S. anthropologists, see Wilson, "Physical Ethnology."

125. Broca, "Sur les caractères des crânes basques de Zaraus (Guipuzcoa)," *Bulletins de la Société d'anthropologie de Paris* 3 (1862) and "Second mémoire sur les caractères des crânes basques de Zaraus (Guipuzcoa), Réponse aux objections de M. Pruner-Bey," *Bulletins de la Société d'anthropologie de Paris* 4 (1863), in *Mémoires d'anthropologie,* vol. 2 (Paris: C. Reinwald, 1874), pp. 1–32; and Topinard, "Mensurations des crânes des grottes de Baye," pp. 1–9.

126. See, for example, H. Ten Kate, "Observations anthropologiques recueillies dans la Guyane et le Venezuela (résumé)," *Revue d'anthropologie,* 3rd ser., 2 (1887): 44–68.

127. See Hecht, "Vigilant Anthropology." Also Léonce Manouvrier, "Considérations présentées par M. L. Manouvrier à l'appui de sa candidature" (1899), AN F^{17}13554 : Collège de France—Demandes de création des chaires; Archives Nationale, Paris.

128. See Hammond, "Anthropology as a Weapon of Social Combat," pp. 124–25.

129. On this split, see Clark, *Social Darwinism in France,* chap. 9; Hammond, "Anthropology as a Weapon of Social Combat"; and Harvey, "Evolutionism Transformed." For more on Mortillet, see Nathalie Richard, "La Revue *L'Homme* de Gabriel de Mortillet: Anthropologie et politique au début de la troisième république," *Société d'Anthropologie de Paris: Bulletins et Mémoires,* n.s., 1 (1989): 231–56.

130. Elizabeth Williams argues that the Broca group studied intelligence rather than the "moral" capacities because it "relieved their science of the weight of postulating the somatic determinism of moral behavior." Williams, *Physical and the Moral,* p. 262.

131. Le Bon, "Variations du volume du cerveau," pp. 60–61; Armand de Quatrefages, "Reprise de la discussion sur la perfectibilité des races," *Bulletins de la Société d'anthropologie de Paris* 1 (1859): 429. See also Perier, "Reprise de la discussion sur la perfectibilité des races," *Bulletins de la Société d'anthropologie de Paris* 1 (1859): 419.

132. Three of the four—cephalic index, brain weight, and cranial capacity—merited separate chapters in Topinard's *Eléments d'anthropologie générale,* along with skin color and body size.

133. On brain size and convolutions, see Claude Bernard, "Des fonctions du cerveau," *Revue des deux mondes* 98 (1872): 373–85, pp. 375–76.

134. Anders A. Retzius, "Mémoire sur les formes du crâne des habitants du Nord" (1842), *Annales des sciences naturelles,* 3rd ser., Zoologie, 6 (1846): 133–71.

135. For a thorough analysis of the cephalic index and the criticisms raised against it, see Claude Blanckaert, "L'Indice céphalique et l'ethnogénie européenne: A. Retzius, P. Broca, F. Pruner-Bey (1840–1870)," *Société d'Anthropologie de Paris: Bulletins et Mémoires,* n.s., 1 (1989): 165–202.

136. See Fee, "Nineteenth-Century Craniology, esp. pp. 426–29.

137. See Topinard, *Eléments,* pp. 402 and 405; and Retzius, "Mémoire sur les formes du crâne," pp. 137–38.

138. Blanckaert, "L'Indice céphalique et l'ethnogénie européenne," esp. pp. 175–82; and Gould, *Mismeasure of Man,* pp. 98–100.

139. Gould has discussed this issue in *Mismeasure of Man,* esp. pp. 86–92.

140. Paul Broca, *Bulletins de la Société d'Anthropologie de Paris,* meeting of July 3, 1879, p. 501; quoted in Adolphe Bloch, "L'Intelligence est-elle en rapport avec le volume du cerveau," *Revue d'anthropologie,* 2nd ser., 8 (1885): 585.

141. Le Bon, "Variation du volume du cerveau," p. 103 (emphasis in original).

142. For an analysis of the way in which the "variability hypothesis" has been applied to women, see Stephanie A. Shields, "The Variability Hypothesis: The History of a Biological Model of Sex Differences in Intelligence," *Signs: Journal of Women in Culture and Society* 7 (1982): 769–97.

143. For more on reactions to Le Bon's article and other efforts in craniometry, see Nye, *Origins of Crowd Psychology,* pp. 33–35.

144. Bloch, "L'intelligence," p. 617.

145. Pierre Gratiolet, "Observations sur la microcéphalie, considérée dans ses rapports avec la question des caractères du genre humain et du parallèle des races," *Société d'Anthropologie de Paris: Bulletins* 1 (1859): 40.

146. Samuel Pozzi, "Du poids du cerveau suivant les races et suivant les individus," *Revue d'anthropologie,* 2nd ser., 1 (1878): 277–85, thoroughly critiqued the idea that there was any absolute connection between brain weight and intelligence. See also M. Alix, "Rapport sur un mémoire de M. H. Wagner, intitulé *Mensuration de la surface du cerveau,*" *Société d'Anthropologie de Paris: Bulletins* 6 (1865): 227–37. On Wagner's work, see Michael Hagner, "Skulls, Brains, and Memorial Culture: On Cerebral Biographies of Scientists in the Nineteenth Century," *Science in Context* 16 (2003): 195–218.

147. Pierre Gratiolet, "Reprise de la discussion sur le volume et la forme du cerveau," *Bulletins de la société d'Anthropologie de Paris* 2 (1861): 238–75, p. 245.

148. Paul Broca, "Reprise de la discussion sur le volume et la forme du cerveau," *Bulletins de la Société d'anthropologie de Paris* 2 (1861): 301–22.

149. Gould makes much the same point in *Mismeasure of Man,* pp. 92–94.

150. See, for example, Broca, "Sur le volume et la forme du cerveau suivant les individus et suivant les races," *Bulletins de la Société d'anthropologie de Paris* 2 (1861): 139–204, in *Mémoires d'anthropologie,* vol. 1, p. 163.

151. See Topinard, *Eléments,* p. 506.

152. Léonce Manouvrier, "L'indice céphalique et la pseudo-sociologie,"*Revue de l'Ecole d'Anthropologie* 9 (1899): 233–59, 280–96. See also Hecht, "Vigilant Anthropology."

153. Georges Pouchet, "Le système nerveux et l'intelligence," *Revue des deux mondes* 93 (1871): 717–48, quote on pp. 736–37.

154. William H. Schneider has analyzed the dissolution of the craniometric project in *Quality and Quantity: The Quest for Biological Regeneration in Twentieth-Century France* (Cambridge: Cambridge University Press, 1990). See also Fee, "Nineteenth-Century Craniology." For a discussion of the general demise and then rebirth of both anthropology and ethnology in France, see Alice L. Conklin, "Civil Society, Science, and Empire in Late Republican France: The Foundation of Paris's Museum of Man," *Osiris* 17 (2002): 255–90.

155. See, for example, the 1859 society discussion concerning the amount of intelligence necessary to fashion prehistoric axes. "Discussion sur les haches du diluvium," *Société d'Anthropologie de Paris: Bulletins* 1 (1859): 68–78.

156. Gustave Le Bon, "Recherches anatomiques et mathématiques sur les lois des variations du volume du cerveau et sur leurs relations avec l'intelligence," *Revue d'anthropologie,* 2nd ser., 2 (1879): 27–104, quote on pp. 83–84.

157. See Schiller, *Paul Broca,* esp. chap. 10; and Young, *Mind, Brain, and Adaptation,* esp. chap. 4. For general background on brain localization in addition to Young, see Edwin Clarke and L. S. Jacyna, *Nineteenth-Century Origins of Neuroscientific Concepts* (Berkeley: University of California Press, 1987), chap. 6; and Anne Harrington, *Medicine, Mind, and the Double Brain: A Study in Nineteenth-Century Thought* (Princeton: Princeton University Press, 1987), esp. chap. 2.

158. See, for example, Paul Broca, "De l'influence de l'éducation sur le volume et la forme de la tête," *Bulletins de la société d'Anthropologie de Paris,* 2nd ser., 7 (1872): 879–96.

159. Ibid., pp. 171–72.

160. See Le Bon, "Variations du volume du cerveau," for an example of the consistent use of polarities to establish connections between craniometric characteristics and intelligence.

161. For more details, see Blanckaert, "L'Indice céphalique et l'ethnogénie européenne." On nineteenth-century ideas about the racial makeup of France, see Blanckaert, "On the Origins of French Ethnology"; and Cohen, *French Encounter with Africans,* chap. 8. The failure of Georges Vacher de Lapouge's vision of the French nation as racially heterogeneous is detailed in Clark, *Social Darwinism in France,* pp. 143–58; and Sean Quinlan, "The Racial Imagery of Degeneration and Depopulation: Georges Vacher de Lapouge and 'Anthroposociology' in Fin-de-Siècle France," *History of European Ideas* 24 (1998): 393–413. From the other side, Hippolyte Taine depicted race in national terms, imagining no important cleavages within the nation. On Taine's approach to race, see François Leger, "L'Idée de race chez Taine," in *L'Idée de race dans la pensée politique française contemporaine,* eds. Pierre Guiral and Emile Temime (Paris: Editions du CNRS, 1977), pp. 89–99.

162. Cohen, *French Encounter with Africans,* p. 216. France's preeminent, though largely unread, espouser of racial doctrine was Arthur de Gobineau in his *Essai sur l'inégalité des races humaines* (1854). On late nineteenth-century French racial anti-Semitism, see Zeev Sternhell, "Le déterminisme physiologique et racial

à la base du nationalisme de Maurice Barrès et de Jules Soury," in *L'Idée de race dans la pensée politique française contemporaine,* pp. 117–38.

163. On solidarism, see J.E.S. Hayward, "The Official Social Philosophy of the Third Republic: Léon Bourgeois and Solidarism," *International Review of Social History* 6 (1961): 19–48; Karen Offen, "Depopulation, Nationalism, and Feminism in Fin-de-Siècle France," *American Historical Review* 89 (1984): 648–76; and Harry Paul, "The Debate over the Bankruptcy of Science in 1895," *French Historical Studies* 3 (1968): 299–327. For a contemporary account, see Léon Bourgeois, *Solidarité* (Paris: Colin, 1896). This is not to deny, however, that there remained a strong racialist element to French thought, and a broadly shared willingness to see other peoples as inferior, whether on biological or cultural grounds. On this point, see Cohen, *French Encounter with Africans,* chap. 8.

164. See Clark, *Social Darwinism in France,* chap. 9.

165. See Philip Nord, *The Republican Moment: Struggles for Democracy in Nineteenth-Century France* (Cambridge: Harvard University Press, 1995), p. 43.

166. Nord, *Republican Moment,* p. 44. Elizabeth Williams argues that many society members sharply differentiated between the fates of other races, depicted as permanently inferior, and that of European peoples. Williams, *Physical and the Moral,* pp. 266–67. See also Clark, *Social Darwinism in France,* chap. 9; and Quinlan, "Racial Imagery of Degeneration and Depopulation." For an important contemporary attack on the use of science to espouse limits on human perfectibility, see Célestin Bouglé, *La Démocratie devant la science: Etudes critiques sur l'hérédité, la concurrence et la différentiation,* 3rd ed. (Paris: Alcan, 1923); and also Bouglé, "Anthropologie et démocratie," *Revue de métaphysique et de morale* 5 (1897): 443–61. Within anthropology proper the key work was Manouvrier, "L'indice céphalique et la pseudo-sociologie."

167. See Yvette Conry, *L'Introduction du darwinisme en France au XIXe siècle* (Paris: Vrin, 1974), p. 45; Pietro Corsi and Paul J. Weindling, "Darwinism in Germany, France, and Italy," in *The Darwinian Heritage,* ed. David Kohn (Princeton: Princeton University Press, 1985), p. 702; and Clark, *Social Darwinism in France,* chap. 9. For more on the history of evolution in France, see chapter 4.

168. On the place of French anthropology within the French colonial project, see Patricia M. E. Lorcin, "Imperialism, Colonial Identity, and Race in Algeria, 1830–1870," *Isis* 90 (1999): 653–79; Lorcin, *Imperial Identities,* esp. chaps. 6 and 7; and, for the post–World War I period, Conklin, "Civil Society." See also Cohen, *French Encounter with Africans,* chap. 8; and Osborne, *Nature, the Exotic,* esp. chaps. 3 and 4.

169. Claude Blanckaert makes this point as well in "On the Origins of French Ethnology," p. 49.

170. However, Michael Osborne has shown that society debates over European acclimatization to other (i.e., colonial) milieus suggested that French overseas efforts might not be biologically sustainable. Osborne, *Nature, the Exotic,* esp. pp. 85–97.

171. See Hecht, "Vigilant Anthropology"; and Robert A. Nye, *Crime, Madness, Politics in Modern France: The Medical Concept of National Decline* (Princeton: Princeton Unniversity Press, 1984), p. 111.

Chapter 4

1. Hippolyte Taine, *De l'intelligence,* 2 vols. (Paris: Hachette, 1870); English translation: *On Intelligence,* 2 vols., trans. T. D. Haye (New York: Holt, 1875); quote from *De l'intelligence,* vol. 1, p. vii.

2. See *Dictionnaire de l'Académie Françoise* (Paris: Jean Baptiste Coignard, 1694), vol. 1, p. 601; and Emile Littré, *Dictionnaire de la langue française* (Paris: Hachette, 1874), vol. 3, pp. 124–25.

3. The *Dictionnaire des fréquences: Vocabulaire littéraire des XIXe et XXe siècles* (Paris: Didier, 1971) shows *intelligence* to have been a fairly popular word throughout the nineteenth century. See vol. 3, p. 1141. Daniel N. Robinson discusses Taine's use of the word intelligence in "Preface to Taine's *On Intelligence,*" in Hippolyte A. Taine, *On Intelligence* (1875; reprint, Washington: University Publications of America, 1977), p. xxii.

4. On the rise of a new interest in human differences, see Leslie S. Hearnshaw, "The Concepts of Aptitude and Capacity," *Revue de synthèse* 89 (1968): 345–54; Frank Manuel, "From Equality to Organicism," *Journal of the History of Ideas* 17 (1956): 54–69; and Robert A. Nye, "Sociology and Degeneration: The Irony of Progress," *Degeneration: The Dark Side of Progress,* eds. J. Edward Chamberlin and Sander L. Gilman (New York: Columbia University Press, 1985), pp. 49–71. I use "philosophy of mind" when the term "psychology" would be anachronistic.

5. For an example of the latter usage, see Paul Valéry, "Remarks on Intelligence" (1925), in *The Outlook of Intelligence,* trans. Denise Folliot and Jackson Mathews (Princeton: Princeton University Press, 1962), pp. 72–88.

6. For background on the regime's troubles, see Sanford Elwitt, *The Making of the Third Republic: Class and Politics in France, 1868–1884* (Baton Rouge: Louisiana State University Press, 1979), chap. 1; Roger Magraw, *France, 1815–1914: The Bourgeois Century* (New York: Oxford University Press, 1986), chaps. 5 and 6; Alain Plessis, *The Rise and Fall of the Second Empire, 1852–1871,* trans. Jonathan Mandelbaum (Cambridge: Cambridge University Press, 1985); Barnett Singer, *Modern France: Mind, Politics, Society* (Seattle: University of Washington Press, 1980); and Gordon Wright, *France in Modern Times: From the Enlightenment to the Present* (New York: W. W. Norton, 1987).

7. Robert D. Anderson, *Education in France, 1848–1870* (Oxford: Clarendon Press, 1975); Katherine Auspitz, *The Radical Bourgeoisie: The Ligue de l'enseignement and the Origins of the Third Republic, 1866–1885* (Cambridge: Cambridge University Press, 1982); D. G. Charlton, *Positivist Thought in France during the Second Empire, 1852–1870* (Oxford: Clarendon Press, 1959); Patrick J. Harrigan, *Mobility, Elites and Education in French Society of the Second Empire* (Waterloo: Wilfred Laurier University Press, 1980); William Logue, *From Philosophy to Sociology: The Evolution of French Liberalism, 1870–1914* (DeKalb: Northern Illinois University Press, 1983); Joseph N. Moody, *French Education since Napoleon* (Syracuse: Syracuse University Press, 1978); Robert J. Smith, *The Ecole Normale Supérieure and the Third Republic* (Albany: SUNY Press, 1982); Theodore Zeldin, *France, 1848–1945,* 2 vols. (Oxford: Clarendon Press, 1973 and 1977); and Craig Zwerling, *The Emergence of the Ecole Normale Supérieure*

as a Centre of Scientific Education in Nineteenth-Century France (New York: Garland, 1990).

8. For more on the general political and cultural history of France during this period, see Auspitz, *Radical Bourgeoisie;* Alfred Cobban, *A History of Modern France,* vols. 2 and 3 (New York: George Braziller, 1965); Jean-Martin Mayeur and Madeleine Rebérioux, *The Third Republic from Its Origins to the Great War, 1871–1914* (Cambridge: Cambridge University Press, 1984); Roger Price, *A Social History of Nineteenth-Century France* (London: Hutchinson, 1987); and Zeldin, *France 1848–1945.*

9. For the place of positivism in nineteenth-century culture, see Charlton, *Positivist Thought in France during the Second Empire;* John Eros, "The Positivist Generation of French Republicanism," *Sociological Review,* n.s., 3 (1955): 255–77; Walter M. Simon, "Auguste Comte's English Disciples," *Victorian Studies* 8 (1964–65): 162–72; Simon, *European Positivism in the Nineteenth Century* (Ithaca: Cornell University Press, 1963); and Simon, "The 'Two Cultures' in Nineteenth-Century France: Victor Cousin and Auguste Comte," *Journal of the History of Ideas* 26 (1965): 45–58.

10. See Auguste Comte, *Cours de philosophie positive* (1830–42), 6 vols., in *Œuvres d'Auguste Comte,* vols. 1–6 (Paris: Editions anthropos, 1968); *Correspondance générale et confessions,* vol. 1 (Paris: Mouton, 1973); and *Ecrits de jeunesse, 1816–1828* (Paris: Archives positivistes, 1970). On Comte, see especially Mary Pickering, *Auguste Comte: An Intellectual Biography* (Cambridge: Cambridge University Press, 1993). Also valuable are Keith Michael Baker, "Closing the French Revolution: Saint-Simon and Comte," in *The Transformation of Political Culture, 1789–1848,* eds. François Furet and Mona Ozouf (Oxford: Pergamon Press, 1989), pp. 323–37; Charlton, *Positivist Thought in France,* esp. chap. 2; John C. Greene, "Biology and Social Theory in the Nineteenth Century: Auguste Comte and Herbert Spencer," in *Critical Problems in the History of Science,* ed. Marshall Clagett (Madison: University of Wisconsin Press, 1959), pp. 419–46; Barbara Haines, "The Inter-Relations between Social, Biological, and Medical Thought, 1750–1850: Saint-Simon and Comte," *British Journal for the History of Science* 11 (1978): 19–35; Diana Postlethwaite, *Making It Whole: A Victorian Circle and the Shape of Their World* (Columbus: Ohio State University Press, 1984), esp. chap. 1; and Simon, *European Positivism in the Nineteenth Century,* chap. 1.

11. See Comte, *Correspondance générale et confessions,* vol. 1, p. 64.

12. See, for example, Comte, 28th Lesson, *Cours de philosophie positive,* vol. 2.

13. In "Closing the French Revolution," Baker discusses Comte's appropriation and transformation of the Enlightenment conception of social science as a means of ending political discord. See also Greene, "Biology and Social Theory in the Nineteenth Century," esp. pp. 420–28.

14. Diana Postlethwaite discusses the influence of Comte on Mill in *Making It Whole,* esp. chap. 1.

15. See Emile Littré, *Auguste Comte et la philosophie positive* (Paris: Hachette, 1864); and Claude Bernard, *Introduction to the Study of Experimental Medicine* (1865; reprint, New York: Dover, 1957). For background, see Charlton, *Positivist Thought in France,* chaps. 4 and 5; Logue, *From Philosophy to Sociology,* esp. chaps. 2 and 3; Harry W. Paul, *From Knowledge to Power: The Rise of the*

Science Empire in France, 1860–1939 (Cambridge: Cambridge University Press, 1985), chaps. 1 and 2; and Simon, *European Positivism in the Nineteenth Century,* chaps. 2–4.

16. From midcentury on, the Ecole Normale Supérieure (ENS) was the principal elite educational institution for the French intelligentsia, especially in the humanities and the emerging social sciences. For further information, see Terry N. Clark, *Prophets and Patrons: The French University and the Emergence of the Social Sciences* (Cambridge: Harvard University Press, 1973); and Smith, *Ecole Normale Supérieure and the Third Republic.* Smith (p. 64) notes that although positivism and other scientistic philosophies were not taught at ENS, they were widely discussed among the students.

17. Logue makes much the same point in *From Philosophy to Sociology,* pp. 98–99 specifically and more generally chaps. 5–8. On the place of positivism in Third Republic politics, see also J. Chastenet, *La république des républicains* (Paris: Hachette, 1954); Eros, "Positivist Generation of French Republicanism"; Mayeur and Rebérioux, *Third Republic from Its Origins to the Great War, 1871–1914;* and Simon, *European Positivism in the Nineteenth Century,* chaps. 5–6.

18. For example, while both Taine and Jules Ferry were ardent positivists, Taine was on the right and Ferry on the left.

19. See, for example, Emile Blanchard, "L'Instruction générale en France: L'observation et l'expérience," *Revue des deux mondes* 95 (1871): 815–45; or Gaston Boissier, "Les méthodes dans l'enseignement secondaire," *Revue des deux mondes* 100 (1872): 683–96.

20. Philosophers of mind adopted a version of positivism even though Comte himself had disparaged psychology as an independent field of study, describing it simply as an offshoot of biology.

21. On Taine see Hans Aarsleff, "Taine and Saussure," in *From Locke to Saussure: Essays on the Study of Language and Intellectual History* (Minneapolis: University of Minnesota Press, 1982), pp. 356–71; Susanna Barrows, *Distorting Mirrors: Visions of the Crowd in Late Nineteenth-Century France* (New Haven: Yale University Press, 1981), esp. chap. 3; Charlton, *Positivist Thought in France,* esp. chap. 7; André Cresson, *Hippolyte Taine: Sa vie, son œuvre* (Paris: Presses universitaires de France, 1951); Robinson, "Preface to Taine's *On Intelligence,*" pp. xxi–xxxviii; Jaap van Ginneken, *Crowds, Psychology, and Politics, 1871–1899* (Cambridge: Cambridge University Press, 1992), esp. chap. 1; and Leo Weinstein, *Hippolyte Taine* (New York: Twayne, 1972).

22. Taine, *De l'intelligence,* vol. 1, pp. 3–4.

23. D. G. Charlton discusses at length to what degree Taine was a positivist and idealist. Taine clearly attempted to integrate the two, and was equally explicit about his rejection of most of Comte's own formulations. See Charlton, *Positivist Thought in France,* pp. 134–50, 154–57.

24. For more on associationism, see chapter 1. Taine was not the first to reintroduce associationist ideas into nineteenth-century French philosophy; Pierre-Maurice Mervoyer did so in 1864, meeting profound indifference. See Mervoyer, *Etude sur l'association des idées* (Douai: Wartelle, 1864).

25. "Taine à M. Jules Soury, 13 août 1873," in Hippolyte Taine, *H. Taine: Sa*

vie et sa correspondance, vol. 3 (Paris: Hachette, 1902–1907), p. 253. See also "Taine à Ernest Renan, 9 septembre 1872," in *H. Taine: Sa vie et sa correspondance,* vol. 3, p. 206.

26. For details on Taine's life, see Barrows, *Distorting Mirrors,* pp. 89–92; and Weinstein, *Hippolyte Taine,* esp. pp. 17–27.

27. On the career pattern typical of a successful intellectual in nineteenth-century France, see Clark, *Prophets and Patrons.*

28. Taine's notoriety was confirmed in 1857 with his publication of *Les philosophes français du XIXe siècle* (Paris: Hachette, 1857), which contained a withering attack on Cousineanism.

29. See Hippolyte Taine, *Histoire de la littérature anglaise,* 4 vols. (Paris: Hachette, 1863–64), esp. the preface; Barrows, *Distorting Mirrors,* p. 90; and Weinstein, *Hippolyte Taine,* esp. chap. 3.

30. For reviews of *De l'intelligence,* see Léon Dumont, "La théorie De l'intelligence d'après M. Taine," *Revue politique et littéraire,* 2nd ser., 3 (1872–73): 1124–36; William James, "Taine's 'On Intelligence,'" *Nation* 15 (1872): 139–41; John Stuart Mill, "*De l'intelligence.* Par H. Taine," *Fortnightly Review* 14 (1870): 121–24; F. Pillon, "*De l'intelligence,* par H. Taine," *Critique philosophique* 2 (1873): 155–60 ; Pillon, "Le rapport des signes et des idées générales selon Condillac et selon M. Taine," *Critique philosophique* 2 (1873): 177–89; Théodule Ribot, "M. Taine et sa psychologie," *Revue philosophique* 4 (1877): 17–46; G. Croom Robertson, "M. Taine on Intelligence," *Nature* 4 (1871): 62–63; Edmond Scherer, "De l'intelligence par Taine," *Le Temps,* 2 août 1870, 10th year, #3443; Jules Soury, "De l'intelligence par M. H. Taine," *Le Temps* 3, 8, 12 août 1873, 13th year, # 4494, 4499, 4503; E. Vacherot, "La nouvelle philosophie en France," *Revue des deux mondes* 88 (1870): 611–41; and Y., "*De l'intelligence* par H. Taine," *Revue critique d'histoire et de littérature* 14 (1873): 226–32.

31. Barrows, *Distorting Mirrors,* pp. 81–82.

32. Taine's principal footnote citations were to Vulpian's *Leçons sur la physiologie du système nerveux* (twenty-three times), Mueller's *Manuel de physiologie* (eighteen times), Helmholtz's *Handbuch der Physiologischen Optik* (fifteen times), Brière de Boismont's *Traité des hallucinations* (thirteen times), Maury's *Du sommeil et des rêves, hallucinations hypnogogiques* (thirteen times), *Annales médico-psychologiques* (twelve times), Baillarger's *Mémoire sur les hallucinations* (twelve times), Bain's *Sense and Intellect* (ten times), Landry's *Traité de paralysies* (ten times), and Longet's *Traité de physiologie* (ten times). Nonetheless, his fellow traveler Théodule Ribot took Taine to task for his failure to be sufficiently empirical, and especially comparative. See Ribot, "M. Taine et sa psychologie," p. 46.

33. Taine, *De l'intelligence,* 3rd ed. (Paris: Hachette, 1878), vol. 1, p. 9.

34. The law of contiguity said that two sensations that came one after the other would be linked together; the law of resemblance stated that two sensations that seemed similar would be linked together.

35. See Taine, *De l'intelligence* (1870), vol. 1, pp. 23–50.

36. Ibid., vol. 2, pp. 261–62; vol. 1, pp. 156–57.

37. Ibid., vol. 1, p. 138.

38. Margery Sabin, "The Community of Intelligence and the Avant-Garde," *Raritan* 4 (1985): 1–25, quote on p. 6.

39. "Littré, Dumas, Pasteur, and Taine," *Edinburgh Review,* reprinted in *Popular Science Monthly* 21 (1882): 667–79, quote on p. 668.

40. See Eros, "Positivist Generation of French Republicanism."

41. See Ian R. Dowbiggin, "French Psychiatry and the Search for a Professional Identity: The Société Médico-Psychologique, 1840–1870," *Bulletin of the History of Medicine* 63 (1989): 331–55; and Anson Rabinbach, *The Human Motor: Energy, Fatigue, and the Origins of Modernity* (Berkeley: University of California Press, 1992), esp. pp. 168–69.

42. Théodule Ribot, *Heredity: A Psychological Study of Its Phenomena, Laws, Causes, and Consequences* (New York: D. Appleton, 1875). Ribot describes some of the opposition he faced in Ribot to Espinas, 14 July 1885, *Revue philosophique* 165 (1975): 157–72, quote on p. 160; and "Lettres de Théodule Ribot à Espinas," *Revue philosophique* 147 (1957): 1–14, 152 (1962): 337–40, 154 (1964): 79–84, 160 (1970): 165–73, 339–48, 165 (1975): 157–72. On Ribot's struggles to gain entrance to French higher education, see John I. Brooks III, *The Eclectic Legacy: Academic Philosophy and the Human Sciences in Nineteenth-Century France* (Newark: University of Delaware Press, 1998), esp. chap. 2. For an example of Ribot's disdain for the eclectic philosophy, see Ribot, "Philosophy in France," *Mind* 2 (1877): 366–86.

43. Ribot to Espinas, 15 March 1873, *Revue philosophique* 147 (1957): 10.

44. See, for example, Charles Waddington, "Séance du 29 juin 1885," *Faculté des Lettres de l'Université de Paris—Registre des Actes et Délibérations, 1881–1888,* p. 233, $AJ^{16}4747$, Archives Nationales (AN), Paris. On Ribot's career, see Serge Nicolas and David Murray, "Théodule Ribot (1839–1916), Founder of French Psychology: A Biographical Introduction," *History of Psychology* 2 (1999): 277–301. For an example of the continuing hostility to positivistic psychology, see D. Mercier, *Les origines de la psychologie contemporaine* (Louvain: Institut Supérieur de Philosophie, 1897).

45. On the intellectual culture of Third Republic France generally, see Barrows, *Distorting Mirrors;* Logue, *From Philosophy to Sociology;* Mayeur and Rebérioux, *Third Republic from Its Origins to the Great War,* esp. chaps. 4 and 11; Robert A. Nye, *Crime, Madness, & Politics in Modern France: The Medical Concept of National Decline* (Princeton: Princeton University Press, 1984); Paul, *From Knowledge to Power,* chaps. 1–2; Smith, *Ecole Normale Supérieure and the Third Republic;* Fritz Ringer, *Fields of Knowledge: French Academic Culture in Comparative Perspective, 1890–1920* (Cambridge: Cambridge University Press, 1992); and George Weisz, *The Emergence of the Modern University in France, 1863–1914* (Princeton: Princeton University Press, 1983), esp. chaps. 3 and 9.

46. For Ribot's reactions to the reviews of *Psychologie anglaise contemporaine,* see Ribot to Espinas, 17 juin 1870, *Revue philosophique* 147 (1957): 5–6.

47. Anatole France, *Le Temps,* 12 mars 1893; as quoted in Weinstein, *Hippolyte Taine,* p. 17.

48. On Ribot, see Claude Bénichou, "Ribot et l'hérédité psychologique," in *L'Ordre des caractères: Aspects de l'hérédité dans l'histoire des sciences de l'homme* (Paris: J. Vrin, 1989), pp. 73–94; Brooks, *Eclectic Legacy,* esp. chap. 2; John I.

Brooks III, "Philosophy and Psychology at the Sorbonne, 1885–1913," *Journal of the History of the Behavioral Sciences* 29 (1993): 123–45; *Centenaire de Th. Ribot: Jubilé de la psychologie scientifique française, 1839–1889–1939* (Agen: Imprimerie moderne, 1939); L. Dugas, *Le philosophe Théodule Ribot* (Paris: Payot, 1924); J. Gasser, "La notion de mémoire organique dans l'œuvre de T. Ribot," *History and Philosophy of the Life Sciences* 10 (1988): 293–313; Nicolas and Murray, "Théodule Ribot (1839–1916)"; Rabinbach, *Human Motor,* pp. 168–78; Maurice Reuchlin, *Histoire de la psychologie* (Paris: Presses universitaires de France, 1978), pp. 60–63; Michael S. Roth, "Remembering Forgetting: *Maladies de la Mémoire* in Nineteenth-Century France," *Representations* 26 (1989): 49–68; and Jacqueline Thirard, "La fondation de la 'Revue philosophique,'" *Revue philosophique* 166 (1976): 401–13.

49. Ribot, "M. Taine et sa psychologie," p. 46.

50. Ribot himself, as John Brooks observes, did more to initiate the new psychology than to practice it. See Brooks, *Eclectic Legacy,* pp. 67–68. On French psychology in the late nineteenth century, see especially Jacqueline Carroy and Régine Plas, "The Origins of French Experimental Psychology: Experiment and Experimentalism," *History of the Human Sciences* 9 (1996): 73–84; and Serge Nicolas, *Histoire de la psychologie française: Naissance d'une nouvelle science* (Paris: In Press Éditions, 2002). See also Joseph Ben-David and Randall Collins, "Social Factors in the Origins of a New Science: The Case of Psychology," *American Sociological Review* 31 (1966): 451–65; Kurt Danziger, "The Origins of the Psychological Experiment as a Social Institution," *American Psychologist* 40 (1985): 133–40; Paul Fraisse, Jean Piaget, and Maurice Reuchlin, *Traité de psychologie expérimentale,* vol. 1, *Histoire et méthode* (Paris: Presses universitaires de France, 1963), pp. 31–38; Maurice Reuchlin, "The Historical Background for National Trends in Psychology: France," *Journal of the History of the Behavioral Sciences* 1 (1965): 115–23; and Björn Sjövall, *The Psychology of Tension* (Stockholm: Norstedts & Göner, 1967).

51. Théodule Ribot, *La psychologie anglaise contemporaine* (1870), 2nd ed. (Paris: Germer Baillière, 1875), pp. 25–26.

52. The most persistent late nineteenth-century cultural worry in France was over the degeneration of the race. See Ian R. Dowbiggin, "Degeneration and Hereditarianism in French Medical Medicine, 1840–1890: Psychiatric Theory as Ideological Adaptation," in *The Anatomy of Madness: Essays in the History of Psychiatry,* eds. William F. Bynum, Roy Porter, and Michael Shepherd (New York: Tavistock, 1985), pp. 188–232; Dowbiggin, *Inheriting Madness: Professionalization and Psychiatric Knowledge in Nineteenth-Century France* (Berkeley: University of California Press, 1991); Jack D. Ellis, *The Physician-Legislators of France: Medicine and Politics in the Early Third Republic, 1870–1914* (Cambridge: Cambridge University Press, 1990); Nye, *Crime, Madness and Politics in Modern France;* Nye, "Sociology and Degeneration"; and William H. Schneider, *Quality and Quantity: The Quest for Biological Regeneration in Twentieth-Century France* (Cambridge: Cambridge University Press, 1990), esp. chaps. 1–3.

53. John Brooks argues that Ribot drew much of his inspiration for his pathological method from English neurophysiologists. See Brooks, *Eclectic Legacy,* pp. 86–88.

54. The first generation of French scientific psychologists, in addition to Ribot, included Alfred Binet, Benjamin Bourdon, Georges Dumas, Pierre Janet, Henri Piéron, Charles Richet, and Edouard Toulouse.

55. Ribot, *La psychologie,* p. 37.

56. See Herbert Spencer, *The Principles of Psychology,* 2nd ed., 2 vols. (New York: D. Appleton, 1877). The second edition of *Principles* in 1870–72 marked a substantial enlargement and revision. The standard biography of Spencer is James D. Y. Peel, *Herbert Spencer: The Evolution of a Sociologist* (New York: Basic Books, 1971), well supplemented by M. W. Taylor, *Man versus the State: Herbert Spencer and Late Victorian Individualism* (Oxford: Clarendon Press, 1992). The best accounts of Spencer's psychology are Postlethwaite, *Making It Whole,* esp. pp. 201–21; Théodule Ribot, *English Psychology* (New York: D. Appleton, 1874), pp. 124–93; Robert J. Richards, *Darwin and the Emergence of Evolutionary Theories of Mind and Behavior* (Chicago: University of Chicago Press, 1987), esp. chaps. 6 and 7; C.U.M. Smith, "Evolution and the Problem of Mind: Part I, Herbert Spencer," *Journal of the History of Biology* 15 (1982): 55–88; and Robert M. Young, *Mind, Brain and Adaptation in the Nineteenth Century: Cerebral Localization and Its Biological Context from Gall to Ferrier* (New York: Oxford University Press, 1990), esp. chap. 5.

57. For Spencer's explanation of the correspondence between organism and environment, see "Part III: General Synthesis," *Principles of Psychology,* vol. 1, pp. 291–392. Richards lucidly analyzes Spencer's evolutionary theory in *Darwin and the Emergence of Evolutionary Theories,* pp. 267–94.

58. Spencer, *Principles of Psychology,* vol. 1, pp. 291–92. See also Smith, "Evolution and the Problem of Mind," esp. pp. 58–62.

59. Spencer, *Principles of Psychology,* vol. 1, p. 387.

60. See Richards, *Darwin and the Emergence of Evolutionary Theories,* pp. 282–84; Taylor, *Man versus the State,* pp. 111–15; and Young, *Mind, Brain and Adaptation,* pp. 179–80.

61. Spencer, *Principles of Psychology,* vol. 1, pp. 462–68.

62. See Greene, "Biology and Social Theory in the Nineteenth Century," p. 434; and Smith, "Evolution and the Problem of Mind," pp. 73–75.

63. Spencer, *Principles of Psychology,* vol. 1, pp. 385, 410. Postlethwaite makes much the same point in *Making It Whole,* pp. 208–209.

64. Spencer, *Principles of Psychology,* vol. 1, pp. 368–69. See also George W. Stocking, "Victorian Cultural Ideology and the Image of Savagery," in *Victorian Anthropology* (New York: Free Press, 1987), pp. 224–28.

65. See, for example, Spencer, *Principles of Psychology,* vol. 1, pp. 388–89.

66. Reuchlin argues that Ribot also drew on Spencer's theories indirectly, through his high regard for John Hughlings Jackson and Jackson's Spencer-derived idea of "dissolution" or mental regression. See Reuchlin, *Histoire de la psychologie,* pp. 62–63. On Jackson, see Young, *Mind, Brain and Adaptation,* esp. chap. 6.

67. Ribot to Espinas, 9 March 1867, *Revue philosophique* 147 (1957): 2.

68. On Spencer's central place in Ribot's intellectual development, see Bénichou, "Ribot et l'hérédité psychologique"; Brooks, *Eclectic Legacy,* esp. chap. 2; Nicolas, *Histoire de la psychologie française,* chap. 4; Reuchlin, "Historical

Background for National Trends," pp. 116–17; and Sjöval, *Psychology of Tension,* p. 35.

69. On the history of evolution in France, see Peter J. Bowler, *The Eclipse of Darwinism: Anti-Darwinian Evolution Theories in the Decades around 1900* (Baltimore: Johns Hopkins University Press, 1983), esp. chap. 5; Yvette Conry, *L'Introduction du darwinisme en France au XIXe siècle* (Paris: Vrin, 1974); Pietro Corsi and Paul J. Weindling, "Darwinism in Germany, France, and Italy," in *The Darwinian Heritage,* ed. David Kohn (Princeton: Princeton University Press, 1985), pp. 683–729; John Farley, "The Initial Reaction of French Biologists to Darwin's *Origin of Species,*" *Journal of the History of Biology* 10 (1977): 275–300; Jacques Roger, "Darwin, Haeckel, et les français," in *De Darwin au darwinisme: Science et idéologie,* ed. Yvette Conry (Paris: Vrin, 1983), pp. 149–65; and Robert E. Stebbins, "France," in *The Comparative Reception of Darwinism,* ed. Thomas F. Glick (Austin: University of Texas, 1972), pp. 117–67.

70. See Toby A. Appel, *The Cuvier-Geoffroy Debate: French Biology in the Decades before Darwin* (Oxford: Oxford University Press, 1987); William Coleman, *Georges Cuvier, Zoologist: A Study in the History of Evolution Theory* (Cambridge: Harvard University Press, 1964); Charles C. Gillispie, "The Formation of Lamarck's Evolutionary Theory," *Archives internationales d'histoire des sciences* 9 (1956): 323–38; and Richards, *Darwin and the Emergence of Evolutionary Theories,* pp. 47–70.

71. See Conry, *L'Introduction du darwinisme en France,* p. 45; Corsi and Weindling, "Darwinism in Germany, France, and Italy," p. 702; Bowler, *Eclipse of Darwinism,* pp. 109–17; Roger, "Darwin, Haeckel, et les français"; and Stebbins, "France," pp. 126–28, 152–55, 158–63.

72. See Clémence Royer, "Préface de la première édition," in *De l'origine des espèces,* trans. Clémence Royer (Paris: Guillaumin et Masson, 1866); Conry, *L'Introduction du darwinisme en France,* pp. 40–41; and Stebbins, "France," pp. 126–28. On Royer, see Sara Joan Miles, "Clémence Royer et *De l'origine des espèces:* Traductrice ou traîtresse?" *Revue de synthèse* 110 (1989): 61–83.

73. See, for example, Darwin's obituary from the Société d'Anthropologie de Paris; "Mort de Ch. Darwin," *Bulletins de la Société d'Anthropologie,* 3rd ser., 5 (1882): 348.

74. Ribot, *La psychologie,* p. 39.

75. Hippolyte Taine, *De l'intelligence,* 5th ed., 2 vols. (Paris: Hachette, 1888), vol. 1, p. 291.

76. Ribot, *La psychologie,* p. 196.

77. The still classic study of the pathological method, especially as employed in France, is Georges Canguilhem, *The Normal and the Pathological,* trans. Carolyn R. Fawcett and Robert S. Cohen (New York: Zone Books, 1991). See also Michel Foucault, *The Birth of the Clinic: An Archaeology of Medical Perception,* trans. A. M. Sheridan Smith (New York: Pantheon Books, 1973), esp. chaps. 6–10; John E. Lesch, *Science and Medicine in France: The Emergence of Experimental Physiology, 1790–1855* (Cambridge: Harvard University Press, 1984); Russell Maulitz, *Morbid Appearances: The Anatomy of Pathology in the Early Nineteenth Century* (Cambridge: Cambridge University Press, 1987); Nye, *Crime, Madness, & Politics,* esp. chap. 5; and Schneider, *Quality and Quantity,* esp. chap. 2.

78. Claude Bernard, "Des fonctions du cerveau," *Revue des deux mondes* 98 (1872): 384.

79. This example is drawn from Taine, *De l'intelligence* (1870), vol. 1, pp. 319–29.

80. Two compilations of Charcot's lectures in English are Jean-Martin Charcot, *Clinical Lectures on Diseases of the Nervous System,* ed. Ruth Harris (London: Tavistock/Routledge, 1991); and Charcot, *Charcot, the Clinician: The Tuesday Lectures,* trans. Christopher G. Goetz (New York: Raven Press, 1987). For two very different contemporary assessments of Charcot, see Léon Daudet, *Souvenirs des milieux littéraires, politiques, artistiques et médicaux* (Paris: Nouvelle Librairie Nationale, 1920), pp. 173–224; and Dr. Levillain, "Charcot et l'Ecole de la Salpêtrière," *Revue encyclopédique* (1894): 108–15. For background on Charcot, see Georges Didi-Huberman, *Invention of Hysteria: Charcot and the Photographic Iconography of the Salpêtrière,* trans. Alisa Hartz (Cambridge: MIT Press, 2003); Henri F. Ellenberger, *The Discovery of the Unconscious: The History and Evolution of Dynamic Psychiatry* (New York: Basic Books, 1970), esp. pp. 89–102 and 171–74; Raymond E. Fancher, *Pioneers of Psychology,* 2nd ed. (New York: W. W. Norton, 1990), esp. pp. 337–47; Christopher G. Goetz, Michel Bonduelle, and Toby Gelfand, *Charcot: Constructing Neurology* (New York: Oxford University Press, 1995); Jan Goldstein, *Console and Classify: The French Psychiatric Profession in the Nineteenth Century* (Cambridge: Cambridge University Press, 1990), esp. chap. 9; Ruth Harris, introduction to Charcot, *Clinical Lectures on Diseases of the Nervous System,* pp. ix–lxviii; Alain Lellouch, "La méthode de J. M. Charcot (1825–1893)," *History and Philosophy of the Life Sciences* 11 (1989): 43–69; and Mark S. Micale, "Charcot and the Idea of Hysteria in the Male: Gender, Mental Science, and Medical Diagnosis in Late Nineteenth-Century France," *Medical History* 34 (1990): 363–411.

81. For a vivid analysis of the Salpêtrière during the nineteenth century, see Mark S. Micale, "The Salpêtrière in the Age of Charcot: An Institutional Perspective on Medical History in the Late Nineteenth Century," *Journal of Contemporary History* 20 (1985): 703–31; quotes from pp. 707, 710.

82. See Fancher, *Pioneers of Psychology,* p. 339.

83. For Charcot's commitment to the hereditary nature of neurological disorders, see Toby Gelfand, "Charcot's Response to Freud's Rebellion," *Journal of the History of Ideas* 50 (1989): 293–307; Harris, introduction; and Micale, "Charcot and the Idea of Hysteria in the Male," esp. pp. 382–411. On Charcot's method in general, see Lellouch, "La méthode de J. M. Charcot."

84. On the clinical-laboratory style, see Kurt Danziger's very perceptive article, "Origins of the Psychological Experiment as a Social Institution." On styles of research in psychology in general, see Danziger, *Constructing the Subject: Historical Origins of Psychological Research* (Cambridge: Cambridge University Press, 1990).

85. On the centrality of the clinical and pathological style in French psychology, see Robert A. Nye, *The Origins of Crowd Psychology: Gustave LeBon and the Crisis of Mass Democracy in the Third Republic* (London: Sage, 1975), p. 31; Carroy and Plas, "Origins of French Experimental Psychology"; Régine Plas, "La psychologie pathologique d'Alfred Binet," in *Les origines de la psychologie sci-*

entifique: Centième anniversaire de "L'Année psychologique" (1894–1994), eds. Paul Fraisse and Juan Segui (Paris: Presses universitaires de France, 1994), pp. 229–245; and Reuchlin, *Histoire de la psychologie,* pp. 76–77.

86. Théodule Ribot, *Les maladies de la mémoire* (Paris: Baillière, 1881); *Les maladies de la volonté* (Paris: Baillière, 1883); and *Les maladies de la personnalité* (Paris: Alcan, 1885). See also Gasser, "La notion de la mémoire organique"; and Roth, "Remembering Forgetting."

87. Pierre Janet, *L'Automatisme psychologique: Essai de psychologie expérimentale sur les formes inférieures de l'activité humaine* (Paris: Alcan, 1889). On Janet, see Brooks, *Eclectic Legacy,* esp. chap. 5; Ellenberger, *Discovery of the Unconscious,* chap. 6; Henri Piéron, "Quelques souvenirs," *Psychologie française* 5 (1960): 82–92; Reuchlin, *Histoire de la psychologie,* pp. 63–66; and Sjöval, *Psychology of Tension.*

88. Hippolyte Taine, "Note sur l'acquisition du langage chez les enfants et dans l'espèce humaine," *Revue philosophique* 1 (1876): 5–23.

89. Ibid., p. 22.

90. Ibid., pp. 22–23.

91. The quoted phrase is from Alfred Binet, *Inédits d'Alfred Binet avec une page autographe hors-texte* (Cahors: Coueslant, 1960), p. 13. Carroy and Plas argue that French experimental psychology had two distinct traditions, the pathological, which they associate with Ribot and Janet, and the Wundtian, which they associate with Binet. See Carroy and Plas, "Origins of French Experimental Psychology." On French laboratory psychology, see also Brooks, "Philosophy and Psychology at the Sorbonne"; Nicolas, *Histoire de la psychologie française;* and Reuchlin, "Historical Background for National Trends." However, the dichotomy being drawn between clinical/pathological and laboratory/experimental styles was by no means clear-cut.

92. For more on Wundt and his influence, see Edwin G. Boring, *A History of Experimental Psychology* (New York: D. Appleton, 1929), esp. chap. 15; Danziger, *Constructing the Subject;* and Robert W. Rieber, ed., *Wilhelm Wundt and the Making of a Scientific Psychology* (New York: Plenum Press, 1980). Theta Wolf notes, however, that Binet, for one, was not completely appreciative of Wundt's efforts. See Theta H. Wolf, *Alfred Binet* (Chicago: University of Chicago Press, 1973), pp. 90–91, 111–12.

93. See Nicolas, *Histoire de la psychologie française;* Jean Beuchet, "La vie et œuvre de Benjamin Bourdon (1860–1943)," *Psychologie française* 6 (1961): 173–81; and Paul Fraisse, "L'Evolution de la psychologie expérimentale," in *Traité de psychologie expérimentale,* vol. 1, esp. pp. 34–38. For a firsthand account of French psychology laboratories at the turn of the century, see Nicolas Vaschide, "L'Enseignement de la psychologie expérimentale en France," *Revue internationale de l'enseignement* 45 (1903): 203–14.

94. Henri Piéron provides a description of the Sorbonne laboratory as of 1894 in "Le laboratoire de psychologie de la Sorbonne," in *Centenaire de Th. Ribot.* Edouard Toulouse and Piéron describe the outfitting of an ideal psychological laboratory in *Technique de psychologie expérimentale,* 2nd ed. (Paris: Octave Doin et Fils, 1911), pp. 6–16. See also Françoise Parot, "La psychologie scientifique française et ses instruments au début du XXe siècle," in *Studies in the History of*

Scientific Instruments, eds. Christine Blondel, Françoise Parot, Anthony Turner, and Mari Williams (London: Rogers Turner Books, 1989), pp. 259–64.

95. Françoise Parot emphasizes both the Wundtian orientation and the disinterest in animal experimentation in "La psychologie scientifique française," esp. pp. 261–62.

96. The analysis in this section is heavily indebted to Kurt Danziger's arguments about the nature of modern psychological research. See Danziger, *Constructing the Subject,* esp. pp. 27–39; Danziger, "Origins of the Psychological Experiment"; and Danziger, "The Positivist Repudiation of Wundt," *Journal of the History of the Behavioral Sciences* 15 (1979): 205–30. The Wundtian style is also characterized in Mitchell G. Ash, "Academic Politics in the History of Science: Experimental Psychology in Germany, 1879–1941," *Central European History* 13 (1980): 255–86; R. Steven Turner, "Helmholtz, Sensory Physiology, and the Disciplinary Development of German Psychology," in *The Problematic Science: Psychology in Nineteenth-Century Thought,* eds. William R. Woodward and Mitchell G. Ash (New York: Praeger, 1982), pp. 147–66; and William R. Woodward, "Wundt's Program for the New Psychology: Vicissitudes of Experiment, Theory, and System," in *Problematic Science,* pp. 167–97.

97. For examples of Bourdon's approach, see Benjamin Bourdon, *La perception visuelle de l'espace* (Paris: Costes, 1902); and Bourdon, "Les sensations," vol. 1, and "La perception," vol. 2, in Georges Dumas, *Traité de psychologie* (Paris: Alcan, 1923). For Toulouse, see Edouard Toulouse, Nicolas Vaschide, and Henri Piéron, *Technique de psychologie expérimentale,* 1st ed. (Paris: Octave Doin, 1904); and Toulouse and Piéron, *Technique,* 2nd ed. (1911).

98. Detailed information on Binet's experimental style during the 1890s can be gained from his extant records in the Piéron papers, 520 AP 44–46, AN. See also Alfred Binet, *The Experimental Psychology of Alfred Binet: Selected Papers,* eds. Robert H. Pollack and Margaret W. Brenner (New York: Springer, 1969); Binet, *Introduction à la psychologie expérimentale* (Paris: Alcan, 1894); *L'Année psychologique* 1 (1894)-7 (1900); and Paul Fraisse, "L'Œuvre d'Alfred Binet en psychologie expérimentale," *Psychologie française* 3 (1958): 105–12.

99. "Assemblée des Professeurs, Chaire de Psychologie expérimentale et Comparée," 10 novembre 1901; F^{17} 13556 20bis, "Chaire de psychologie expérimentale et comparée," AN.

100. On the importance of induced versus naturally occurring phenomena, see Carroy and Plas, "Origins of French Experimental Psychology."

101. Binet in 1909 argued explicitly for the novelty of the Parisian approach to experimental psychology in "Le bilan de la psychologie en 1908," *L'Année psychologique* 15 (1909): v–xii.

102. See Binet, "SHARP.—Individual Psychology. A Study in Psychological Method," *L'Année psychologique* 6 (1899): 583–93, quote on p. 583.

103. Carroy and Plas, "Origins of French Experimental Psychology," p. 81.

104. Alfred Binet and Victor Henri, "Le développement de la mémoire visuelle chez les enfants," *Revue générale des sciences pures et appliquées* 5 (1894): 162–69, quote on p. 167.

105. See Alfred Binet and Victor Henri, "La psychologie individuelle," *L'Année psychologique* 2 (1895): 411–65; Binet and Henri, "Le développement de la

mémoire visuelle chez les enfants"; "Recherches sur le développement de la mémoire visuelle des enfants," *Revue philosophique* 37 (1894): 348–50; and Binet, "Psychologie individuelle—La description d'un objet," *L'Année psychologique* 3 (1896): 296–332.

106. See, for example, Binet, "The Perception of Lengths and Numbers in Some Small Children," *Revue philosophique* 30 (1890): 68–81, in *The Experimental Psychology of Alfred Binet,* p. 85; and Theta H. Wolf, "The Emergence of Binet's Conception and Measurement of Intelligence: A Case History of the Creative Process," *Journal of the History of the Behavioral Sciences* 5 (1969): 113–34, 207–37, esp. pp. 117–21.

107. See Binet, "La mesure en psychologie individuelle," *Revue philosophique* 46 (1898): 113–23, quote on p. 113; Binet and Henri, "Le développement de la mémoire visuelle," p. 165; and Jules de la Vaissière, *Eléments de la psychologie expérimentale* (Paris: Beauchesne, 1912), esp. pp. 323–25.

108. Binet and Henri, "La psychologie individuelle," p. 411. For a similar sentiment, see Toulouse, Vaschide, and Piéron, *Technique de psychologie expérimentale* (1904), p. 252.

109. See Binet and Henri, "La psychologie individuelle," p. 435; and Binet, "Perception of Lengths and Numbers," p. 85.

110. Binet and Henri, "Le développement de la mémoire visuelle," p. 166.

111. Alfred Binet and Nicolas Vaschide, "Historique des recherches sur les rapports de l'intelligence avec la grandeur et la forme de la tête," *L'Année psychologique* 5 (1898): 245–98.

112. Nicolas Vaschide and Madelaine Pelletier, *Recherches expérimentales sur les signes physiques de l'intelligence* (LaChapelle-Montligeon: Imprimerie de Montligeon, 1904), p. 10.

113. See Binet and Vaschide, "Historique des recherches," pp. 294–98.

114. By Binet, see his series of articles in *L'Année psychologique* 7 (1900): 314–429 and 8 (1901): 341–89; as well as "Questions de technique céphalométrique d'après M. Bertillon," *L'Année psychologique* 10 (1904): 139–41; and "Les signes physiques de l'intelligence chez les enfants," *L'Année psychologique* 16 (1910): 1–30. See also Vaschide and Pelletier, *Recherches expérimentales sur les signes physiques de l'intelligence;* Théodore Simon, "Recherches anthropométriques sur 223 garçons anormaux âgés de 8 à 23 ans," *L'Année psychologique* 6 (1899): 191–247; and Simon, "Recherches céphalométriques sur les enfants arriérés de la colonie de Vaucluse," *L'Année psychologique* 7 (1900): 430–89.

115. Binet, "Recherches sur la technique de la mensuration de la tête vivante," *L'Année psychologique* 7 (1900): 314.

116. See Binet, "Recherches sur la technique de la mensuration"; Simon, "Recherches céphalométriques sur les enfants arriérés"; and Simon, "Le développement du corps et de la tête chez les enfants anormaux," *Bulletin de la Société libre pour l'étude psychologique de l'enfant* 2 (1901–1902): 109–13.

117. See Binet, "Recherches de céphalométrie sur des enfants d'élite et arriérés des écoles primaires de Paris," *L'Année psychologique* 7 (1900): 412–29. Stephen Jay Gould discusses Binet's "flirtation" with and ultimate rejection of craniometry in *The Mismeasure of Man* (New York: W. W. Norton, 1981), pp. 146–48. For a similar analysis of craniometry that then proposes "direct measures of intelli-

gence" by means of measuring sensory acuity, see Jules-Jean Van Biervliet, "La mesure de l'intelligence," *Journal de psychologie* 1 (1904): 225–35.

118. Vaschide and Pelletier describe much the same procedure in *Recherches expérimentales sur les signes physiques,* p. 27.

119. Monique Vial has come to a similar conclusion in "Les débuts de l'enseignement spécial en France; les revendications qui ont conduit à la loi du 15 avril 1909 créant les classes et écoles de perfectionnement," in *Les institutions de l'éducation spécialisée* (Paris: INRP, 1985), pp. 23–66.

120. Binet, "Recherches de céphalométrie," p. 412.

121. Vaschide and Pelletier, *Recherches expérimentales sur les signes physiques,* p. 33.

122. Binet, *Inédits d'Alfred Binet,* p. 31.

123. See Edouard Toulouse, *Enquête médico-psychologique sur les rapports de la supériorité intellectuelle avec la névropathie. Emile Zola* (Paris: Société d'Editions scientifique, 1896); and "Lettres de Edouard Toulouse à Emile Zola, 1895–1901," papers of Emile Zola, B.N. mss. n.a.fr. 24524 fols. 116–54.

124. See, for example, Charles Chabot, "L'Ecole sur mesure," *Revue pédagogique* 57 (1910): 1–22.

125. See Toulouse, *Enquête médico-psychologique,* pp. x–xi.

126. See Raymond E. Fancher, *The Intelligence Men: Makers of the I.Q. Controversy* (New York: W. W. Norton, 1985), pp. 63–65.

127. See Alfred Binet and L. Henneguy, "Observations et expériences sur le calculateur J. Inaudi," *Revue philosophique* 34 (1892): 204–20; "Les grandes mémoires: Résumé d'une enquête sur les joueurs d'échecs," *Revue des deux mondes* 117 (1893): 826–59; Binet and Jean-Martin Charcot, "Un calculateur de type visuel," *Revue philosophique* 35 (1893): 590–94; Binet and L. Henneguy, *La psychologie des grandes calculateurs et joueurs d'échecs* (Paris: Hachette, 1894); Binet and Jacques Passey, "La psychologie des auteurs dramatiques," *Revue philosophique* 37 (1894): 228–40; Binet and Passey, "Etudes de psychologie sur les auteurs dramatiques," *L'Année psychologique* 1 (1894): 60–118; Binet, "M. François de Curel (Notes psychologiques)," *L'Année psychologique* 1 (1894): 119–73; and Binet, "La création littéraire. Portrait psychologique de M. Paul Hervieu," *L'Année psychologique* 10 (1903): 1–62.

128. Alfred Binet, *La suggestibilité* (Paris: Schleicher, 1900), pp. 119–20, quoted in Fancher, *Intelligence Men,* pp. 62–63.

129. Binet, "La création littéraire," p. 3.

130. Binet and Henneguy, *La psychologie des grandes calculateurs et joueurs d'échecs,* p. 24.

131. For an extended and rich analysis of the phenomenon of choreographing multiple identities, as she has put it, see Charis Cussins, "Ontological Choreography: Agency through Objectification in Infertility Clinics," *Social Studies of Science* 26 (1996): 575–610.

132. For a related example, see Toulouse, *Enquête médico-psychologique;* and Toulouse, "Lettres de Edouard Toulouse à Emile Zola, 1895–1901."

133. Binet, "The Perception of Lengths and Numbers"; "Children's Perceptions," *Revue philosophique* 30 (1890): 582–611; "Studies on Movements in

Some Young Children," *Revue philosophique* 29 (1890): 297–309; and "Imageless Thought," *Revue philosophique* 55 (1903): 138–52; all in *The Experimental Psychology of Alfred Binet,* pp. 79–92, 93–126, and 207–21; and *L'Etude expérimentale de l'intelligence* (Paris: Schleicher Frères & Cie, 1903). On the high regard for *L'Etude expérimentale,* see Nikolai Kostyleff, *La crise de la psychologie expérimentale* (Paris: Alcan, 1911), p. 64; Jacques Larguier des Bancels, "L'Œuvre d'Alfred Binet," in *Centenaire de Th. Ribot,* p. 159; La Vaissière, *Eléments de la psychologie expérimentale,* p. 17; M. Michotte, "Discours de M. Michotte," in *Centenaire de Th. Ribot,* p. 54; Henri Piéron, "Préface," in François-Louis Bertrand, *Alfred Binet et son œuvre* (Paris: Alcan, 1930), p. iv; and Wolf, *Alfred Binet,* p. 117.

134. Binet, *L'Etude expérimentale,* pp. 299–300.

135. Theta Wolf and Raymond Fancher have argued that it is misleading to consider *L'Etude expérimentale* a study of intelligence, because we would now say it examines personality differences. However true currently, at the turn of the century "intelligence" still *could* denote differences in kind rather than simply in degree. For their positions, see Wolf, *Alfred Binet,* p. 116; and Fancher, *Intelligence Men,* p. 65.

136. Maurice Reuchlin has developed this theme in "La psychologie différentielle"; see also Wolf, "Emergence of Binet's Conception and Measurement of Intelligence."

137. See Alfred Binet and Théodore Simon, "Sur la nécessité d'établir un diagnostic scientifique des états inférieurs de l'intelligence"; "Méthodes nouvelles pour le diagnostic du niveau intellectuel des anormaux"; and "Application des méthodes nouvelles au diagnostic du niveau intellectuel chez des enfants normaux et anormaux d'hospice et d'école primaire"; all in *L'Année psychologique* 11 (1905): 163–336. For English translations, see *The Development of Intelligence in Children,* trans. Elizabeth Kite (Baltimore: Williams & Wilkins, 1916). On the composition of the commission, see Louette, "Commission des anormaux," *Bulletin de la Société libre pour l'étude psychologique de l'enfant* 4 (1903–1904): 406–407, quote on p. 406; Monique Vial, "Les débuts de l'enseignement spécial en France, les instances politiques nationales et la création des classes et écoles de perfectionnement: Les artisans du projet de loi (1904–1909)," *Travaux du CRESAS* (1982): 7–150; and Joseph Chaumié, "Arrêté du 4 octobre 1904," in *La Commission Bourgeois, 1904–1905: Documents pour l'histoire de l'éducation spécialisée,* eds. Monique Vial and Marie-Anne Hugon (Paris: Éd. du CTNERHI, 1998), pp. 35–36.

138. Binet and Simon, "Upon the Necessity of Establishing a Scientific Diagnosis of Inferior States of Intelligence," in *Development of Intelligence in Children,* p. 9. Louette summarized the commission's proposals in "Commission des anormaux," p. 407.

139. See "Travaux de la commission ministérielle pour les enfants anormaux," *Bulletin de la Société libre pour l'étude psychologique de l'enfant* 6 (1905–1906): 57–61, quote on p. 61; and Binet, "Commission des anormaux," *Bulletin de la Société libre pour l'étude psychologique de l'enfant* 4 (1903–1904): 407–408, quote on p. 408.

140. Binet and Simon, "Necessity of Establishing a Scientific Diagnosis," p. 10.

141. As a model, Binet may have looked to the work of Dr. Blin, a colleague of Simon's at Vaucluse, and Blin's student Henri Damaye, who in the early 1900s developed a questionnaire for assessing an individual's global intellectual level that was also based on grading responses to general knowledge questions. Binet aided them in this project and reviewed the result in his presentation of the Binet-Simon scale. Wolf suggests that they may have been especially influential in suggesting to Binet that he drop analyzing intelligence as a collection of independent faculties, and instead treat it clinically as a complex whole. See Binet and Simon, "Necessity of Establishing a Scientific Diagnosis," pp. 27–36; and Wolf, *Alfred Binet,* esp. pp. 173–75.

142. Binet and Simon, "Application of the New Methods to the Diagnosis of Intellectual Level," in *Development of Intelligence in Children,* pp. 99–101, 145–46.

143. See Félix Larrivé, *Education, traitement, et assistance des enfants anormaux* (Isère: Etablissement médical de Meyzieux, 1903), p. 1.

144. "Diagnosis" was also a taxonomic term employed since Linnaeus to refer to identifying characteristics that can be used to separate individuals into categories.

145. Binet and Simon, "Application of the New Methods," p. 140. Indeed, Simon recollected fifty years later that focus on the feebleminded was critical to the development of the scale. See Y. Jenger, "Rencontre avec Docteur Simon," *L'Education nationale* 1 (1958): 1–3.

146. Binet and Simon, "New Methods for the Diagnosis of the Intellectual Level of the Abnormal," in *Development of Intelligence in Children,* pp. 41, 42.

147. Binet's admonitions about scoring are found especially on pp. 69, 123–39, and in his case studies, pp. 139–81, in "Application of the New Methods."

148. Binet and Simon, "Méthodes nouvelles pour le diagnostic," p. 243.

149. Theta Wolf discusses this sense of the term in "Emergence of Binet's Conception," p. 220.

150. See Vaney, "Combien existe t-il en France d'enfants arriérés de l'intelligence?" *Bulletin de la Société libre pour l'étude psychologique de l'enfant* 6 (1905–1906): 94–97, quote on p. 96.

151. Binet and Simon, "Necessity of Establishing a Scientific Diagnosis," pp. 11, 14, 24.

152. Ibid., p. 28 (emphasis added).

153. See ibid., "The Development of Intelligence in the Child," *L'Année psychologique* 11 (1905): 1–90 in *Development of Intelligence in Children,* pp. 184–235.

154. On expanding the target population for the Binet-Simon, see Wolf, *Alfred Binet,* p. 190.

155. Binet and Simon, "Development of Intelligence in the Child," p. 229 (emphasis added).

156. Binet vacillated between acknowledging the machinelike quality of the scale and then repudiating these very same qualities. See Binet and Simon, "Development of Intelligence in the Child," pp. 239–40; and Alfred Binet, *Les idées modernes sur les enfants* (Paris: Flammarion, 1909), pp. 136–37.

157. Binet, "New Investigations Upon the Measure of the Intellectual Level Among School Children," *L'Année psychologique* 17 (1911): 145–201, in *Development of Intelligence in Children,* pp. 274–329.

158. For examples of Binet's emphasis on the need for experimenter judgment, see Binet, "New Investigations," esp. pp. 295–96.

159. See, for example, ibid., *Les idées modernes sur les enfants,* p. 137.

160. Ibid., "New Investigations," pp. 297–98.

161. Stephen Jay Gould, Theta Wolf, and Read D. Tuddenham emphasize that Binet never committed to viewing intelligence as a single mental faculty, preferring to see intelligence as a complex of mental functions expressed in a set of externalized behaviors. Nevertheless, as constructed, the Binet-Simon scale *did* produce a singular measurement and was designed to be broadly administered. See Gould, *Mismeasure of Man,* pp. 150–52; Wolf, "Emergence of Binet's Conception," pp. 235–36; and Tuddenham, "The Nature and Measurement of Intelligence," in *Psychology in the Making,* ed. Leo Postman (New York: Knopf, 1962), pp. 469–525, quote on p. 490.

162. On the development of the notion of "normal" and its relation to both pathology and statistics, see Canguilhem, *Normal and the Pathological;* Michel Foucault, *Discipline and Punish: The Birth of the Prison,* trans. Alan Sheridan (New York: Vintage, 1979); Ian Hacking, *The Taming of Chance* (Cambridge: Cambridge University Press, 1990), esp. chaps. 19 and 21; Elizabeth Lunbeck, *The Psychiatric Persuasion: Knowledge, Gender, and Power in Modern America* (Princeton: Princeton University Press, 1994); and Theodore Porter, *The Rise of Statistical Thinking, 1820–1900* (Princeton: Princeton University Press, 1986).

163. Binet, "New Investigations," p. 281.

164. See ibid., "Development of Intelligence in the Child," p. 184 below; and Morlé, "L'Influence de l'état social sur le dégrée de l'intelligence des enfants," *Bulletin de la Société libre pour l'étude psychologique de l'enfant* 12 (1911–12): 8–15.

165. See Binet, "New Investigations," p. 277.

166. Gaby Netchine-Grynberg similarly argues that only with Binet did the various discrete categories of mental inferiority get connected both to one another and to the normal; see Gaby Netchine-Grynberg, "De l'idiotie a la débilité mentale, ou les étapes de l'abstraction nécessaire," in *Les débilités mentales,* ed. René Zazzo (Paris: Colin, 1979), esp. pp. 71–85.

167. See, for example, V. Vaney, "Nouvelles méthodes de mesure applicables au degré d'instruction des élèves," *L'Année psychologique* 11 (1905): 146–62. On the general phenomenon of the regularization of French society during this period, see Eugen Weber, *Peasants into Frenchmen: The Modernization of Rural France, 1870–1914* (Stanford: Stanford University Press, 1976); and for education, Monique Vial, "La création du perfectionnement en 1909," in *Intégration ou marginalisation? Aspects de l'éducation spécialisée* (Paris: INRP, 1984), pp. 47–73.

168. Reuchlin has commented on Binet's shift in orientation in "La psychologie différentielle," esp. p. 390.

169. Alfred Binet and Théodore Simon, "Méthodes nouvelles pour diagnostiquer l'idiotie, l'imbécillité et la débilité mentale," *Atti del V Congresso internazionale di psicologia* (Rome: Foranzi, 1905), p. 508.

170. Binet, "Préface," *L'Année psychologique* 14 (1908): v–vi, quote on p. v.

171. Ibid., "Le bilan de psychologie en 1910," *L'Année psychologique* 17 (1911): v–xi, quote on p. x.

172. On technologies as mediators, see M. Norton Wise, "Mediating Machines," *Science in Context* 2 (1988): 77–113.

173. Gould makes much the same point in *Mismeasure of Man,* pp. 152–54.

174. Binet, *Les idées modernes sur les enfants,* p. 141. See also Binet and Simon, "New Methods for the Diagnosis of Intellectual Level," p. 37.

175. For background on Binet, see Fancher, *Intelligence Men,* esp. chap. 2; and Wolf, *Alfred Binet,* esp. chaps. 1–2. What little remains of Binet's correspondence is scattered in a number of collections, although most of these are located in the Salle des manuscrits, Bibliothèque Nationale (BN), Paris. See the papers of Edmond Goncourt, BN mss. n.a.fr. 22452 fols. 266–70; Louis Havet, BN mss. n.a.fr. 24488 fols. 55–63; and Gaston Paris, BN mss. n.a.fr. 24432 fols. 125–29. In addition, see the papers of Jacques Larguier des Bancels, Bibliothèque Cantonale, Lausanne; Marian Bejat, "Une correspondance inédite d'Alfred Binet," *Revue roumaine des sciences sociales: Série de psychologie* 10 (1966): 199–212; the papers of Edouard Claparède in Geneva; and the Binet materials in the Piéron papers, 520 AP 44–46, AN. Other important secondary works on Binet include Bertrand, *Alfred Binet et son œuvre;* Carroy and Plas, "Origins of French Experimental Psychology"; Serge Nicolas, "Alfred Binet et *L'Année psychologique* d'après une correspondance inédite," *L'Année psychologique* 97 (1997): 665–99; Nicolas, *Histoire de la psychologie française;* Nicolas and Ludovic Ferrand, "Alfred Binet and Higher Education," *History of Psychology* 5 (2002): 264–83; Edith J. Varon, "Alfred Binet's Concept of Intelligence," *Psychological Review* 43 (1936): 32–58; and Wolf, "Emergence of Binet's Conception."

176. Alfred Binet, *Les altérations de la personnalité* (Paris: Alcan, 1892), esp. pp. 67–68; and Binet, *La suggestibilité.* For an overview of the incident, see Jacqueline Carroy, *Hypnose, suggestion et psychologie: L'Invention du sujets* (Paris: Presses universitaires de France, 1991); Ellenberger, *Discovery of the Unconscious;* Goetz, Bonduelle, and Gelfand, *Charcot,* chap. 6; Serge Nicolas, "L'Ecole de la Salpêtrière en 1885," *Psychologie et histoire* 1 (2000): 165–207; and Wolf, *Alfred Binet,* chap. 2.

177. On the founding of the journal, see Serge Nicolas, Juan Segui, and Ludovic Ferrand, "*L'Année psychologique:* History of the Founding of a 100-Year-Old French Journal," *History of Psychology* 3 (2000): 44–61. See also E. Chapuis, "*L'Année psychologique* dans la correspondance de Jean Larguier des Bancels," *L'Année psychologique* 97 (1997): 643–63; and Nicolas, "Alfred Binet et *L'Année psychologique.*"

178. On the activity at the Sorbonne laboratory, see Binet's extant records in the Piéron papers, 520 AP 44–46, AN. A layout of the laboratory is included in file #2, p. 2, 520 AP 45. Piéron provides reminiscences about the early years of the laboratory, though from a perspective decidedly hostile to Binet, in "Le laboratoire de psychologie de la Sorbonne" and "Discours d'Henri Piéron pour le 75e anniversaire du laboratoire de psychologie de la Sorbonne," *L'Année psychologique* 65 (1965): 6–15. For a secondary account of the laboratory, see Wolf, *Alfred Binet,* esp. pp. 12–15, 85–115.

179. On Binet's attempt in 1901 to gain the chair being vacated by Ribot at the Collège de France, see the records of the Collège for 10 and 14 novembre 1901,

19 and 22 janvier 1902, and 13 février 1902 in F^{17} 13556 20bis, AN; and letter to Paul Passy (4 juillet 1901) in the papers of Louis Havet. Pierre Janet was elected in a close vote. Binet discussed his desire to obtain Janet's post at the Sorbonne in a letter to Havet (11 mai 1902); Binet again lost out, this time to Georges Dumas. For the debate in the Faculty of Letters of the University of Paris, see "Conseil de la Faculté, 15 mars 1902," *Faculté des Lettres de l'Université de Paris—Registre des actes et déclarations, 1900–1906,* pp. 65–69, AJ^{16} 4749, AN. Wolf analyzes both of these rejections in *Alfred Binet,* pp. 22–28. See also Nicolas and Ferrand, "Alfred Binet and Higher Education."

180. On Binet and the Société libre, see Wolf, *Alfred Binet,* pp. 21–22, 160–72, and 283–326.

181. On this general point, see Wolf, *Alfred Binet,* pp. 25–28. Robert Nye illuminates the continued importance of salon-style culture in Third Republic France in his discussion of the monthly banquets organized by Ribot and Le Bon in 1892, the Déjeuner des XX, and of Le Bon's own weekly banquets, started in 1901, "Les Déjeuners du Mercredi." See Nye, *Origins of Crowd Psychology,* pp. 30, 83–85; and also Ringer, *Fields of Knowledge.*

182. Michotte makes much the same point in "Discours de M. Michotte," pp. 48–54.

183. On Piéron's selection to be director, see Piéron, "Le laboratoire de psychologie de la Sorbonne"; and "Discours d'Henri Piéron."

184. On this point, see Nicolas and Ferrand, "Alfred Binet and Higher Education"; and Wolf, *Alfred Binet.*

185. On this phenomenon generally, see John W. Burrow, *The Crisis of Reason: European Thought, 1848–1914* (New Haven: Yale University Press, 2000); H. Stuart Hughes, *Consciousness and Society: The Reorientation of European Social Thought, 1890–1930* (New York: Knopf, 1958); Harry Paul, "The Debate over the Bankruptcy of Science in 1895," *French Historical Studies* 3 (1968): 299–327; and Pierre-Olivier Walzer, *Le XXe siècle I, 1896–1920,* vol. 1 (Paris: Arthaud, 1975), esp. pp. 121–35.

186. Paul Bourget, *Le Disciple* (1889; reprint, Paris: A. Lemerre, 1899).

187. Maurice Blondel, *L'Action: Essai d'une critique de la vie et d'une science de la pratique* (Paris: Alcan, 1893); and Ferdinand Brunetière, *La Science et le religion: Réponse à quelques objections* (Paris: Firmin-Didot, 1895). For background, see Paul, "Debate over the Bankruptcy of Science."

188. For more on Bergson, see Hughes, *Consciousness and Society,* esp. pp. 113–25.

189. See Henri Bergson, *Creative Evolution* (1907), trans. Arthur Mitchell (New York: Holt, 1911).

190. On this event, see Nicolas and Ferrand, "Alfred Binet and Higher Education"; and Nicolas, *Histoire de la psychologie française.*

191. Kostyleff, *La crise de la psychologie expérimentale,* p. 47.

192. See, for example, Georges Dwelshauvers, *La psychologie française contemporaine* (Paris: Alcan, 1920), p. 113.

193. For an excellent examination of the power of physicians in Third Republic France, see Ellis, *Physician-Legislators of France.*

194. See Zeldin, *France, 1848–1945,* vol. 2, pp. 193–95.

195. On this point, see Vial, "Les débuts de l'enseignement spécial en France; les revendications," esp. pp. 53–58; and "Les débuts de l'enseignement spécial en France; les instances politiques," pp. 111–32.

196. For more on the system of "talent," see chapter 2.

197. On this phenomenon, see Zeldin, *France 1848–1945;* and, for a more theoretical take, Pierre Bourdieu, *Distinction: A Social Critique of the Judgment of Taste,* trans. Richard Nice (Cambridge: Harvard University Press, 1987).

198. Theodore Porter, *Trust in Numbers: The Pursuit of Objectivity in Science and Public Life* (Princeton: Princeton University Press, 1995), chap. 6.

199. The two most important scholars of the development of special education in France are Gaby Netchine-Grynberg and Monique Vial. For examples of Netchine-Grynberg's contributions, see "De l'idiotie a la débilité mentale," pp. 53–86; and "De l'idiotie au handicap: La marche vers l'indétermination," *Raison présente* 65 (1983): 51–63. For Vial, in addition to those already cited, see especially *Les enfants anormaux à l'école* (Paris: Colin, 1990); and Marie-Ann Hugon, Jacqueline Gateaux, and Monique Vial, "Les enfants des classes de perfectionnement (1907–1950)," in *Intégration ou marginalisation? Aspects de l'éducation spécialisée* (Paris: INRP, 1984), pp. 75–104. Also valuable are Marc Barthélémy, *Histoire de l'enseignement spécial en France, 1760–1990* (Cergy-Saint-Christophe: Editions dialogues, 1996); Michel Chauvière, "Pour une histoire de l'éducation spéciale: Les interfaces des années 1940," in *The Making of Frenchmen: Current Directions in the History of Education in France, 1679–1979,* eds. Donald N. Baker and Patrick J. Harrigan (Waterloo: Historical Reflections Press, 1980), pp. 601–15; Yves Pelicier and Guy Thullier, "Pour une histoire de l'éducation des enfants idiots en France (1830–1914)," *Revue historique* 261 (1979): 99–130; and Philippe Raynaud, "L'Education spécialisée en France, 1882–1982," *Esprit* 5 (1982): 76–99 and 7/8 (1982): 104–26. On Binet's contribution, see Plas, "La psychologie pathologique d'Alfred Binet."

200. See, for example, "Les enfants anormaux," *Manuel général de l'instruction primaire* 35 (1900): 215; "Les anormaux," *Journal des instituteurs* 44 (1899–1900): 793–94; Ch. Drouard, *Les Ecoles urbaines: Education des enfants normaux, des enfants anormaux et des adolescents. Œuvres circumscolaires et postscolaires* (Paris: Belin Frères, 1902), pp. 367–79; Levistre, *Rapport de la commission spéciale sur la Création d'Ecoles pour les Enfants anormaux et les indisciplines, Supplément au Bulletin de la Société Pédagogique des Directeurs et Directrices d'Ecoles publiques de la Ville de Paris* (Paris: J Couturier, 1904); and Jean Philippe and G. Paul-Boncour, "Vrai et faux anormaux scolaires," *Revue pédagogique* 45 (1904): 441–52.

201. For one of many examples, see "Les enfants anormaux," *Manuel général de l'instruction primaire.*

202. "L'Instruction et l'éducation des anormaux," *Journal des instituteurs* 44 (1899–1900): 45–46, quote on p. 46; and *Rapport de la Commission Spéciale,* p. 2.

203. Schneider, *Quality and Quantity,* esp. chap. 3.

204. On this general phenomenon, see Linda L. Clark, *Social Darwinism in*

France (Tuscaloosa: University of Alabama Press, 1984), esp. chaps. 3, 4, 8, 9; Nye, *Crime, Madness, and Politics in Modern France,* esp. chaps. 4–5; Daniel Pick, *Faces of Degeneration: A European Disorder, c.1848–c.1918* (Cambridge: Cambridge University Press, 1989), chaps. 2–4; and Schneider, *Quality and Quantity.*

205. See, for example, "Les anormaux," *Journal des instituteurs;* Larrivé, *Education, traitement et assistance des enfants anormaux;* and Désiré Bourneville, "Classification établie par M. le Docteur Bourneville dans sa lettre du 5 novembre 1896, adressé à M. le Directeur de l'Enseignement primaire du Département de la Seine," in *La Commission Bourgeois (1904–1905),* pp. 65–72.

206. See Nye, *Crime, Madness, & Politics,* chaps. 4 and 5; and Léon Bourgeois, *Solidarité* (Paris: A. Colin, 1896).

207. Chaumié, "Arrêté du 4 octobre 1904," p. 35.

208. For a list of Commission members, see Chaumié, "Arrêté du 4 octobre 1904," pp. 35–36.

209. For the final report, including all of the excerpts provided, see "Travaux de la Commission ministérielle pour les enfants anormaux," *Bulletin de la Société libre pour l'étude psychologique de l'enfant* 6 (1905–1906): 57–61. For the subcommittee report drafted by Binet, see Alfred Binet, "Rapport au nom de la Sous-Commission chargée d'étudier les solutions à proposer en faveur des arriérés et des instables," April 1905, in *La Commission Bourgeois,* pp. 245–63.

210. "Lois et règlements relatifs à l'éducation des enfants arriérés, Loi du 15 avril 1909," in *Bulletin de la Société libre pour l'étude psychologique de l'enfant* 11 (1910–11): 123–25.

211. "Décret relatif aux conditions d'obtention du certificat d'aptitude à l'enseignement des enfants arriérés," 25 August 1909, in *Bulletin de la Société libre pour l'étude psychologique de l'enfant* 11 (1910–11): 126.

212. See Louette, "Commission des anormaux," p. 406.

213. See Zeldin, *France, 1848–1945,* vol. 2, p. 200.

214. See Nye, *Crime, Madness, and Politics in Modern France,* chaps. 1 and 2; and, for a broad overview, Canguilhem, *Normal and the Pathological.*

215. See, for example, Louis Gobron, "Législation des établissements spéciaux aux enfants arriérés," *Revue pédagogique,* n.s., 57 (1910): 563–79, esp. pp. 564, 566, 575, 577; and [Alfred Binet and Théodore Simon], "Les classes pour les enfants arriérés," *Bulletin de la Société libre pour l'étude psychologique de l'enfant* 11 (1910–11): 51–150, pp 56–59.

216. V. Vaney, "L'Examen pédagogique des arriérés," *Bulletin de la Société libre pour l'étude psychologique de l'enfant* 9 (1908–09): 99–100.

217. See, for example, [Binet and Simon], "Les classes pour les enfants arriérés," pp 65–66.

218. See Binet, "Development of Intelligence in the Child," esp. pp. 254–59.

219. See Porter, *Trust in Numbers,* pp. 142–45.

220. Vial, "Les débuts de l'enseignement spécial en France; les instances politiques," p. 127.

221. Ibid., "La création du perfectionnement," pp. 58–64.

222. Roger Cousinet, "Intelligence," in *Nouveau dictionnaire de pédagogie et instruction primaire,* ed. Ferdinand Buisson (Paris: Hachette, 1911), p. 862.

Chapter 5

1. G. F. Stout and James Mark Baldwin, "Intellect (or Intelligence)," in *Dictionary of Philosophy and Psychology,* vol. 1 (1901; reprint, New York: Macmillan, 1918), p. 558.

2. Samuel C. Kohs, "The Binet-Simon Measuring Scale for Intelligence: An Annotated Bibliography," *Journal of Educational Psychology* 5 (1914): 215–24, 279–90, 335–46; and Kohs, "An Annotated Bibliography of Recent Literature on the Binet-Simon Scale (1913–1917)," *Journal of Educational Psychology* 8 (1917): 425–38, 488–502, 559–65, 609–18.

3. For a revealing contemporary discussion of the meaning(s) of intelligence, see Lightner Witmer, "On the Relation of Intelligence to Efficiency," *Psychological Clinic* 9 (1915): 61–86. For a provocative overview, see Daniel C. Calhoun, *The Intelligence of a People* (Princeton: Princeton University Press, 1973), esp. chap. 1.

4. Theodore M. Porter, *Trust in Numbers: The Pursuit of Objectivity in Science and Public Life* (Princeton: Princeton University Press, 1995); Dorothy Ross, *The Origins of American Social Science* (Cambridge: Cambridge University Press, 1991); and Robert H. Wiebe, *The Search for Order, 1877–1920* (New York: Hill and Wang, 1967).

5. Mike Hawkins, *Social Darwinism in European and American Thought, 1860–1945: Nature as Model and Nature as Threat* (Cambridge: Cambridge University Press, 1997).

6. Mark B. Adams, ed., *The Wellborn Science: Eugenics in Germany, France, Brazil, and Russia* (New York: Oxford University Press, 1990); Mark H. Haller, *Eugenics* (New Brunswick: Rutgers University Press, 1963); Daniel J. Kevles, *In the Name of Eugenics: Genetics and the Uses of Human Heredity* (Berkeley: University of California Press, 1985); Nancy Leys Stepan, *"The Hour of Eugenics": Race, Gender, and Nation in Latin America* (Ithaca: Cornell University Press, 1991); and Alexandra Minna Stern, *Eugenic Nation: Faults and Frontiers of Better Breeding in Modern America* (Berkeley: University of California Press, 2005).

7. Although Leila Zenderland points out that rival systems of classification meant that asylum populations looked heterogeneous to most early twentieth-century observers. See Leila Zenderland, *Measuring Minds: Henry Herbert Goddard and the Origins of American Mental Testing* (Cambridge: Cambridge University Press, 1998).

8. On the early twentieth-century history of American psychology, see Edwin G. Boring, "The Influence of Evolutionary Theory upon American Psychological Thought," in *History, Psychology, and Science* (New York: Wiley, 1963), pp. 159–84; Thomas M. Camfield, "The Professionalization of American Psychology, 1870–1917," *Journal of the History of the Behavioral Sciences* 9 (1973): 66–75; Hamilton Cravens, *The Triumph of Evolution: American Scientists and the Heredity-Environment Controversy, 1900–1941* (Philadelphia: University of Pennsylvania Press, 1978); Kurt Danziger, *Constructing the Subject: Historical Origins of Psychological Research* (Cambridge: Cambridge University Press, 1990); R.G.A. Dolby, "Transmission of Two New Scientific Disciplines from Europe to North America in the Late Nineteenth Century," *Annals of Science* 34 (1977):

287–310; Ernest R. Hilgard, *Psychology in America: A Historical Survey* (San Diego: Harcourt Brace Jovanovich, 1987); Jill G. Morawski and Gail A. Hornstein, "Quandary of the Quacks: The Struggle for Expert Knowledge in American Psychology, 1890–1940," in *The Estate of Social Knowledge,* eds. JoAnne Brown and David K. vanKeuren (Baltimore: Johns Hopkins University Press, 1991), pp. 106–33; John M. O'Donnell, *The Origins of Behaviorism: American Psychology, 1870–1920* (New York: New York University Press, 1985); and Daniel J. Wilson, *Science, Community, and the Transformation of American Philosophy, 1860–1930* (Chicago: University of Chicago Press, 1990).

9. See Zenderland's excellent analysis in *Measuring Minds* chaps. 5–8; and Matthew Frye Jacobson, *Barbarian Virtues: The United States Encounters Foreign Peoples at Home and Abroad, 1876–1917* (New York: Hill and Wang, 2000), pp. 163–72.

10. Similar arguments emphasizing the importance of the pathological to the development of the intelligence scale have been made by Nikolas Rose and Leila Zenderland. See Nikolas Rose, *The Psychological Complex: Psychology, Politics and Society in England, 1869–1939* (London: Routledge & Kegan Paul, 1985); Rose, "Calculable Minds and Manageable Individuals," *History of the Human Sciences* 1 (1988): 179–200; and Zenderland, *Measuring Minds.*

11. See American Psychological Association, *A Survey of Psychological Investigations with Reference to Differentiations between Psychological Experiments and Mental Tests* (Swarthmore: Committee on the Academic Status of Psychology, 1916).

12. E. L. Youmans, "On the Scientific Study of Human Nature," in *The Culture Demanded By Modern Life; A Series of Addresses and Arguments on the Claims of Scientific Education* (New York: D. Appleton, 1867), pp. 375–76, 377, 378–94. Even Emerson acknowledged the place of science in understanding human nature; see Ralph Waldo Emerson, "Natural History of Intellect" (1870), in *The Complete Works of Ralph Waldo Emerson,* vol. 12: *Natural History of Intellect, and Other Papers* (Boston: Houghton Mifflin, 1904).

13. Morton G. White casts this as a turn from formalism to embrace of the dynamic quality of social life; see *Social Thought in America: The Revolt Against Formalism* (New York: Viking Press, 1949). See also Paul J. Croce, *Science and Religion in the Era of William James.* Volume 1: *The Eclipse of Certainty, 1820–1880* (Chapel Hill: University of North Carolina Press, 1995); Thomas L. Haskell, *The Emergence of Professional Social Science* (Urbana: University of Illinois Press, 1977); John Higham, "Hanging Together: Divergent Unities in American History," *Journal of American History* 41 (1974): 5–28; Louis Menand, *The Metaphysical Club: A Story of Ideas in America* (New York: Farrar, Straus & Giroux, 2001); and Wiebe, *Search for Order.*

14. On the social impacts of Darwinism, see Robert C. Bannister, *Social Darwinism: Science and Myth in Anglo-American Thought* (Philadelphia: Temple University Press, 1979); Donald C. Bellomy, "'Social Darwinism' Revisited," *Perspectives in American History,* n.s., 1 (1984): 1–129; Paul F. Boller, *American Thought in Transition: The Impact of Evolutionary Naturalism, 1865–1900* (Chicago: Rand McNally, 1969); Carl N. Degler, *In Search of Human Nature: The Decline and*

Revival of Darwinism in American Social Thought (New York: Oxford University Press, 1991), esp. chaps. 1–2; Hawkins, *Social Darwinism;* and Richard Hofstadter, *Social Darwinism in American Thought* (Boston: Beacon Press, 1955).

15. See Youmans, "Preface" and "Introduction," *Culture Demanded By Modern Life.* For a similar critique of the classical curriculum, see John Fiske, "Considerations on University Reform," *Atlantic Monthly* 19 (1867): 451–65.

16. On the changes in American higher education, see Julie A. Reuben, *The Making of the Modern University: Intellectual Transformation and the Marginalization of Morality* (Chicago: University of Chicago Press, 1996); Fredrick Rudolph, *The American College and University: A History* (New York: Vintage, 1962), esp. chaps. 12–20; and Laurence R. Veysey, *The Emergence of the American University* (Chicago: University of Chicago Press, 1965).

17. On the spread of scientific education in general in the antebellum period, see Rudolph, *American College and University,* pp. 221–40.

18. On the secularization of postbellum American society, see Boller, *American Thought in Transition;* George M. Fredrickson, *The Inner Civil War: Northern Intellectuals and the Crisis of the Union* (Urbana: University of Illinois Press, 1993); John Higham, "The Reorientation of American Culture in the 1890's," in *Writing American History: Essays on Modern Scholarship* (Bloomington: Indiana University Press, 1970), pp. 73–102; Kevles, *In the Name of Eugenics,* esp. chaps. 1–7; James T. Kloppenberg, *Uncertain Victory: Social Democracy and Progressivism in European and American Thought, 1870–1920* (New York: Oxford University Press, 1986); Menand, *Metaphysical Club;* Ross, *Origins of American Social Science;* Wiebe, *Search for Order;* and Wilson, *Science, Community, and the Transformation of American Philosophy.*

19. Owen Hannaway and David Hollinger have both emphasized the widespread belief in, in Hollinger's words, "the moral efficacy of scientific practice." See Owen Hannaway, "The German Model of Chemical Education in America: Ira Remsen at Johns Hopkins (1876–1913)," *Ambix* 23 (1976): 145–64; and David A. Hollinger, "Inquiry and Uplift: Late Nineteenth-Century American Academics and the Moral Efficacy of Scientific Practice," in *The Authority of Experts,* ed. Thomas L. Haskell (Bloomington: Indiana University Press, 1984), pp. 142–56. For a good example, see J. McKeen Cattell, "Science, Education, and Democracy," *Science,* n.s., 39 (1914): 154–64.

20. Louis Agassiz, "Prof. Agassiz on the Origin of Species," *American Journal of Science and Arts,* 2nd ser., 30 (1860): 154; Agassiz, "Evolution and the Permanence of Type," *Atlantic Monthly* 33 (1874): 92–101; and John Bascom, "Darwin's Theory of the Origin of Species," *American Presbyterian Review,* n.s., 3 (1871): 349–79.

21. Bascom, "Darwin's Theory," pp. 372, 379.

22. Edward Drinker Cope, *The Origin of the Fittest: Essays on Evolution* (London: Macmillan, 1887); John Fiske, *Darwinism, and Other Essays* (New York: Macmillan, 1879); Asa Gray, *Darwiniana: Essays and Reviews Pertaining to Darwinism* (New York: D. Appleton, 1876); and Chauncey Wright, "Evolution and Its Explanation," "The Logic of Biology," and "The Meaning of Accident," in *The Philosophical Writings of Chauncey Wright: Representative Selections,* ed. Edward H. Madden (New York: Liberal Arts Press, 1958), pp. 28–38. For back-

ground, see Croce, *Science and Religion in the Era of William James,* chaps. 4–6; Menand, *Metaphysical Club,* esp. chaps. 6, 9.

23. See, for example, Cope, *Origin of the Fittest,* pp. 281, 378–404.

24. See, for example, Lambert Adolphe Jacques Quetelet, *A Treatise on Man, and the Development of His Faculties* (1835; reprint, Gainesville: Scholars' Facsimiles & Reprints, 1969).

25. Ian Hacking, *The Taming of Chance* (Cambridge: Cambridge University Press, 1990), chaps. 13, 21; Theodore M. Porter, *The Rise of Statistical Thinking, 1820–1900* (Princeton: Princeton University Press, 1986), chaps. 4–5; and Stephen M. Stigler, *The History of Statistics: The Measurement of Uncertainty Before 1900* (Cambridge: Harvard University Press, 1986), chap. 5.

26. See, for example, E. B. Tylor, "Quetelet on the Science of Man," *Popular Science Monthly* 1 (1872): 45–55 (reprinted from *Nature*). On Quetelet in America, see Menand, *Metaphysical Club,* chap. 8.

27. Emile Durkheim, *Suicide: A Study in Sociology* (1897), trans. John A. Spaulding and George Simpson (Glencoe: Free Press, 1951). See also Hacking, *Taming of Chance,* chap. 20; and Porter, *Rise of Statistical Thinking,* pp. 68–69.

28. Francis Galton, *Hereditary Genius: An Inquiry into its Laws and Consequences* (London: Macmillan, 1869); and Galton, *Natural Inheritance* (London: Macmillan, 1889). On Galton, see Ruth Schwartz Cowan, "Nature and Nurture: The Interplay of Biology and Politics in the Work of Francis Galton," *Studies in the History of Biology* 1 (1977): 133–208; Cowan, *Sir Francis Galton and the Study of Heredity in the Nineteenth Century* (New York: Garland, 1985); Solomon Diamond, "Francis Galton and American Psychology," in *Psychology: Theoretical-Historical Perspectives,* eds. Robert W. Rieber and Kurt Salzinger (New York: Academic Press, 1980), pp. 43–55; Raymond E. Fancher, "Alphonse de Candolle, Francis Galton, and the Early History of the Nature-Nurture Controversy," *Journal of the History of the Behavioral Sciences* 19 (1983): 341–52; Fancher, *The Intelligence Men: Makers of the IQ Controversy* (New York: W. W. Norton, 1985), pp. 18–40; D. W. Forrest, *Francis Galton: The Life and Work of a Victorian Genius* (London: Elek, 1974); Nicholas W. Gillham, *A Life of Sir Francis Galton: From African Exploration to the Birth of Eugenics* (New York: Oxford University Press, 2001); Kevles, *In the Name of Eugenics,* esp. chap. 1; Porter, *Rise of Statistical Thinking,* esp. pp. 128–46, 270–314; and Stigler, *History of Statistics,* chap. 8.

29. See William James, "Great Men, Great Thoughts, and the Environment," *Atlantic Monthly* 46 (1880): 441–59; James, "The Importance of Individuals" (1890), in *The Will to Believe and Other Essays in Popular Philosophy* (1897; reprint, Cambridge: Harvard University Press, 1979), pp. 190–95; John Fiske, "Sociology and Hero-Worship: An Evolutionist's Reply to Dr. James," *Atlantic Monthly* 47 (1881): 75–84; Grant Allen, "The Genesis of Genius," *Atlantic Monthly* 47 (1881): 371–81; Allen, "Idiosyncrasy," *Popular Science Monthly* 24 (1884): 387–403; and Allen, "Genius and Talent," *Popular Science Monthly* 34 (1889): 341–56.

30. For an important response to the "Negro problem," see W.E.B. Du Bois, *The Philadelphia Negro: A Social Study* (1899; reprint, Philadelphia: University of Pennsylvania Press, 1996). On Du Bois's study, see Mia Bay, "'The World Was Thinking Wrong About Race': *The Philadelphia Negro* and Nineteenth-Century

Science," in *W.E.B. Du Bois, Race, and the City: The Philadelphia Negro and Its Legacy,* eds. Michael Katz and Thomas J. Sugrue (Philadelphia: University of Pennsylvania Press, 1998), pp. 41–59. See also Michele Mitchell, *Righteous Propagation: African Americans and the Politics of Racial Destiny after Reconstruction* (Chapel Hill: University of North Carolina Press, 2004).

31. On responses to immigration and imperialism, see Henry Cabot Lodge, "The Distribution of Ability in the United States," *Century Magazine* 42 (1891): 687–94; and "The Restriction of Immigration," *North American Review* 152 (1891): 27–35. For background, see Gail Bederman, *Manliness & Civilization: A Cultural History of Gender and Race in the United States, 1880–1917* (Chicago: University of Chicago Press, 1995); John Higham, *Strangers in the Land: Patterns of American Nativism, 1860–1925* (New Brunswick: Rutgers University Press, 1994); Matthew Frye Jacobson, *Whiteness of a Different Color: European Immigrants and the Alchemy of Race* (Cambridge: Harvard University Press, 1998), chaps. 2, 6; and Jacobson, *Barbarian Virtues.*

32. See William Graham Sumner, *What Social Classes Owe to Each Other* (New York: Harper, 1883).

33. See, for example, Grant Allen, "Plain Words on the Woman Question," *Popular Science Monthly* 36 (1889): 170–81; and Edward H. Clarke, *Sex in Education: Or, A Fair Chance for the Girls* (Boston: Osgood, 1873). On evolutionary theory and women, see Louise M. Newman, ed., *Men's Ideas/Women's Realities: Popular Science, 1870–1915* (New York: Pergamon Press, 1985), esp. pp. 1–16, 54–68; and Cynthia Eagle Russett, *Sexual Science: The Victorian Construction of Womanhood* (Cambridge: Harvard University Press, 1989). For the British case, see Katharina Rowold, ed., *Gender & Science: Late Nineteenth-Century Debates on the Female Mind and Body* (Bristol: Thoemmes Press, 1996).

34. See Charlotte Perkins Gilman, *Women and Economics: A Study of the Economic Relation Between Men and Women as a Factor in Social Evolution* (Boston: Small, Maynard, 1899).

35. See, for example, Nancy Cohen, *The Reconstruction of American Liberalism, 1865–1914* (Chapel Hill: University of North Carolina Press, 2002), chaps. 5–7; Higham, *Strangers in the Land,* chap. 6; Kevles, *In the Name of Eugenics,* chaps. 4–7; Jeffrey Sklansky, *The Soul's Economy: Market Society and Selfhood in American Thought, 1820–1920* (Chapel Hill: University of North Carolina Press, 2002); and Wiebe, *Search for Order,* chaps. 6–7.

36. See, for example, Lawrence A. Cremin, *The Transformation of the School* (New York: Vintage, 1964), esp. chaps. 4, 6; and Ross, *Origins of American Social Science,* esp. chap. 5. On this general phenomenon, see Michel Foucault, *Discipline and Punish: The Birth of the Prison,* trans. Alan Sheridan (New York: Vintage, 1979); and Foucault, *The Order of Things: An Archaeology of the Human Sciences,* trans. Alan Sheridan (New York: Vintage, 1973).

37. For an insightful overview, see Sklansky, *Soul's Economy,* chap. 5. On the relations between Darwinism and the study of the mind, see Robert J. Richards, *Darwin and the Emergence of Evolutionary Theories of Mind and Behavior* (Chicago: University of Chicago Press, 1987). For a contemporary critique of the shift toward biology, see "Hereditary Genius," *Catholic World* 11 (1870): 721–32.

38. G. Stanley Hall, "The New Psychology," *Andover Review* 3 (1885): 120–35, 239–48, quote on pp. 247–48.

39. John O'Donnell, David Leary, and Paul Croce have provided excellent accounts of how concern with religion and science influenced American psychology. See David E. Leary, "Telling Likely Stories: The Rhetoric of the New Psychology," *Journal of the History of the Behavioral Sciences* 23 (1987): 315–31; O'Donnell, *Origins of Behaviorism,* esp. chaps. 3–7; and Croce, *Science and Religion in the Era of William James.* See also Deborah J. Coon, "Salvaging the Self in a World Without Soul: William James's *The Principles of Psychology,*" *History of Psychology* 3 (2000): 83–103.

40. O'Donnell rightly points out that there were actually many continuities between the old mental philosophy and new psychology. See O'Donnell, *Origins of Behaviorism,* esp. chap. 1; and Russell D. Kosits, "Of Faculties, Fallacies, and Freedom: Dilemma and Irony in the Secularization of American Psychology," *History of Psychology* 7 (2004): 340–66.

41. For lively characterizations of James's early career, see Croce, *Science and Religion in the Era of William James;* and Menand, *Metaphysical Club,* esp. chaps. 4–6, 9, 13. There is an extensive secondary literature on William James. Other important works include Daniel W. Bjork, *The Compromised Scientist: William James in the Development of American Psychology* (New York: Columbia University Press, 1983); Howard M. Feinstein, *Becoming William James* (Ithaca: Cornell University Press, 1984); Bruce J. Kuklick, *The Rise of American Philosophy, Cambridge, Massachusetts, 1860–1930* (New Haven: Yale University Press, 1977), esp. chaps. 9–11, 14–17; Gerald E. Myers, *William James: His Life and Thought* (New Haven: Yale University Press, 1986); O'Donnell, *Origins of Behaviorism,* esp. chap. 6; Richards, *Evolutionary Theories of Mind and Behavior,* esp. chap. 9; and Wilson, *Transformation of American Philosophy.*

42. See O'Donnell, *Origins of Behaviorism;* and Wilson, *Transformation of American Philosophy.*

43. Although a very influential text was William B. Carpenter, *The Principles of Mental Physiology, with Their Applications to the Training and Discipline of the Mind, and the Study of Its Morbid Conditions* (New York: D. Appleton, 1875).

44. O'Donnell, *Origins of Behaviorism,* p. 97. On James and Darwinism, see esp. Richards, *Evolutionary Theories of Mind and Behavior,* chap. 9.

45. O'Donnell argues that this appointment was seen as a way to stave off materialism; see O'Donnell, *Origins of Behaviorism,* pp. 102–105.

46. William James, "Lowell Lectures on 'The Brain and the Mind' (1878)" in *Manuscript Lectures* (Cambridge: Harvard University Press, 1988), pp. 16–43.

47. Ibid., "A Plea for Psychology as a 'Natural Science,'" *Philosophical Review* 1 (1892): 146–53; and *The Principles of Psychology,* 2 vols. (New York: Henry Holt, 1890).

48. See ibid., "Lowell Lectures," p. 29; and Coon, "Salvaging the Self."

49. For background on the early American psychologists, see Raymond E. Fancher, *Pioneers of Psychology,* 2nd ed. (New York: W. W. Norton, 1990), pp. 258–74; Geraldine Jonçich Clifford, *Edward L. Thorndike: The Sane Positivist* (Middletown: Wesleyan University Press, 1968); O'Donnell, *Origins of Behavior-*

ism; Richards, *Evolutionary Theories of Mind and Behavior,* chap. 10; Dorothy Ross, *G. Stanley Hall: The Psychologist as Prophet* (Chicago: University of Chicago Press, 1972); and Wilson, *Transformation of American Philosophy.* For a subtle analysis of the process of making a scientific persona in the new psychology, see Francesca Bordogna, "Scientific Personae in American Psychology: Three Case Studies," *Studies in the History and Philosophy of the Biological and Biomedical Sciences* 36 (2005): 95–134.

50. O'Donnell argues that a patina of experimental experience was for most American psychologists more desirable than extensive commitment to the laboratory. See O'Donnell, *Origins of Behaviorism,* esp. chap. 8. For the debate between James and Ladd on the ways to make psychology scientific, see George Trumbull Ladd, "Psychology as So-Called 'Natural Science,'" *Philosophical Review* 1 (1892): 24–53; Ladd, "Is Psychology a Science?" *Psychological Review* 1 (1894): 392–95; and James's response, "A Plea for Psychology as a 'Natural Science,'" *Philosophical Review* 1 (1892): 146–53.

51. See, for example, *Studies from the Yale Psychological Laboratory* 1 (1893)-5 (1897), and, in particular, E. W. Scripture, "Elementary Course in Psychological Measurements," *Studies* 4 (1896): 89–141. On Titchener and structural psychology in America, see Bjork, *Compromised Scientist,* chap. 4; Hilgard, *Psychology in America,* pp. 73–79; and O'Donnell, *Origins of Behaviorism,* pp. 9–11. On introspection in America, see Deborah J. Coon, "Standardizing the Subject: Experimental Psychologists, Introspection, and the Quest for a Technoscientific Ideal," *Technology and Culture* 34 (1993): 757–83.

52. On functionalism, see Hilgard, *Psychology in America,* pp. 79–91; and O'Donnell, *Origins of Behaviorism,* pp. 11–14 and chaps. 9–11.

53. Kurt Danziger, "The Origins of the Psychological Experiment as a Social Institution," *American Psychologist* 40 (1985): 133–40, quote on p. 137; and Danziger, "Statistical Method and the Historical Development of Research Practice in American Psychology," in *The Probabilistic Revolution,* vol. 2, eds. Lorenz Krüger, Gerd Gigerenzer, and Mary S. Morgan (Cambridge: MIT Press, 1987), pp. 35–47. See also O'Donnell, *Origins of Behaviorism,* pp. 159–78.

54. James, "A Plea for Psychology as a 'Natural Science,'" p. 148.

55. O'Donnell, *Origins of Behaviorism,* chap. 9. For an example of the importance functionalists placed on the practical, see G. Stanley Hall, "Mental Science," *Science,* n.s., 20 (1904): 481–90.

56. O'Donnell, *Origins of Behaviorism,* p. 173.

57. James, *Principles of Psychology,* vol. 2, p. 695.

58. E. W. Scripture, *The New Psychology* (New York: Charles Scribner's Sons, 1897), p. 494.

59. William James, "Mental Evolution in Man, by George J. Romanes (1889)," *Cambridge Tribune,* April 20, 1889, p. 5, in *Essays, Comments, and Reviews* (Cambridge: Harvard University Press, 1987), p. 416. On James's skepticism about craniometry, see "Lowell Lectures," p. 35.

60. See, for example, Linus W. Kline, "Suggestions Toward a Laboratory Course in Comparative Psychology," *American Journal of Psychology* 10 (1899): 399–430; Edward L. Thorndike, "Animal Intelligence: An Experimental Study of

the Associative Processes in Animals," *The Psychological Review, Monograph Supplements* 2 (1898); and Thorndike, "Some Experiments on Animal Intelligence," *Science,* n.s., 7 (1898): 818–24. For background, see Robert Boakes, *From Darwin to Behaviourism: Psychology and the Minds of Animals* (Cambridge: Cambridge University Press, 1984).

61. See, for example, Charles Darwin, *Expression of the Emotions in Man and Animals* (New York: D. Appleton, 1873); C. Lloyd Morgan, *Animal Life and Intelligence* (Boston: Ginn, 1891); Morgan, "Instinct and Intelligence in Animals," *Nature* 57 (1898): 326–30; Edward L. Thorndike, *Animal Intelligence: An Experimental Study of the Associative Processes in Animals* (New York: Macmillan, 1898); John B. Watson, *Behavior: An Introduction to Comparative Psychology* (New York: Holt, 1914); and Robert M. Yerkes, *The Mental Life of Monkeys and Apes: A Study of Ideational Behavior* (New York: Holt, 1916).

62. See Kosits, "Of Faculties, Fallacies, and Freedom."

63. James, "Lowell Lectures," p. 42.

64. See, for example, "Faculty," in *Dictionary of Philosophy and Psychology,* vol. 1, p. 369.

65. E. B. Titchener to Francis Galton; Sept. 15, 1894; Folder 328, Francis Galton Papers, Archives of University College London, London.

66. For more on "brass-instrument" psychology practices, see Coon, "Standardizing the Subject."

67. Hall, "New Psychology," p. 126.

68. See O'Donnell, *Origins of Behaviorism,* pp. 229–30.

69. James McKeen Cattell, "Mental Tests and Measurements," *Mind* 15 (1890): 373–81.

70. See especially Michael M. Sokal, "James McKeen Cattell and the Failure of Anthropometric Testing, 1890–1901," in *The Problematic Science: Psychology in Nineteenth-Century Thought,* eds. William R. Woodward and Mitchell G. Ash (New York: Praeger, 1982), pp. 322–45; and James Allen Young, "Height, Weight, and Health: Anthropometric Study of Human Growth in Nineteenth-Century American Medicine," *Bulletin of the History of Medicine* 53 (1979): 214–43.

71. On Cattell, see Bjork, *Compromised Scientist,* chap. 5; Fancher, *Intelligence Men,* pp. 44–49; O'Donnell, *Origins of Behaviorism,* esp. pp. 31–35; Michael M. Sokal, introduction to James McKeen Cattell, *An Education in Psychology: James McKeen Cattell's Journal and Letters from Germany and England, 1880–1888,* ed. Sokal (Cambridge: MIT Press, 1981), pp. 1–18; Sokal, "James McKeen Cattell and the Failure of Anthropometric Testing"; Sokal, "James McKeen Cattell and Mental Anthropometry, Nineteenth-Century Science and Reform and the Origins of Psychological Testing," in *Psychological Testing and American Society, 1890–1930,* ed. Sokal (New Brunswick: Rutgers University Press, 1987), pp. 21–45; and Read D. Tuddenham, "The Nature & Measurement of Intelligence," in *Psychology in the Making,* ed. Leo Postman (New York: Knopf, 1963), pp. 469–525, esp. pp. 476–81.

72. On Wundtian psychology, see especially Kurt Danziger, "The Positivist Repudiation of Wundt," *Journal of the History of the Behavioral Sciences* 15 (1979): 205–30; and Danziger, *Constructing the Subject,* pp. 27–39. See also William R.

Woodward, "Wundt's Program for the New Psychology: Vicissitudes of Experiment, Theory, and System," in *Problematic Science,* pp. 167–97. For Wundt's American students, see Ludy T. Benjamin, et al., "Wundt's American Doctoral Students," *American Psychologist* 47 (1992): 123–31.

73. Sokal, "Cattell and the Failure of Anthropometric Testing," p. 327.

74. Galton, *Hereditary Genius.*

75. See, for example, ibid., "Measurement of Character," *Popular Science Monthly* 25 (1884): 732–39. On Galton's statistical style and its relation to psychology, see especially Danziger, *Constructing the Subject.*

76. Cattell, "Mental Tests and Measurements," p. 373, note.

77. Danziger has discussed the American adoption of the Galtonian style in "Origins of Psychological Experiment," and "Statistical Method in American Psychology."

78. James McKeen Cattell and Livingston Ferrand, "Physical and Mental Measurements of the Students of Columbia University," *Psychological Review* 3 (1896): 618–48, 620, 623.

79. Joseph Jastrow, "A Study in Mental Statistics," *New Review* 5 (1891): 559–68, quote on p. 563. See also Jastrow, "Some Anthropometric and Psychologic Tests on College Students," *American Journal of Psychology* 4 (1892): 420–28.

80. Mary Whiton Calkins, "Community of Ideas of Men and Women," *Psychological Review* 3 (1896): 426–31. See also Cordelia C. Nevers, "Dr. Jastrow on Community of Ideas of Men and Women," *Psychological Review* 2 (1895): 363–67; and Jastrow's response to Nevers, "Community of Ideas of Men and Women," *Psychological Review* 3 (1896): 68–71.

81. On this general issue, see Danziger, *Constructing the Subject;* and Porter, *Trust in Numbers.*

82. R. Meade Bache, "Reaction Time with Reference to Race," *Psychological Review* 2 (1895): 475–86; Anna Tolman Smith, "A Study in Race Psychology," *Popular Science Monthly* 50 (1896): 354–60; W. Townsend Porter, "The Physical Basis of Precocity and Dullness," *Transactions of the Academy of Science of St. Louis* 6 (1892–1894): 161–81; and J. Allen Gilbert, "Researches on the Mental and Physical Development of School-Children," *Studies from the Yale Psychological Laboratory* 2 (1894): 40–100.

83. On Porter's study, see Young, "Height, Weight, and Health," pp. 236–38.

84. Porter, "Physical Basis of Precocity and Dullness," p. 162.

85. For a contemporary critique of Porter's study, see Franz Boas, "On Dr. William Townsend Porter's Investigation of the Growth of School Children of St. Louis," *Science* 1 (1895): 225–30.

86. Gilbert, "Mental and Physical Development of School-Children," pp. 54, 71.

87. Gilbert only differentiated his population by mental level for certain of the tests, when measuring the child's weight, height, lung capacity, reaction-time, discrimination time, and time memory. Studies on children were also conducted by Edmund C. Sanford and Harry K. Wolfe; see Sokal, "James McKeen Cattell and Mental Anthropometry," pp. 30–32.

88. See the report on Jastrow's paper "Popular Tests of Mental Capacity" in "Physical and Mental Tests," *Psychological Review* 5 (1898): 172–79, quotes on pp. 172, 173.

89. See, for example, the remarks of Baldwin and Cattell in "Physical and Mental Tests," pp. 175–79.

90. See Sokal, "James McKeen Cattell and Mental Anthropometry," pp. 37–39; also Tuddenham, "Nature & Measurement of Intelligence," pp. 478–81.

91. Stella E. Sharp, "Individual Psychology: A Study in Psychological Method," *American Journal of Psychology* 10 (1899): 329–91; and Clark Wissler, "The Correlation of Mental and Physical Tests," *Psychological Review. Monograph Supplements* 3 (1901).

92. See Wissler, "Correlation of Mental and Physical Tests," pp. 42–62.

93. Sokal, "James McKeen Cattell and Mental Anthropometry," p. 38.

94. See, for example, Issac Ray, *A Treatise on the Medical Jurisprudence of Insanity* (London: G. Henderson, 1839), p. 3.

95. Cattell and Ferrand, "Physical and Mental Measurements," p. 623.

96. Jastrow, "Physical and Mental Tests," pp. 174–75.

97. Zenderland also emphasizes the importance of the periphery in the development of mental testing in America. See *Measuring Minds,* esp. introduction, chaps. 3–4.

98. Henry H. Goddard, *European Diary,* 1908; folder AA4(1), box M33.1, Henry Herbert Goddard Papers, Archives of the History of American Psychology, University of Akron, Akron, Ohio (hereafter cited as Goddard Papers).

99. See chapter 4.

100. On Goddard and the Binet-Simon scales, see Henry H. Goddard, introduction to Alfred Binet and Théodore Simon, *The Development of Intelligence in School Children,* trans. Elizabeth S. Kite (Baltimore: Williams & Wilkins, 1916), pp. 5–8; Goddard, "In the Beginning," *Understanding the Child* 3 (1933): 2–6; Fancher, *Intelligence Men,* esp. pp. 105–115; Stephen Jay Gould, *The Mismeasure of Man* (New York: W. W. Norton, 1981), pp. 158–74; Tuddenham, "Nature & Measurement of Intelligence," pp. 490–92; Leila Zenderland, "The Debate Over Diagnosis: Henry Herbert Goddard and the Medical Acceptance of Intelligence Testing," in *Psychological Testing and American Society,* pp. 46–74, esp. pp. 60–63; and Zenderland, *Measuring Minds.*

101. Goddard, "Introduction," p. 5.

102. Ibid., "Four Hundred Feeble-Minded Children Classified by the Binet Method," *Pedagogical Seminary* 17 (1910): 387–97, quote on p. 389. For an extensive discussion of the reasons Goddard adopted the Binet test, see Zenderland, *Measuring Minds,* chap. 3; and Kevles, *In the Name of Eugenics,* pp. 76–80.

103. Although somewhat critical of the Binet scale, Leonard P. Ayers noted that it was winning "rapid and widespread use among hundreds of practical teachers and workers with children" because "the tests 'work' successfully when applied." See Ayers, "The Binet-Simon Measuring Scale for Intelligence: Some Criticisms and Suggestions," *Psychological Clinic* 5 (1911): 187–96, quotes on pp. 193–94. For more on the initial response to the Binet scale, see Zenderland, *Measuring Minds,* chap. 4.

104. Goddard provides some figures on the dissemination of the Binet-Simon scale in "Introduction." See also Lewis M. Terman, "A Report of the Buffalo Conference on the Binet-Simon Tests of Intelligence," *Pedagogical Seminary* 20 (1913): 549–54.

105. Goddard, "Four Hundred Feeble-Minded Children." As Zenderland rightly argues, Goddard's goal was to fuse Victorian moral concerns with the rigors of science; see Zenderland, *Measuring Minds.*

106. Lewis M. Terman and H. G. Childs, "A Tentative Revision and Extension of the Binet-Simon Measuring Scale of Intelligence," *Journal of Educational Psychology* 3 (1912): 61–74, 133–43, 198–208, 277–89. See also Terman, "The Binet-Simon Scale for Measuring Intelligence," *Psychological Clinic* 7 (1911): 199–206.

107. Edmund B. Huey, "The Binet Scale for Measuring Intelligence and Retardation," *Journal of Educational Psychology* 1 (1910): 435–44, quote on p. 435. See also Edmund B. Huey, "The Present Status of the Binet Scale of Tests for the Measurement of Intelligence," *Psychological Bulletin* 9 (1912): 160–68; and Fred Kuhlmann, "The Results of Grading Thirteen Hundred Feeble-Minded Children with the Binet-Simon Tests," *Journal of Educational Psychology* 4 (1913): 261–68. For general background, see Henry L. Minton, *Lewis M. Terman: Pioneer in Psychological Testing* (New York: New York University Press, 1988), esp. chap. 5.

108. For a survey of the early contributions to Binet testing, see Huey, "Present Status of the Binet Scale of Tests for the Measurement of Intelligence."

109. See, for example, Fred Kuhlmann's remarks circa 1910, in an untitled, undated, note marked "COPY" in possession of Goddard; file: correspondence Miscellaneous, box M614, Goddard Papers. See also Alice C. Strong, "Three Hundred Fifty White and Colored Children Measured by the Binet-Simon Measuring Scale of Intelligence: A Comparative Study," *Pedagogical Seminary* 20 (1913): 485–515; and J. E. Wallace Wallin, "Human Efficiency," *Pedagogical Seminary* 18 (1911): 74–84.

110. Ayres, "Binet-Simon Measuring Scale for Intelligence."

111. Lewis M. Terman, *The Measurement of Intelligence: An Explanation of and a Complete Guide for the Use of the Stanford Revision and Extension of the Binet-Simon Intelligence Scale* (Boston: Houghton Mifflin, 1916). For Terman's marginal comments on the Binet-Simon scales, see Alfred Binet and Theodore Simon, *The Development of Intelligence in School Children,* marginal notes by Lewis M. Terman (Nashville: Williams Printing Co., 1980).

112. The testing community was divided up to the outbreak of World War I between advocates of the Binet approach and those, such as Yerkes, who championed a point-scale approach. For Yerkes' critique of Binet-style instruments, see Robert M. Yerkes, "The Binet Versus the Point Scale Method of Measuring Intelligence," *Journal of Applied Psychology* 1 (1917): 111–22. For background on Yerkes, see James Reed, "Robert M. Yerkes and the Mental Testing Movement," in *Psychological Testing and American Society,* pp. 75–94.

113. Lewis M. Terman, "Trails to Psychology," in *A History of Psychology in Autobiography,* vol. 2, ed. Carl Murchison (Worcester: Clark University Press, 1932), p. 324.

114. Edwin D. Starbuck to Goddard; 1 August 1912; correspondence S, box M615, Goddard Papers. See also Edward B. Titchener, "Anthropometry and Experimental Psychology," *Philosophical Review* 2 (1893): 187–92.

115. Goddard to Starbuck; 6 August 1912; correspondence S, box M615, Goddard Papers. Huey emphasized the practicality of the tests for making assessments in the schools in "Binet Scale for Measuring Intelligence and Retardation."

116. Goddard to J. S. Woodward, President, and members of the Board of Trustees of the Carnegie Institute; 1 June 1908; file AA-4, box M33, Goddard Papers.

117. Bureau of Juvenile Research, Ohio Board of Administration; undated; file X, box M35.1, Goddard Papers. For statistics on the number of and institutions for the feebleminded in the United States, see George E. Johnson, "Contribution to the Psychology and Pedagogy of Feeble-Minded Children," *Pedagogical Seminary* 3 (1894): 246–301.

118. Johnson lists a number of different systems for classifying the feebleminded, only some of which were clearly hierarchical, in "Psychology and Pedagogy of Feeble-Minded Children," pp. 253–57. Martin W. Barr, physician at the Pennsylvania Training School for Feeble-Minded Children in Elwyn, made clear that he used a system based on intelligence deficits existing in degrees. See Barr, "The Training of Mentally Deficient Children," *Popular Science Monthly* 53 (1898): 531–35; and "Mental Defectives and the Social Welfare," *Popular Science Monthly* 54 (1899): 746–59. On the use of intelligence in consigning individuals to asylums, see Steven A. Gelb, "'Not Simply Bad and Incorrigible': Science, Morality, and Intellectual Deficiency," *History of Education Quarterly* 29 (1989): 359–79; James W. Trent, *Inventing the Feeble Mind: A History of Mental Retardation in the United States* (Berkeley: University of California Press, 1994); and Zenderland, *Measuring Minds,* chap. 3.

119. See, for example, John C. Bucknill and Daniel H. Tuke, *A Manual of Psychological Medicine: Containing the History, Nosology, Description, Statistics, Diagnosis, Pathology, and Treatment of Insanity* (Philadelphia: Blanchard and Lea, 1858), esp. pp. 101–20, 508; and Edouard Seguin, *Idiocy: And Its Treatment by the Physiological Method* (1866; reprint, New York: Teachers College, 1907). For background on treatment of the feebleminded, see Leo Kanner, *A History of the Care and Study of the Mentally Retarded* (Springfield: Charles Thomas, 1964); Peter L. Taylor and Leland V. Bell, *Caring for the Retarded in America: A History* (Westport: Greenwood Press, 1984); and Trent, *Inventing the Feeble Mind.*

120. For more on Goddard's task of elucidating confusing diagnoses, see Zenderland, "Debate Over Diagnosis"; and *Measuring Minds.* For Goddard's views on feeblemindedness, see Henry H. Goddard, *Feeble-Mindedness: Its Causes and Consequences* (New York: Macmillan, 1916).

121. See, for example, Leonard P. Ayers, *Laggards in Our Schools: A Study of Retardation and Elimination in City School Systems* (New York: Charities Publication Committee, 1909).

122. Alexander Johnson, "Concerning a Form of Degeneracy," *American Journal of Sociology* 4 (1899): 326–34, quote on p. 334.

123. See, for example, Cattell and Ferrand, "Physical and Mental Measurements."

124. See, for example, Henry H. Goddard, *The Research Department: What it is Doing, What It Hopes to Do* (Vineland: The Training School, 1914) in file F. M. Club, box M37, Goddard Papers.

125. See, for example, Walter Fernald to Goddard; 2 January 1915; correspondence C-D, box M614, Goddard Papers. For an analysis of the complicated ways in which testing practices were adapted to specific institutional settings, see

Michael A. Rembis, "'I Ain't Been Reading While on Parole': Experts, Mental Tests, and Eugenic Commitment Law in Illinois, 1890–1940," *History of Psychology* 7 (2004): 225–47, esp. pp. 233–41.

126. Henry H. Goddard, contribution to "Mentality Tests: A Symposium," *Journal of Educational Psychology* 7 (1916): 229–40, 278–86, 348–60.

127. Zenderland examines in detail early attempts by Goddard and others to promote intelligence testing as a way of addressing a variety of social pathologies. See Zenderland, *Measuring Minds,* esp. chaps. 4–8.

128. Alfred Binet, *L'Etude expérimentale de l'intelligence* (Paris: Schleicher Frères & Cie, 1903).

129. No title; no date; p. 2; file: Case History, box M614, Goddard Papers.

130. Goddard himself conducted one of the first large-scale research projects on normal children, during which he concluded that "the Binet Scale was wonderfully accurate." See Goddard, "Two Thousand Normal Children Measured by the Binet Measuring Scale of Intelligence," *Pedagogical Seminary* 18 (1911): 232–59.

131. On Terman, see Paul Davis Chapman, *Schools as Sorters: Lewis M. Terman, Applied Psychology, and the Intelligence Testing Movement, 1890–1930* (New York: New York University Press, 1988), esp. chap. 1; Fancher, *Intelligence Men,* pp. 132–45; Gould, *Mismeasure of Man,* pp. 174–92; Minton, *Lewis Terman;* and Henry L. Minton, "Lewis M. Terman and Mental Testing: In Search of the Democratic Ideal," in *Psychological Testing and American Society,* pp. 95–112.

132. Elizabeth Lunbeck, *The Psychiatric Persuasion: Knowledge, Gender, and Power in Modern America* (Princeton: Princeton University Press, 1994).

133. See, for example, George V. Dearborn, "The Criteria of Mental Abnormality," *Psychological Review* 5 (1898): 505–10.

134. On Terman's background, see Minton, *Lewis Terman.*

135. Lewis M. Terman, "Genius and Stupidity: A Study of Some of the Intellectual Processes of Seven 'Bright' and Seven 'Stupid' Boys," *Pedagogical Seminary* 13 (1906): 307–73.

136. Ibid., pp. 313, 310, 372.

137. Many psychologists, including Binet and Goddard, however, did vacillate between seeing degrees of feeblemindedness as arrested stages of normal development and as constituting their own specific mental manifestations. See, for example, Florence Mateer, "The Diagnostic Fallibility of Intelligence Ratios," *Pedagogical Seminary* 25 (1918): 369–92; and Witmer, "Relation of Intelligence to Efficiency," pp. 72–74.

138. Many critics of testing focused particularly on this feature. See, for example, J. Victor Haberman, "The Intelligence Examination and Evaluation: A Study of the Child's Mind (Second Report)," *Psychological Review* 23 (1916): 352–79, 484–500.

139. Charles Spearman, "'General Intelligence,' Objectively Determined and Measured," *American Journal of Psychology* 15 (1904): 201–93. On Spearman, see Bernard Norton, "Charles Spearman and the General Factor in Intelligence: Genesis and Interpretation in the Light of Sociopersonal Considerations," *Journal of the History of the Behavioral Sciences* 15 (1979): 142–54; and A. D. Lovie and

P. Lovie, "Charles Spearman, Cyril Burt, and the Origins of Factor Analysis," *Journal of the History of the Behavioral Sciences* 29 (1993): 308–21. On mental testing in England, see Rose, *Psychological Complex,* chap. 5; Gillian Sutherland, *Ability, Merit, and Measurement: Mental Testing and English Education, 1880–1940* (New York: Oxford University Press, 1984); and Adrian Wooldridge, *Measuring the Mind: Education and Psychology in England, c.1860–1990* (Cambridge: Cambridge University Press, 1994).

140. Spearman, "General Intelligence," p. 284 (emphasis in original).

141. On the early reception of the Binet tests by the American psychology community, see Zenderland, *Measuring Minds,* pp. 235–60.

142. Harry D. Kitson, contribution to "Mentality Tests: A Symposium," p. 279.

143. Clara Schmitt, "The Binet-Simon Tests of Mental Ability. Discussion and Criticism," *Pedagogical Seminary* 19 (1912): 186–200; quotes on pp. 188, 196.

144. See Terman, "Report of the Buffalo Conference on the Binet-Simon Tests of Intelligence," pp. 549–50. Frank N. Freeman expressed a similar worry in 1916; see contribution to "Mentality Tests: A Symposium," p. 234.

145. Lightner Witmer, "Clinical Records," *Psychological Clinic* 9 (1915): 1–17, quote on p. 2. See also Frances Porter, "Difficulties in the Interpretation of Mental Tests—Types and Examples," *Psychological Clinic* 9 (1915): 140–58, 167–80; and William Healy, *The Individual Delinquent: A Text-Book of Diagnosis and Prognosis for All Concerned in Understanding Offenders* (Boston: Little Brown, 1917), chaps. 6–7.

146. See, for example, Edward L. Thorndike, Wilfrid Lay, and P. R. Dean, "The Relation of Accuracy in Sensory Discrimination to General Intelligence," *American Journal of Psychology* 20 (1909): 364–69. On Thorndike, see Jonçich, *Sane Positivist,* esp. chap. 12.

147. See Edward L. Thorndike, *Educational Psychology,* vol. III: *Mental Work and Fatigue and Individual Differences and Their Causes* (New York: Teachers College, 1913), part II; and Thorndike, *Animal Intelligence.* For Thorndike's mature views, including his description of his CAVD (completion, arithmetic, vocabulary, and directions) test of intelligence, see Thorndike, et al., *Measurement of Intelligence* (New York: Teachers College, 1927). See also Jonçich, *Sane Positivist,* pp. 368–70.

148. Although in his eugenic writings Thorndike seemed comfortable positing a unitary "determiner of intellect." Edward L. Thorndike, "Eugenics: With Special Reference to Intellect and Character," *Popular Science Monthly* 83 (1913): 125–38.

149. See Carl E. Seashore, "The Binet-Simon Tests," *Journal of Educational Psychology* 3 (1912): 50; and Haberman, "Intelligence Examination and Evaluation."

150. W. H. Pyle, contribution to "Mentality Tests: A Symposium," p. 284.

151. Huey, "Present Status of the Binet Scale of Tests for the Measurement of Intelligence," p. 167.

152. See, for example, Carl C. Brigham's detailed analysis of the Binet tests: "Two Studies in Mental Tests," *Psychological Monographs* 24 (1917): 1–254; and Thomas H. Haines, contribution to "Mentality Tests: A Symposium," pp. 234–37.

153. Thomas H. Haines, "Program of the New Ohio Bureau of Juvenile Research: Its Aims and Possibilities," reprint from *Ohio State Medical Journal* (January 1915), p. 1, in file DD, box M37, Goddard Papers. Fred Kuhlmann, contribution to "Mentality Tests: A Symposium," p. 280; see also Kuhlmann, "Binet and Simon's System for Measuring the Intelligence of Children," *Journal of Psycho-Asthenics* 15 (1911): 79–92.

154. Clara Harrison Town, "The Binet-Simon Scale and the Psychologist," *Psychological Clinic* 8 (1912): 239–44; and Martha Adler, "Mental Tests Used as a Basis for the Classification of School Children," *Journal of Educational Psychology* 5 (1914): 22–28, quote on p. 25. See also William H. Winch, "Binet's Mental Tests: What They Are, and What We Can Do With Them," *Child-Study* 6 (1913): 113–17.

155. Terman's "unselected" did not mean random, but rather "representative," and representative meant that all of the schools chosen were "middle-class," racially homogeneous, and largely full of native born. See Terman, *Measurement of Intelligence,* pp. 52–53; and Lewis M. Terman et al., *The Stanford Revision and Extension of the Binet-Simon Scale for Measuring Intelligence* (Baltimore: Warwick & York, 1917), p. 29.

156. William Stern, *The Psychological Methods of Testing Intelligence,* trans. Guy M. Whipple (Baltimore: Warwick & York, 1914), p. 80; and Terman, *The Intelligence of School Children* (Boston: Houghton Mifflin, 1919), pp. 8–9.

157. Terman, *Intelligence of School Children,* pp. 9, 10.

158. Yerkes argued for his point-scale approach over the IQ method that Terman adopted in part because "the point-scale method has the merit of indicating directly the rate, or annual increments of intellectual growth." Robert M. Yerkes, "Methods of Expressing Results of Measurements of Intelligence: Coefficient of Intelligence," *Journal of Educational Psychology* 7 (1916): 593–606.

159. Terman, *Measurement of Intelligence,* pp. 149–50; 260–62; 151–52; 281–90, 310–13.

160. Terman and Childs, "Tentative Revision and Extension," p. 282.

161. Terman, *Intelligence of School Children,* p. 10.

162. Ibid., *Measurement of Intelligence,* p. 166.

163. See Witmer, "Relation of Intelligence to Efficiency," pp. 68–69, 75.

164. Terman, *Measurement of Intelligence,* pp. 157–59, 181–82, 215–16.

165. Ibid., p. 197.

166. I am indebted to Sheila Jasanoff for this point.

167. However, one consistent critic of the Stanford revision was J. E. Wallace Wallin; see "Preliminary Impressions of the Stanford Revision of the Binet-Simon Scale," *Psychological Clinic* 12 (1918): 1–15.

168. Arthur S. Otis, "Some Logical Aspects of the Binet Scale," *Psychological Review* 23 (1916): 129–79; criteria on p. 179.

169. Walter F. Dearborn, "The Measurement of Intelligence," *Psychological Bulletin* 14 (1917): 221–24; quote on p. 221. Ethel D. Whitmire, "Intelligence Tests *versus* Teacher's Estimate," *Psychological Clinic* 13 (1920): 197–98.

170. Henry H. Goddard to Abraham Flexner, 5 April 1917; correspondence with General Education Board, File 8, Box 2, Lewis M. Terman Papers, Stanford University Archives, Palo Alto, CA.

171. Yerkes to Terman, 14 July 1916; Folder 907, Box 46, Robert M. Yerkes Papers, Yale University Archives, New Haven, CT.

172. In the point-scale method, all subjects answered all of the questions, and received varying amounts of points depending on the quality of their answers. The points were then tallied to give an overall total, which could be compared to norms for their group. See Yerkes, "Binet Versus Point Scale Method," p. 113 (emphasis in original); and especially Robert M. Yerkes, James W. Bridges, and Rose S. Hardwick, *A Point Scale for Measuring Mental Ability* (Baltimore: Warwick & York, 1915).

173. On Terman's response, see Terman to Yerkes, 25 October 1916; Folder 907, Box 46, Yerkes Papers; and Terman, contribution to "Mentality Tests: A Symposium," pp. 348–51.

174. Zenderland, *Measuring Minds,* p. 250.

175. Gail Hornstein notes that there was little disagreement among psychologists about the possibility of quantifying intelligence. See Gail A. Hornstein, "Quantifying Psychological Phenomena: Debates, Dilemmas, and Implications," in *The Rise of Experimentation in American Psychology,* ed. Jill Morawski (New Haven: Yale University Press, 1988), pp. 1–34.

176. For one applied psychologist's view, see Walter V. Bingham, "Mentality Testing of College Students," *Journal of Applied Psychology* 1 (1917): 38–45.

177. See Kevles, *In the Name of Eugenics,* esp. chaps. 1–7; Jacobson, *Barbarian Virtues,* chap. 4; Lawrence B. Goodheart, "Rethinking Mental Retardation: Education and Eugenics in Connecticut, 1818–1917," *Journal of the History of Medicine and Allied Sciences* 59 (2004): 90–111; Wendy Kline, *Building a Better Race: Gender, Sexuality, and Eugenics from the Turn of the Century to the Baby Boom* (Berkeley: University of California Press, 2001), pp. 1–59; Edward J. Larson, *Sex, Race, and Science: Eugenics in the Deep South* (Baltimore: Johns Hopkins University Press, 1995), esp. chaps. 1–3; Martin S. Pernick, *The Black Stork: Eugenics and the Death of "Defective" Babies in American Medicine and Motion Pictures Since 1915* (New York: Oxford University Press, 1996); and Stern, *Eugenic Nation.*

178. See National Conference on Race Betterment, *Proceedings of the First National Conference on Race Betterment* (Battle Creek: Race Betterment Foundation, 1914), pp. 604–19. On immigration and testing, see Henry H. Goddard, "The Binet Tests in Relation to Immigration," *Journal of Psycho-Asthenics* 18 (1913): 104–107; E. K. Sprague, "Mental Examination of Immigrants," *Survey* 31 (1914): 466–68; and Clifford Kirkpatrick, *Intelligence and Immigration* (Baltimore: Williams & Wilkins, 1926). For details on Goddard's experiment, see Steven A. Gelb, "Henry H. Goddard and the Immigrants, 1910–1917: The Studies and Their Social Context," *Journal of the History of the Behavioral Sciences* 22 (1986): 324–32; John T. E. Richardson, "Howard Andrew Knox and the Origins of Performance Testing on Ellis Island, 1912–1916," *History of Psychology* 6 (2003): 143–70; Zenderland, *Measuring Minds,* pp. 266–81; and Jacobson, *Barbarian Virtues,* pp. 163–68.

179. See "Court Bars Binet Test," *New York Times,* July 19, 1916, p. 7; and "No Psychology in Law," *Literary Digest* 53 (1916): 405–406. The Goff incident is also mentioned by Daniel Kevles, Theodore Porter, and Leila Zenderland. See

Daniel J. Kevles, "Testing the Army's Intelligence: Psychologists and the Military in World War I," *Journal of American History* 55 (1968): 566; Kevles, *In the Name of Eugenics,* p. 80; Porter, *Trust in Numbers,* p. 209; and Zenderland, *Measuring Minds,* p. 253.

180. "Scorns the Aid of Science," *New York Times,* July 20, 1916, p. 10.

181. See O'Donnell, *The Origins of Behaviorism,* pp. 212–26; and, on the marketing of intelligence, Stephen Petrina, "The 'Never-To-Be-Forgotten Investigation': Luella W. Cole, Sidney L. Pressey, and Mental Surveying in Indiana, 1917–1921," *History of Psychology* 4 (2001): 245–71.

182. I thank David Hollinger for this point. See John Erskine, "The Moral Obligation to be Intelligent," in *The Moral Obligation to be Intelligent and Other Essays* (New York: Duffield, 1915).

183. John Dewey, "The Need for a Recovery of Philosophy," in *Creative Intelligence: Essays in the Pragmatic Attitude* (New York: Henry Holt, 1917), p. 63.

Chapter 6

1. This chapter is a revised version of "Army Alpha, Army Brass, and the Search for Army Intelligence," *Isis* 84 (1993): 278–309.

2. On the various roles of psychology in World War I, see Thomas M. Camfield, "Psychologists at War: The History of American Psychology and the First World War" (Ph.D. dissertation: University of Texas, 1969); Stephen Jay Gould, *The Mismeasure of Man* (New York: W. W. Norton, 1981), pp. 192–233; Daniel J. Kevles, "Testing the Army's Intelligence: Psychologists and the Military in World War I," *Journal of American History* 55 (1968): 565–81; Franz Samelson, "World War I Intelligence Testing and the Development of Psychology," *Journal of the History of the Behavioral Sciences* 13 (1977): 274–82; Samelson, "Putting Psychology on the Map: Ideology and Intelligence Testing," in *Psychology in Social Context,* ed. Allan R. Buss (New York: Irvington Publishers, 1979), pp. 103–68; Richard T. von Mayrhauser, "The Manager, the Medic, and the Mediator: The Clash of Professional Psychological Styles and the Wartime Origins of Group Mental Testing," in *Psychological Testing and American Society, 1890–1930,* ed. Michael M. Sokal (New Brunswick: Rutgers University Press, 1987), pp. 128–57; von Mayrhauser, "Making Intelligence Functional: Walter Dill Scott and Applied Psychological Testing in World War I," *Journal of the History of the Behavioral Sciences* 25 (1989): 60–72; and von Mayrhauser, "The Practical Language of American Intellect," *History of the Human Sciences* 4 (1991): 371–93.

3. Yerkes to Braisted, 5 May 1917 (emphasis added); File: A.P.A.: Committee on Psychological Examination of Recruits; Institutions, Associations, Individuals—April 1917; National Research Council Papers, Washington, D.C. (hereafter cited as NRC Papers).

4. Sheila Jasanoff has described such a process of mutual constitution as "co-production." See, for example, Jasanoff, "The Idiom of Co-Production," in *States of Knowledge: The Co-Production of Science and Social Order* (London: Routledge, 2004); and "Science, Politics, and the Recognition of Expertise at EPA," *Osiris* 7 (1992): 195–217.

5. See L.-Ch. Bonne, *Explication de la loi du 27 juillet 1872 sur le recrutement*

de l'armée (Paris: Ch. Delagrave et Cie, 1872) for the text of the law, and for the debates over it, *Historiques des diverses lois sur le recrutement depuis la révolution jusqu'à nos jours* (Paris: Imprimerie nationale, 1902).

6. "Ex-Trooper," *The French Army from Within* (New York: George H. Doran, 1914), p. 176.

7. For the General Orders, see G.O. No. 128 of July 12, 1906, in U.S. Adjutant General's Office, *General Orders and Circulars. War Department—1906* (Washington: Government Printing Office, 1907). Although the efficiency report form went through a number of versions between the late 1890s and 1917, it retained the basic flavor of the 1895 version. For a continuous run of efficiency report forms from 1895 to 1917, see file #1100-Frederick G. Lawton, Box 13, RG 94: Office of the Adjutant General-Document Files, National Archives, Washington, D.C. (hereafter cited as AGO, pre-1917).

8. See, for example, General Orders No. 130, paragraphs 857–860, of July 16, 1906, in U.S. AGO, *General Orders and Circulars. 1906.*

9. For examples of pre-1917 officer efficiency reports, see John S. Battle, #1112, Box 13; Sylvester Bonnaffon 3rd, #267109, Box 1170; Francis P. Casey, #21234, Box 148; Frederick G. Lawton, #1100, Box 13; Paul Murray, #1831514, Box 6677; and William Townsend, #1488113, Box 5678; all AGO, pre-1917.

10. See efficiency reports for 1903 and 1916 on Sylvester Bonnaffon 3rd in file #267109-Sylvester Bonnaffon 3rd, Box 1170, AGO, pre-1917.

11. While the 1904 efficiency report form asked for an assessment of "Intelligence and judgment shown in instructing, drilling, and handling men," this seems always to have been read as asking how well the officer handled enlisted men.

12. Examples of the new efficiency report form are published in U.S. Adjutant General's Office, *The Personnel System of the United States Army,* vols. 1–2 (Washington: Government Printing Office, 1919). For a history of the form's development, see von Mayrhauser, "Practical Language of American Intellect."

13. For the official history of the army testing project during World War I, see Robert M. Yerkes, *Psychological Examining in the United States Army,* Volume XV of Memoirs of the National Academy of Sciences (Washington: Government Printing Office, 1921).

14. There is no one comprehensive source for efficiency reports and other ratings of personnel during World War I. For some examples, see File: 201.6, Box 304; File: 220.12 (11-8-24 to 5-13-17), Box 416; and File: 220.81 Honorable Discharges-1917, Box 460; all in RG 407: Adjutant General's Office—Central Decimal Files 1917–1925, National Archives, Washington, D.C. (hereafter cited as AGO, 1917–1925).

15. Colonel H. O. Williams, Inspector General, Memorandum to the Adjutant General of the Army, 22 July 1918; File: 201.6 (8-20-18) to (7-22-18), Box 304, AGO, 1917–1925.

16. See for example Samelson, "World War I Intelligence Testing," pp. 274–76; Thomas M. Camfield, "The Professionalization of American Psychology," *Journal of the History of the Behavioral Sciences* 9 (1973): 66–75; and for general background, Dorothy Ross, *The Origins of American Social Science* (Cambridge: Cambridge University Press, 1991).

17. For the size of the army, see U.S. Adjutant General's Office, *The Personnel System,* vol. 1, chap. 2; and Russell F. Weigley, *History of the United States Army*

(Bloomington: Indiana University Press, 1984), pp. 598–99. For more information on the pre–World War I army, see John W. Chambers, *To Raise an Army: The Draft Comes to Modern America* (New York: Free Press, 1987), esp. chaps. 1–5; and Robert H. Ferrell, *Woodrow Wilson and World War I, 1917–1921* (New York: Harper & Row, 1985), esp. pp. 14–15. On West Point, see Roger H. Nye, "The United States Military Academy in an Era of Educational Reform, 1900–1925" (Ph.D. dissertation: Columbia University, 1968). Kevles makes a similar point in "Testing the Army's Intelligence," p. 567.

18. John W. Chambers discusses the reorganization of military recruiting along scientific principles in *To Raise an Army.*

19. Newton D. Baker, in *Personnel* 1 (8/21/1918): 1–4; Folder 1785, Box 94, Yerkes Papers, Yale University Archives, New Haven, CT (hereafter cited as Yerkes Papers).

20. For more on the army's response to its rapid growth during the war, see Fred D. Baldwin, "The American Enlisted Man in World War I" (Ph.D. dissertation: Princeton University, 1964); Chambers, *To Raise an Army,* pp. 232–34; John G. Clifford, *The Citizen Soldiers: The Plattsburg Training Camp Movement, 1913–1920* (Lexington: University Press of Kentucky, 1972); Edward M. Coffman, *The Hilt of the Sword: The Career of Peyton C. March* (Madison: University of Wisconsin Press, 1968); Jack C. Lane, *Armed Progressive: General Leonard Wood* (San Rafael: Presidio Press, 1978); and Timothy K. Nenninger, *The Leavenworth Schools and the Old Army: Education, Professionalism and the Officer Corps of the United States Army, 1881–1918* (Westport: Greenwood Press, 1978).

21. See Bruce White, "The American Military and the Melting Pot in World War I," in *The Military in America: From the Colonial Era to the Present,* ed. Peter Karsten (New York: Free Press, 1986).

22. For more on Yerkes, see Donna Haraway, *Primate Visions: Gender, Race, and Nature in the World of Modern Science* (New York: Routledge, 1989), esp. pp. 59–83; James Reed, "Robert M. Yerkes and the Mental Testing Movement," in *Testing and American Society, 1890–1930,* pp. 75–94; and von Mayrhauser, "The Manager, the Medic, and the Mediator."

23. Robert M. Yerkes, *Plan for the Psychological Examining of Recruits to Eliminate the Mentally Unfit;* File: American Psychological Association: Committee on Psychological Examination of Recruits; Institutions, Associations, Individuals—April 1917; NRC Papers. See also Yerkes' entries for April 26 and April 29, 1917, in his *War Diary;* Folder 2663, Box 171, Yerkes Papers.

24. Yerkes, *Plan,* pp. 1–2.

25. On the Yerkes-Bridges Point Scale, see Robert M. Yerkes, James W. Bridges, and Rose S. Hardwick, *A Point Scale for Measuring Mental Ability* (Baltimore: Warwick & York, 1915); and Yerkes, "The Binet versus the Point Scale Method of Measuring Intelligence," *Journal of Applied Psychology* 1 (1917): 111–22.

26. See Brigham to Yerkes, 26 March 1917; File: Committee on Psychology: Subcommittee on Incapacity, Reeducation, Vocational Training; Executive Committee—1917; NRC Papers; and Yerkes' report of his trip in Yerkes to Hale, 16 April 1917; File: Committee on Psychology: General; Executive Committee—April, 1917; NRC Papers.

27. Yerkes was supported by Joseph P. Byers, executive secretary of the Committee on Provision for the Feebleminded. Byers's group provided $700, and

Byers himself lobbied the secretary of war in favor of recruit screening. See Byers to the secretary of war, 10 April 1917; Byers to Yerkes, 19 April 1917; Byers to Yerkes, 28 April 1917; and Yerkes to Byers, 4 May 1917; all in File: APA: Committee on Psychological Examination of Recruits; Institutions, Associations, Individuals—April 1917; NRC Papers.

28. See Yerkes, *Psychological Examining,* p. 299.

29. Braisted to Yerkes, 8 May 1917; File: APA: Committee on Psychological Examination of Recruits; Institutions, Associations, Individuals—April 1917; NRC Papers.

30. See, for example, Angell to Hale, 23 May 1917; File: Committee on Psychology: General; Executive Committee—April 1917; NRC Papers; Maj. King to Yerkes, 27 May 1917; Dr. Fernberger to Yerkes, 24 June 1917; and "Report of Dr. Yerkes," penciled date of May 30, 1917, but which may refer to a meeting mentioned in Yerkes' "War Diary" as taking place on May 1; last three in File: APA: Committee on Psychological Examination of Recruits; Institutions, Associations, Individuals—April 1917; NRC Papers.

31. Yerkes, *Psychological Examining,* p. 301.

32. See ibid., pp. 299–305.

33. In the period 1900–1916 there was actually little explicit concern with the issue of "validity," and psychologists generally spoke of the "reliability" of a measuring instrument, without, however, any formalized techniques for assessing it. See Richard T. von Mayrhauser, "The Mental Testing Community and Validity: A Prehistory," *American Psychologist* 47 (1992): 244–53, although I disagree with a number of his conclusions.

34. Yerkes, *Psychological Examining,* p. 305.

35. See, for example, Kelly to Yerkes, 27 August 1917; File: 254, Box 14, Series 4, James Earl Russell Papers, Special Collections, Teachers College, Columbia University, New York, NY; and Goddard to Chase, 26 November 1917; File: Correspondence C-D, Box M614, Henry H. Goddard Papers, Archives of the History of Psychology, University of Akron, Akron, Ohio (hereafter cited as Goddard Papers).

36. McComas to Yerkes, 2 July 1917; File: APA: Committee on Psychological Examination of Recruits; Institutions, Associations, Individuals—July 1917; NRC Papers.

37. Thorndike to Yerkes, 23 July 1917; File: APA: Committee on Psychological Examination of Recruits; Institutions, Associations, Individuals—July 1917; NRC Papers.

38. See Robert M. Yerkes, "Description of Plan for Psychological Military Service"; July 16, 1917; File: APA: Committee on Psychological Examination of Recruits; July 1917; Institutions, Associations, Individuals; NRC Papers; and Yerkes to McComas, 9 July 1917; File: APA: Committee on Psychological Examination of Recruits; Institutions, Associations, Individuals—July 1917; NRC Papers.

39. For more information, see Yerkes, *Psychological Examining,* pp. 313, 314.

40. Ibid., p. 316.

41. For more on the differences between Yerkes and Thorndike, see von Mayrhauser, "Manager, the Medic, and the Mediator." See also Yerkes, "Binet versus the Point Scale Method," p. 116.

42. Yerkes, *Psychological Examining,* p. 305.

43. Ibid., pp. 316 (emphasis added), 317.

44. For Thorndike's mature views on intelligence, see Edward L. Thorndike, E. O. Bregman, M. V. Cobb, and Ella Woodyard, *The Measurement of Intelligence* (New York: Teachers College, 1926). See also Thorndike, *Education: A First Book* (New York: Macmillan, 1912), esp. pp. 102–16; and Thorndike, *Educational Psychology,* vols. 1–3 (New York: Teachers College, 1913–14). In addition, see Geraldine Jonçich Clifford, *Edward L. Thorndike: The Sane Positivist* (Middletown: Wesleyan University Press, 1984), chaps. 14–17 and esp. pp. 368–70.

45. Memo from the Surgeon General of the Army to the Chief of Staff; 21 August 1917; in Yerkes, *Psychological Examining,* p. 12.

46. See memo from Yerkes to the Surgeon General, 20 October 1917; Subject: Weekly Diary; Folder 1751, Box 91, Yerkes Papers.

47. Memo from Major General A. Cronkhite, the Commanding General, 80th Division, Camp Lee, to the Adjutant General; 10 November 1917; File: 10716-29: Sanitary Corps—Medical Dept., Box 561, RG 165: War College Division—General Correspondence 1903–1919, National Archives, Washington, D.C. (hereafter cited as WCD, 1903–1919).

48. For Shaw (16 November 1917) and Birmingham (7 December 1917), see Yerkes, *Psychological Examining,* pp. 19–24.

49. Memo from John J. Bradley, Colonel, General Staff, for and in the absence of John F. Morrison, Major General, Director of Training, War College Division, to Chief of Staff; 24 December 1917; File: 10195-4: Classification and Selection of Personnel for U.S. Army, Box 542, WCD, 1903–1919.

50. See Lt. Bingham, "Report of Company Commanders"; File: 10716–29: Sanitary Corps—Medical Dept., Box 561, WCD, 1903–1919.

51. Memo from Surgeon General Gorgas to the Adjutant General; 3 January 1918; Group #12, File: 702 (2-7-22) to (6-25-18), Box 1090, AGO, 1917–1925.

52. See memo from Yerkes to the Surgeon General, 20 October 1917; Subject: Weekly Diary; Folder 1751, Box 91, Yerkes Papers.

53. Yerkes, *Psychological Examining,* p. 327.

54. See ibid., pp. 328–30, 338, and 540–41.

55. Ibid., p. 341.

56. For Terman's views on intelligence, see Lewis M. Terman, *The Measurement of Intelligence* (Boston: Houghton Mifflin, 1916), esp. pp. 42–46; and Terman, "Intelligence and Its Measurement: A Symposium," *Journal of Educational Psychology* 12 (1921): 127–33. See also Henry L. Minton, *Lewis M. Terman: Pioneer in Psychological Testing* (New York: New York University Press, 1988), esp. pp. 46–51.

57. Although the actions of Scott and Yerkes tended in many ways to reinforce one another, there was little love lost between them. Scott had parted ways with Yerkes in April 1917. See von Mayrhauser, "The Manager, the Medic, and the Mediator."

58. See letter from E. L. Thorndike to Robert M. Yerkes, 18 July 1917; File: American Psychological Association: Committee on Psychological Examination of Recruits, July 1917; Institutions, Associations, Individuals; NRC Papers.

59. Letter from Yerkes to Prof. E. L. Thorndike, 2 April 1918; File: Surgeon General's Office: Division of Psychology: Psychological Examination of Recruits: General, 1918; Agencies and Departments: War; NRC Papers.

60. Letter from Robert M. Yerkes to Lewis M. Terman; 6 November 1917 (emphasis added); File: Committee on Psychology: Subcommittee on Methods for Psychological Examination of Recruits, 1917; Executive Committee; NRC Papers.

61. Yerkes, *Psychological Examining,* p. 327.

62. Memo from Yerkes to the Surgeon General; Subject: Weekly Diary, 26 January 1918; pp. 25–26, Folder 1751, Box 91, Yerkes Papers.

63. For monthly data on the number of psychological examinations given, see Yerkes, *Psychological Examining,* p. 100.

64. Memo from Yerkes to Camp Division Surgeons; 24 January 1918; Yerkes, *Psychological Examining,* p. 58.

65. See memo from Brigadier General Henry Jervey, Asst. Chief of Staff, to Adjutant General, 18 February 1918; Group #2; and Jervey to Acting Secretary of War, 6 April 1918; Group #5; both in File: Psychological Tests, Box 148, #1150, RG 165: Office of the Chief of Staff—Correspondence 1918–1921, National Archives, Washington, D.C. (hereafter cited as OCS, 1918–1921).

66. See Lytle Brown, Brigadier General, War Plans Division; memorandum to the Chief of Staff, 6 May 1918; File: 10716-5: Sanitary Corps—Medical Dept., Box 561, WCD, 1903–1919.

67. Letter sent to all Commanding Officers by the Adjutant General, quoted in memo from Yerkes to the Chief of Staff, 12 September 1918; in Yerkes, *Psychological Examining,* p. 43.

68. A large number of these reports are collected in Group #1, File: 322.3 (7-10-18) to (1-4-18), Box 698, AGO, 1917–1925.

69. See G. H. Dorr, memorandum to Assistant Secretary of War, 10 June 1918; File: 702—Psychological (7-31-18), Box 1856, AGO, 1917–1925; and Col. R. J. Burt, memorandum to Chief of Staff, 18 June 1918; File: 10195-25: Classification and Selection of Personnel for U.S. Army, Box 542, WCD, 1903–1919.

70. See Dorr, memorandum to Assistant Secretary of War, 10 June 1918, pp. 11 (emphasis Dorr's), 13.

71. See Frederick P. Keppel, memorandum for the Secretary of War, 24 May 1918; Group #12, File: 702—Psychological (2-7-22) to (6-25-18), Box 1090, AGO, 1917–1925; Brigadier General Lytle Brown, memorandum for the Chief of Staff, 25 July 1918; #13; and Brigadier General Henry Jervey, memorandum for the Chief of Staff, 29 July 1918; #15; both in File: Psychological Tests, Box 148, #1150, OCS, 1918–1921.

72. See General Orders No. 74, August 14, 1918; File: 10716-29: Sanitary Corps—Medical Dept., Box 561, WCD, 1903–1919.

73. See Brown, memorandum for the Chief of Staff, 31 October 1918; and Jervey, memorandum for the Adjutant General, 6 November 1918; both in File: 10195-40 Classification and Selection of Personnel for U.S. Army, Box 542, WCD, 1903–1919.

74. Yerkes, *War Diary,* p. 119; Folder 2663, Box 171, Yerkes Papers.

75. For copies of all of the army tests, including their directions and scoring instructions, see Yerkes, *Psychological Examining,* pp. 123–95. The examples in the text are taken from the Alpha test, p. 157.

76. Terman, *Measurement of Intelligence,* pp. 224, 327.

77. See letter from Robert M. Yerkes to Surgeon General William C. Gorgas, 23 July 1917; File: American Psychological Association: Committee on Psychological Examination of Recruits, July 1917; Institutions, Associations, Individuals; NRC Papers.

78. Yerkes, *Psychological Examining,* p. 424.

79. Yerkes, memorandum to all Camp Psychological Examiners, 12 November 1918; File: War: SGO: Division of Psychology: Psychological Examination Results: Reports: Monthly—1918, Agencies and Departments, NRC Papers.

80. "Division of Psychology, Third Monthly Report; August, 1918"; September 3, 1918; Folder 1755, p. 175, Box 92, Yerkes Papers.

81. Letter from George F. Arps to Yerkes, 27 August 1918; Folder 1761, pp. 17–22, Box 92, Yerkes Papers.

82. For a rich collection of reactions to the intelligence testing program see File: 702-Psychological (7-31-18), Box 1856, AGO, 1917–1925; also File: 10716-29: Sanitary Corps—Medical Dept., Box 561, WCD, 1903–1919.

83. For an early meeting where the issue of territorial encroachment was discussed, see "Report of Dr. Yerkes" (May 30, 1917 [date penciled in]); File: American Psychological Association: Committee on Psychological Examination of Recruits; Institutions, Associations, Individuals—April 1917; NRC Papers.

84. See esp. File: 702-Psychological (7-31-18), Box 1856, AGO, 1917–1925.

85. Captain Norbarue Berkeley, Commanding Officer, Battery "E," 313th Field Artillery, Camp Lee; Memorandum to Regimental Adjutant, 13 December 1917; Exhibit F: Camp Lee Questionnaire Responses, File: 702-Psychological (7-31-18), Box 1856, AGO, 1917–1925.

86. See, for example, G. H. Dorr, Office of the Assistant Secretary, War Department; Memorandum on Psychological Tests, Camp Lee—General Adelbert Cronkhite; File: 702-Psychological (7-3-18), Box 1856, AGO, 1917–1925.

87. Colonel Robin S. Welsh, Commanding Officer, 314th Field Artillery, 155th Field Artillery Brigade, Camp Lee, Memorandum to Chief Psychological Examiner, Camp Lee, 22 December 1917; Exhibit F: Camp Lee Questionnaire Responses, File: 702-Psychological (7–31–18), Box 1856, AGO, 1917–1925.

88. First Lieutenant A. R. P[?], Commanding Officer, 41st Company, 11th Training Battalion, 155th Depot Brigade, Camp Lee, Memorandum to Commanding Officer, 11th Battalion, Depot Brigade, Camp Lee, 11 December 1917; Exhibit F: Camp Lee Questionnaire Responses, File: 702—Psychological (7-31-18), Box 1856, AGO, 1917–1925.

89. Statistics drawn from *The Medical Department of the United States Army in the World War,* vol. 1, *The Surgeon General's Office* (Washington: Government Printing Office, 1923), p. 402. See also Samelson, "Putting Psychology on the Map," pp. 142–45.

90. See Robert M. Yerkes, memorandum to the Surgeon General, 5 July 1918; Folder 1754, pp. 1–12, Box 92, Yerkes Papers.

91. Samelson, "Putting Psychology on the Map," p. 145; see also *Personnel System,* vol. 1, pp. 311–13.

92. For information on the new tests for illiterates see Group #31, File: Psychological Tests, Box 148, #1150, OCS, 1918–1921; and *Recruit Psychological Examination for Illiterates and Non-English-Speaking Citizens and Aliens,*

stamped November 21, 1919; Group #4, File: 702 (2-7-22 to 6-25-18), Box 1090, AGO, 1917–1925. On the tests for prisoners, see Group #44, File: Prisoners, Box 145, #1150, OCS, 1918–1921.

93. Special Regulations No. 65, "Physical examination for entrance into the army by voluntary enlistment or by induction under the Selective Service law, 1918," was formally adopted as the second edition of Form 75, *Standards of Physical Examination for the Use of Local Boards, District Boards, and Medical Advisory Boards under the Selective-Service Regulations* (Washington: Government Printing Office, 1918), issued by the Office of the Provost Marshall General in 1918. See pp. 37–38 for the addition of "moron" to the list of "disqualifying defects" and for its definition.

94. For an explanation of the modification of regulations, see Brigadier General Lytle Brown, Director of War Planning Division, Memorandum to Chief of Staff, 13 June 1919; Group #30, File: Psychological Tests, Box 148, #1150, OCS, 1918–1921.

95. See *Regulations Governing Physical Examinations,* Form No. 11, Office of the Provost Marshall General (Washington: Government Printing Office, 1917), p. 11; and *Manual of Instructions for Medical Advisory Boards,* Form 64, Office of the Provost Marshall General (Washington: Government Printing Office, 1918), pp. 13–14.

96. See Kevles, "Testing the Army's Intelligence," pp. 578–80.

97. For basic background on the French military during the Third Republic, see Richard D. Challener, *The French Theory of the Nation in Arms* (New York: Columbia University Press, 1955); Robert A. Doughty, *The Seeds of Disaster: The Development of French Army Doctrine, 1919–1939* (Hamden: Archon Books, 1985); Raoul Girardet, *La société militaire de 1815 à nos jours* (Paris: Perrin, 1998); Paul-Marie de La Gorce, *The French Army: A Military-Political History,* trans. Kenneth Douglas (New York: George Braziller, 1963); David B. Ralston, *The Army of the Republic: The Place of the Military in the Political Evolution of France, 1871–1914* (Cambridge: MIT Press, 1967); and H. L. Wesseling, *Soldier and Warrior: French Attitudes toward the Army and War on the Eve of the First World War* (Westport: Greenwood Press, 2000).

98. See Articles 17, 19, and 20 in Bonne, *Explication de la loi du 27 juillet 1872,* pp. 35–43, 38–39, 34.

99. See, for example, Article 20 of the law of July 15, 1889, in Jean d'Estournelles de Constant, *Lois et règlements sur l'enseignement primaire et sur les différents services de l'enfance qui ne dépendent pas du Ministère de l'Instruction publique* (Paris: Imprimerie nationale, 1890), p. 651; and François Roussel, *La Nouvelle législation du recrutement de l'armée: Analyse de Lois et Règlements en vigueur au 1er avril 1891, accompagnée des textes les plus importants* (Paris: Marchal et Billard, 1891), pp. 7–8.

100. See Bonne, *Explication de la loi du 27 juillet 1872,* pp. 49–50.

101. Constant, *Lois et règlements sur l'enseignement primaire,* p. 650.

102. Roussel, *Nouvelle législation du recrutement de l'armée,* p. 371.

103. "Ex-Trooper," *French Army from Within,* p. 44.

104. See, for example, Louis Catrin, *L'Aliénation mentale dans l'armée* (Paris: J. Rueff, 1901); Paul Chavigny, "La Débilité mentale considérée spécialement au

point de vue du service militaire.—Son expertise médico-légale," *Annales d'hygiène publique et de médecine légale,* 4th ser., 11 (1909): 393–443; Adrien-Pierre-Léon Granjux, "L'Aliénation mentale dans l'armée," *Bulletin médical* 15 (1902): 179–84; Granjux, "L'Aliénation mentale dans l'armée au point de vue clinique et médico-légal," *Revue neurologique* 17 (1909): 1026–32; René Jude, *Les Dégénérés dans les bataillons d'Afrique* (Vannes: B. Le Beau, 1907); A.-J. Rayneau, "L'Aliénation mentale dans l'armée," *Revue neurologique* 17 (1909): 1032–43; and J. Simonin, "Les Dégénérés dans l'armée: origine.—caractères.—prophylaxie," *Annales d'hygiène publique et de médecine légale,* 4th ser., 11 (1909): 32–52.

105. See, for example, Rayneau, "L'Aliénation mentale dans l'armée."

106. Emmanuel Régis, discussion comment contained in Rayneau, "L'Aliénation mentale dans l'armée," p. 1041.

107. See, for example, Doughty, *The Seeds of Disaster,* chaps. 1 and 2; and Ralston, *Army of the Republic,* p. 311. For an engaging description of the life of the French trooper, which shows just how completely defined the training was, see Lionel Decle, *Trooper 3809: A Private Soldier of the Third Republic* (New York: Charles Scribner's Sons, 1899).

108. [Hubert Lyautey], "Du rôle social de l'officier," *Revue des deux mondes* 212 (1891): 443–59, quote on p. 455.

109. Alfred Binet and Théodore Simon, "Sur la nécessité d'une méthode applicable au diagnostic des arriérés militaires," *Annales médico-psychologique* 68 (1910): 123–36, summarized in *L'Année psychologique* 17 (1910): 475–80. On this episode, see William H. Schneider, "After Binet: French Intelligence Testing, 1900–1950," *Journal of the History of the Behavioral Sciences* 28 (1992): 111–32, quote on p. 115.

110. Théodore Simon, discussion comment contained in "Sur la nécessité d'une méthode applicable au diagnostic des arriérés militaires," p. 135.

111. J. Simonin, "Essai des Tests psychiques scolaires pour apprécier l'Aptitude intellectuelle au Service militaire," *Revue neurologique* 17 (1909): 1043–48.

112. Ibid., pp. 1046–47, 1048.

113. Binet and Simon, "Sur la nécessité d'une méthode applicable au diagnostic des arriérés militaires," pp. 124–27.

114. Jacques Roubinovitch, discussion comment contained in "Sur la nécessité d'une méthode applicable au diagnostic des arriérés militaires," p. 129.

115. Chavigny, "La Débilité mentale considérée spécialement au point de vue du service militaire," p. 422.

116. On this system of selection, see Girardet, *La société militaire de 1815 à nos jours,* chap. 5.

117. Capitaine d'Arbeux, *L'Officier contemporaine: La démocratisation de l'armée (1899–1910)* (Paris: Bernard Grasset, 1911), pp. 119–20, note.

118. See Capitaine E. Vincent, *Du service à trois ans: Son application générale, ses conséquences et ses avantages* (Paris: Librairie militaire de J. Dumaine, 1882), pp. 30–32. For a general overview, see Girardet, *La société militaire de 1815 à nos jours,* pp. 140–43.

119. Ralston, *Army of the Republic,* p. 268.

120. On André and the *affaire des fiches* see, for example, Girardet, *La société militaire de 1815 à nos jours,* chap. 7; La Gorce, *French Army,* chap. 4; and Ral-

ston, *Army of the Republic,* chap. 6. For general background, see Jean-Marie Mayeur and Madeleine Rebérioux, *The Third Republic from Its Origins to the Great War, 1871–1914,* trans. J. R. Foster (Cambridge: Cambridge University Press, 1987), chaps. 8 and 9.

121. See Lyautey, "Du rôle social de l'officier," p. 454.

122. On the réveil national, see Girardet, *La société militaire de 1815 à nos jours,* chaps. 6 and 8; Ralston, *Army of the Republic,* chap. 7; and Wesseling, *Soldier and Warrior,* chap. 2.

123. See M. Norton Wise and Crosby Smith, "Measurement, Work and Industry in Lord Kelvin's Britain," *Historical Studies in the Physical Sciences* 17 (1986): 147–73, quote on p. 172.

124. See, for example, Captain Ashley Williams, Commanding Officer, Company E, 320th Infantry, Camp Lee, Memorandum, 11 December 1917; Exhibit F: Camp Lee Questionnaire Responses, File: 702-Psychological (7-31-18), Box 1856, AGO, 1917–1925.

125. See, for example, Colonel E. W. Markham, Commanding Officer, 303rd Engineers, Camp Dix, Memorandum to Chief Psychological Examiner, 78th Division, 22 November 1917; File: 10716-29: Sanitary Corps—Medical Dept., Box 561, WCD, 1903–1919.

126. Captain Henry H. Burdick, Company L, 318th Infantry, Camp Lee, Memorandum, 8 December 1917; Exhibit F: Camp Lee Questionnaire Responses, File: 702-Psychological (7-31-18), Box 1856, AGO, 1917–1925.

127. See, for example, Dorr, Memorandum to Assistant Secretary of War, 10 June 1918; and Burt, Memorandum to Chief of Staff, 18 June 1918.

128. See Dorr, Memorandum to Assistant Secretary of War, 10 June 1918, p. 11.

129. See chapter 5.

Chapter 7

1. Henry W. Holmes, "The Responsibility of the College and University in a Democracy," *School and Society* 16 (1922): 144–52, quote on pp. 145–46.

2. It is important to emphasize, however, that this ideology was rarely given free rein in France or America's imperial/colonial projects. On the American approach, see Gail Bederman, *Manliness & Civilization: A Cultural History of Gender and Race in the United States, 1880–1917* (Chicago: University of Chicago Press, 1995). On the ideology of French colonialism, see Alice Conklin, *A Mission to Civilize: The Republican Idea of Empire in France and West Africa, 1895–1930* (Stanford: Stanford University Press, 1997).

3. Theodore M. Porter, *Trust in Numbers: The Pursuit of Objectivity in Science and Public Life* (Princeton: Princeton University Press, 1995), chap. 6; see also Walter R. Sharp, *The French Civil Service: Bureaucracy in Transition* (New York: Macmillan, 1931); and Ezra N. Suleiman, *Elites in French Society: The Politics of Survival* (Princeton: Princeton University Press, 1978).

4. This is true, at least, for France proper. The place of intelligence in the discourse of French colonialism has yet to be explored.

5. On this general phenomenon, see Bruno Latour, "Drawing Things Together," in *Representation in Scientific Practice,* eds. Michael Lynch and Steve Woolgar (Cambridge: MIT Press, 1990), pp. 19–68; and Porter, *Trust in Numbers,* chap. 2.

6. Hornell Hart's analysis of the popular periodical press, for example, reveals that "intelligence testing" went from a topic of negligible presence in 1905–1909 to one in 1922–1924 that exceeded in frequency all other scientific topics, and then maintained popularity throughout the 1920s and 1930s. Hornell Hart, "Changing Social Attitudes and Interests," in *Recent Social Trends in the United States,* vol. 1 (New York: McGraw-Hill, 1933).

7. See Porter, *Trust in Numbers,* chap. 7.

8. See Georges Dwelshauvers, *La psychologie française contemporaine* (Paris: Alcan, 1920), p. 128. William H. Schneider has provided a perceptive analysis of the concept of intelligence and the use of intelligence tests in France during the period 1920s–1940s in "After Binet: French Intelligence Testing, 1900–1950," *Journal of the History of the Behavioral Sciences* 28 (1992): 111–32. My discussion in this section is greatly indebted to his insights.

9. For critiques of quantitative methods, see Dwelshauvers, *La psychologie française,* esp. p. 127; and Henri Piéron, *Le développement mental et l'intelligence* (Paris: Alcan, 1929), esp. p. 77. On the complexity of mind, see Benjamin Bourdon, *L'intelligence* (Paris: Alcan, 1926), esp. pp. 235–36; Dwelshauvers, *La psychologie française,* esp. p. 131; and Piéron, *Principles of Experimental Psychology* (1927), trans. J. B. Miner (New York: Harcourt, Brace, and Co., 1929), esp. pp. 176–79.

10. François Parot, "Psychology in the Human Sciences in France, 1920–1940: Ignace Meyerson's Historical Psychology," *History of Psychology* 3 (2000): 104–21. See also Serge Nicolas, *Histoire de la psychologie française: Naissance d'une nouvelle science* (Paris: In Press, 2002), esp. chaps. 8–10.

11. Parot, "Psychology in the Human Sciences in France, 1920–1940," p. 113.

12. Schneider, "After Binet."

13. Alice Descoeudres, *Le développement de l'enfant de deux à sept ans: Recherches de psychologie expérimentale* (1930; reprint, Neuchatel: Delachaux & Niestlé, 1946). See also Henri Delacroix, "Les opérations intellectuelles," in *Traité de psychologie,* vol. 2, ed. Georges Dumas (Paris: Alcan, 1923–24).

14. Descoeudres, *Le développement de l'enfant de deux à sept ans,* p. 18.

15. Marcel Foucault, *La mesure de l'intelligence chez les écoliers* (Paris: Librairie Delagrave, 1933), p. 97. See also Foucault, *Observations et expériences de psychologie scolaire* (Paris: Presses universitaires de France, 1923).

16. Ibid., *Mesure de l'intelligence,* pp. 17–19. Note the similarity between these tasks and those commonly used in American intelligence tests. See also ibid., pp. 13, 70. Although in *Observations et expériences de psychologie scolaire,* Foucault did conclude that "il n'y a pas de limite à la variété des formes intellectuelles." Foucault, *Observations,* p. 118.

17. On this point, see Schneider, "After Binet." See also André Lalande, "La psychologie, ses divers objets et ses méthodes," in *Traité de psychologie,* vol. 1, p. 48.

18. Henri Piéron, "Theoretical and Practical Aspects of Intelligence," *British Journal of Psychology* 22 (1931–32): 353–58, quote on p. 356.

19. Madame Henri Piéron, "Un test d'intelligence pour l'orientation professionnelle. Son étalonnage," *L'Année psychologique* 27 (1926): 174–202, quote on p. 186. See also Henri Piéron, *Le développement mental et l'intelligence;* and Piéron, "Theoretical and Practical Aspects of Intelligence."

20. Piéron, *Principles of Experimental Psychology,* pp. 178–79.

21. Bourdon, *L'intelligence,* p. 250; and Jean-Marie Lahy, "L'intelligence et les classes sociales: Essai d'une définition objective de l'intelligence," *Journal de psychologie normale et pathologique* 32 (1935): 543–601, quote on p. 590.

22. Lahy, "L'intelligence et les classes sociales," p. 585.

23. Edouard Claparède, "The Nature of General Intelligence (II)," *British Journal of Psychology* 14 (1923–24): 236–42, quote on p. 237.

24. Jeanne Monnin, "Quelques données sur les formes de l'intelligence," *L'Année psychologique* 35 (1934): 118–46, quote on p. 121. See also Piéron, *Principles of Experimental Psychology,* p. 177; and "Theoretical and Practical Aspects of Intelligence," pp. 357–58.

25. Yet another approach to intelligence pursued in the French-speaking world was Jean Piaget's work on the stages of intellectual development. Piaget spent time in Paris in the early 1920s administering Binet-Simon tests under the tutelage of Simon. See, for example, Jean Piaget, "Les traits principaux de la logique de l'enfant," *Journal de psychologie normale et pathologique* 21 (1924): 48–101; *La naissance de l'intelligence chez l'enfant* (Neuchatel: Delachaux & Niestlé, 1936); and *The Psychology of Intelligence,* trans. Malcolm Piercy and D. E. Berlyne (London: Routledge & Kegan Paul, 1950). For background, see Fernando Vidal, *Piaget Before Piaget* (Cambridge: Harvard University Press, 1994).

26. See Schneider, "After Binet."

27. Institut national d'études démographiques, *Le Niveau intellectuel des enfants d'âge scolaire,* vols. 1 and 2 (Paris: Presses universitaires de France, 1950 and 1954).

28. Albert Challand, *La Mesure de l'intelligence* (Saint-Louis: Editions "Alsatia," 1927), p. 90.

29. See Delacroix, "Les opérations intellectuelles"; Descoeudres, *Développement de l'enfant;* and Schneider, "After Binet." For Heuyer's approach, see Georges Heuyer, *Enfants anormaux et délinquants juvéniles* (Paris: G. Steinhal, 1914); and "L'examen médico-psychologique des enfants délinquants," *La prophylaxie mentale* 3 (1927): 298–304.

30. R. Duthil, "Les tests collectifs d'intelligence et leurs applications," *Revue pédagogique,* n.s., 87 (1925): 260–70 ; quote on p. 260. See also p. 269.

31. Henri Piéron, "L'utilisation des tests et les examens," in *Deux conférences d'éducation pédagogique* (Paris: Syndicat national des instituteurs et institutrices, 1934), pp. 15, 14.

32. Foucault, *Mesure de l'intelligence,* pp. 96–97.

33. Gustave Le Bon, *Psychologie de l'éducation* (Paris: Flammarion, 1918), p. 271, as quoted in Challand, *Mesure de l'intelligence,* p. 33, note.

34. Henri Piéron, Mme. Henri Piéron, and Henri Laugier, "Etude critique de la valeur sélective du certificat d'études et comparaison de cet examen avec une

épreuve par tests," in *Etudes docimologiques sur le perfectionnement des examens et concours* (Paris: Conservatoire national des arts et métiers, 1934), p. 13.

35. For more on testing for the purposes of occupational placement, see Maurice Reuchlin, "Naissance de la psychologie appliquée," in *Traité de psychologie appliquée,* vol. 1 (Paris: Presses universitaires de France, 1971), pp. 11–52; William H. Schneider, "Henri Laugier, the Science of Work and the Workings of Science in France, 1920–1940," *Cahiers pour l'Histoire du CNRS* 5 (1989): 7–34; and Schneider, "The Scientific Study of Labor in Interwar France," *French Historical Studies* 17 (1991): 410–46.

36. See Madame Piéron, "Un test d'intelligence pour l'orientation professionnelle,"p. 174.

37. Schneider, "Scientific Study of Labor in Interwar France," pp. 433–35.

38. Lahy, "L'intelligence et les classes sociales," p. 595.

39. Nonetheless French psychologists' vision of how psychological/psychotechnical explorations could contribute to society could be vast. See H. Laugier, E. Toulouse, and D. Weinberg, *Biotypologie et aptitudes scolaires* (Paris: Conservatoire national des arts et métiers, 1934), pp. 243–64; and Lahy, "L'intelligence et les classes sociales,"p. 600.

40. In the United States, Johnson O'Connor and others were also developing an approach to career guidance that focused on extensive individual profiles and relegated intelligence testing to a minor role. See, for example, O'Connor, *Psychometrics: A Study of Psychological Measurements* (Cambridge: Harvard University Press, 1934).

41. Challand, *Mesure de l'intelligence,* p. 93.

42. For an example of the publicity the army testing program received, see "Secret Mind Tests of the Army," *New York Times,* February 16, 1919, p. 69.

43. See Stanley Coben, *Rebellion Against Victorianism: The Impetus for Cultural Change in 1920s America* (New York: Oxford University Press, 1991); John M. Jordan, *Machine-Age Ideology: Social Engineering & American Liberalism* (Chapel Hill: University of North Carolina Press, 1994); James T. Kloppenberg, *Uncertain Victory: Social Democracy and Progressivism in European and American Thought, 1870–1920* (New York: Oxford University Press, 1986); Nell I. Painter, *Standing at Armageddon: The United States, 1877–1919* (New York: W. W. Norton, 1987); Edward A. Purcell, Jr., *The Crisis of Democratic Theory: Scientific Naturalism and the Problem of Value* (Lexington: University Press of Kentucky, 1973); Daniel T. Rodgers, *Atlantic Crossings: Social Politics in a Progressive Age* (Cambridge: Harvard University Press, 1998); Christine Stansell, *American Moderns: Bohemian New York and the Creation of a New Century* (New York: Metropolitan Books, 2000); and Robert H. Wiebe, *The Search for Order, 1877–1920* (New York: Hill and Wang, 1967).

44. See T. J. Jackson Lears, *No Place of Grace: Antimodernism and the Transformation of American Culture, 1880–1920* (New York: Pantheon Books, 1981).

45. Clarence S. Yoakum and Robert M. Yerkes, *Army Mental Tests* (New York: Holt, 1920); and Carl C. Brigham, *A Study of American Intelligence* (Princeton: Princeton University Press, 1923).

46. See, for example, Roderick Nash, *The Nervous Generation: American Thought, 1917–1930* (Chicago: Rand McNally, 1971); and Painter, *Standing at Armageddon.*

47. See, for example, "Despite Years, What is Your Mental Age?" *Washington Post,* November 6, 1921, p. 71.

48. Cornelia James Cannon, "American Misgivings," *The Atlantic Monthly* (February 1922): 145–57, p. 154.

49. George B. Cutten, "The Reconstruction of Democracy," *School and Society* 16 (1922): 479.

50. Lothrop Stoddard, *The Revolt Against Civilization: The Menace of the Under Man* (New York: Charles Scribner's Sons, 1925), p. 106.

51. On this general outlook in turn-of-the-century America, see Bederman, *Manliness & Civilization;* and Matthew P. Guterl, *The Color of Race in America, 1900–1940* (Cambridge: Harvard University Press, 2001), esp. chap. 1.

52. William McDougall, *Is America Safe for Democracy?* (New York: Charles Scribner's Sons, 1921), pp. 168; 51–58.

53. Daniel J. Kevles, *In the Name of Eugenics: Genetics and the Uses of Human Heredity* (Berkeley: University of California Press, 1986), pp. 72, 76; Bederman, *Manliness & Civilization;* Louise Michele Newman, *White Women's Rights: The Racial Origins of Feminism in the United States* (New York: Oxford University Press, 1999); and Martin S. Pernick, *The Black Stork: Eugenics and the Death of 'Defective' Babies in American Medicine and Motion Pictures Since 1915* (New York: Oxford University Press, 1996).

54. Joseph Kinmont Hart, "The Progress of Science and the Fate of Democracy," *School and Society* 9 (1919): 249–59. See also Hart, *The Discovery of Intelligence* (New York: Century, 1924).

55. Frederick Jackson Turner, *The Frontier in American History* (New York: Holt, Rinehart and Winston, 1920). For background, see Kerwin Klein, *Frontiers of Historical Imagination: Narrating the European Conquest of Native America, 1890–1990* (Berkeley: University of California Press, 1997).

56. Hart, "The Progress of Science and the Fate of Democracy," pp. 256–57.

57. On the social sciences in America, see Robert C. Bannister, *Sociology and Scientism: The American Quest for Objectivity, 1880–1940* (Chapel Hill: University of North Carolina Press, 1987); Mary O. Furner, *Advocacy & Objectivity: A Crisis in the Professionalization of American Social Science, 1865–1905* (Lexington: University Press of Kentucky, 1975); Matthew Hale, *Human Science and Social Order: Hugo Münsterberg and the Origins of Applied Psychology* (Philadelphia: Temple University Press, 1980); Thomas L. Haskell, *The Emergence of Professional Social Science: The American Social Science Association and the Nineteenth-Century Crisis of Authority* (Urbana: University of Illinois Press, 1977); Dorothy Ross, *The Origins of American Social Science* (Cambridge: Cambridge University Press, 1991); Helene Silverberg, ed., *Gender and American Social Science: The Formative Years* (Princeton: Princeton University Press, 1998); and Mark C. Smith, *Social Science in the Crucible: The American Debate over Objectivity and Purpose, 1918–1941* (Durham: Duke University Press, 1994).

58. For examples, see Jacob A. Riis, *How the Other Half Lives: Studies Among the Tenements of New York* (New York: Charles Scribner's Sons, 1890); and *The Pittsburgh Survey,* 6 vols. (New York: Charities Publication Committee, 1909–14). On this phenomenon, see Coben, *Rebellion Against Victorianism;* Maurine W. Greenwald and Margo Anderson, eds., *Pittsburgh Surveyed: Social Science*

and Social Reform in the Early Twentieth Century (Pittsburgh: University of Pittsburgh Press, 1996); Elizabeth Lunbeck, *The Psychiatric Persuasion: Knowledge, Gender, and Power in Modern America* (Princeton: Princeton University Press, 1994); Rodgers, *Atlantic Crossings;* Ross, *The Origins of American Social Science,* esp. chaps. 9, 10; and Smith, *Social Science in the Crucible.*

59. On science and the American war effort, see Thomas M. Camfield, "Psychologists at War: The History of American Psychology and the First World War" (Ph.D. dissertation: University of Texas, 1969); Daniel J. Kevles, *The Physicists: The History of a Scientific Community in Modern America,* 2nd ed. (Cambridge: Harvard University Press, 1995); and Robert M. Yerkes, *The New World of Science: Its Development During the War* (New York: The Century Co., 1920).

60. For a related approach from the vantage point of biology, see Albert E. Wiggam, "The New Decalogue of Science: An Open Letter from the Biologist to the Statesman," *The Century Magazine* 103 (1922): 643–50.

61. For a strikingly similar post–World War II argument, see Karl Popper, *The Open Society and Its Enemies* (Princeton: Princeton University Press, 1950).

62. It was also in the immediate postwar period that another approach to the psychological, Freudianism, spread widely especially among the intellectual and upper middle classes in the United States. On this phenomenon, see Mari Jo Buhle, *Feminism and Its Discontents: A Century of Struggle with Psychoanalysis* (Cambridge: Harvard University Press, 1998); John C. Burnham, *Psychoanalysis and American Medicine: 1894–1918; Medicine, Science, and Culture* (New York: International Universities Press, 1967); Burnham, "The New Psychology: From Narcissism to Social Control," in *Change and Continuity in Twentieth-Century America: The 1920s,* eds. John Braeman, Robert Bremner, and David Brody (Columbus: Ohio State University Press, 1968); Eric Caplan, *Mind Games: American Culture and the Birth of Psychotherapy* (Berkeley: University of California Press, 1998); Nathan G. Hale, Jr., *The Rise and Crisis of Psychoanalysis in the United States: Freud and the Americans, 1917–1985* (New York: Oxford University Press, 1995); and Eli Zaretsky, *Secrets of the Soul: A Social and Cultural History of Psychoanalysis* (New York: Knopf, 2004).

63. See, for example, "Secret Mind Tests of the Army."

64. On psychology in the United States during the 1920s, see Ernest R. Hilgard, *Psychology in America: A Historical Survey* (San Diego: Harcourt Brace Jovanovich, 1987). For a trenchant analysis of an early example of making testing into a profession, see Stephen Petrina, "The 'Never-to-Be-Forgotten Investigation': Luella W. Cole, Sidney L. Pressey, and Mental Surveying in Indiana, 1917–1921," *History of Psychology* 4 (2001): 245–71.

65. For the public interest in testing after the war, see Robert M. Yerkes, Letter to Abraham Flexner 17 January 1919, folder 3223, box 308, General Education Board Collection, Rockefeller Archive Center, N. Tarrytown, NY (Rockefeller Archives).

66. Robert M. Yerkes and Lewis M. Terman, Office of the Surgeon General, Letter to General Education Board, 23 January 1919, General Education Board Documents of Record, vol. VIII, 1919, box 21; General Education Board Collection, Rockefeller Archives; "Elementary School Intelligence Examination Board Minutes," First meeting, March 28–29, 1919; "Elementary School Intelligence Examination Board Minutes," Second meeting, April 29–May 2, 1919, both in

Lewis M. Terman Papers, Box 12, Folder 14, Stanford University Archives, Stanford University, Palo Alto, CA (Terman Papers).

67. See Guy M. Whipple, "The National Intelligence Tests," *Journal of Educational Research* 4 (1921): 16–31; and also Paul D. Chapman, *Schools as Sorters: Lewis M. Terman, Applied Psychology, and the Intelligence Testing Movement, 1890–1930* (New York: New York University Press, 1988), pp. 77–82.

68. B. R. Buckingham, "The School as a Selective Agency," *Journal of Educational Research* 3 (1921): 139. See also C. Magie Campbell, Phipps Psychiatric Clinic, the Johns Hopkins Hospital, to Abraham Flexner, 4 February 1919; folder 3223, box 308, r.g. 1.2, file 697: Mental Measurements, 1917–1942, General Education Board Papers, Rockefeller Archives.

69. On the modern American high school, see David L. Angus and Jeffrey E. Mirel, *The Failed Promise of the American High School, 1890–1995* (New York: Teachers College Press, 1999); Lawrence A. Cremin, *American Education: The Metropolitan Experience, 1876–1980* (New York: Harper & Row, 1988); David Nasaw, *Schooled to Order: A Social History of Public Schooling in the United States* (New York: Oxford University Press, 1979); and David B. Tyack, *The One Best System: A History of American Urban Education* (Cambridge: Harvard University Press, 1974). On mental testing in the school system, see Chapman, *Schools as Sorters;* Paula S. Fass, "The IQ: A Cultural and Historical Framework," *American Journal of Education* 88 (1980): 431–58; Daniel P. Resnick, "History of Educational Testing," in *Ability Testing: Uses, Consequences, and Controversies,* eds. Alexandra K. Wigdor and Wendell R. Garner (Washington: National Academy Press, 1982); and Stephen S. Williams, "From Polemics to Practice: IQ Testing and Tracking in the Detroit Public Schools and Their Relationship to the National Debate" (Ph.D. dissertation: University of Michigan, 1986).

70. A. E. Winship to Lewis Terman, 2 August 1919; Box 1, File 35, Terman Papers. See also J. P. Lichtenberger, "Social Significance of Mental Levels," *American Sociological Society. Meeting. Papers and Proceedings* 15 (1920): 102–24.

71. Robert M. Yerkes, "The Mental Rating of School Children," typescript attached to letter from Yerkes to Abraham Flexner; 29 January 1919, folder 3223, box 308, General Education Board Collection, Rockefeller Archives.

72. Guy M. Whipple, "Educational Determinism: A Discussion of Professor Bagley's Address at Chicago," *School and Society* 15 (1922): 602.

73. Yerkes, "Mental Rating of School Children."

74. See, for example, Illinois State Federation of Labor, *Weekly News Letter,* July 26, 1924, as quoted in George S. Counts, *School and Society in Chicago* (New York: Harcourt, Brace, 1928), p. 187.

75. For more on Bagley, see Diane Ravitch, *Left Back: A Century of Failed School Reform* (New York: Simon & Schuster, 2000).

76. William C. Bagley, "Educational Determinism: Or Democracy and the I.Q.," *School and Society* 15 (1922): 373–84 (italics in original).

77. For a full statement of Bagley's arguments, see William C. Bagley, *Determinism in Education: A Series of Papers on the Relative Influence of Inherited and Acquired Traits in Determining Intelligence, Achievement, and Character* (Baltimore: Warwick & York, 1925), especially "Some Hypotheses and Provisional Conclusions."

78. Walter Lippmann, "The Mental Age of Americans," "The Mystery of the 'A' Men," "The Reliability of Intelligence Tests," "The Abuse of the Tests," "Tests of Hereditary Intelligence," "A Future for the Tests," *New Republic* 32 (1922): 213–15, 246–48, 275–77, 297–98, 328–30; and 33 (1923): 9–11.

79. Ibid., "The Abuse of the Tests."

80. However, for a more jaundiced appreciation of mass democracy produced at almost the same time, see Walter Lippmann, *Public Opinion* (New York: Harcourt, Brace, 1922).

81. Ibid., "The Great Confusion: A Reply to Mr. Terman," *New Republic* 33 (1923): 146.

82. John Dewey, "Individuality Equality and Superiority," *New Republic* 33 (1922): 61–63; and Dewey, "Mediocrity and Individuality," *New Republic* 33 (1922): 35–37.

83. See ibid., "Mediocrity and Individuality," pp. 63, 61.

84. For elaboration on Dewey's views, see John Dewey, *Democracy and Education* (1916; reprint, New York: Free Press, 1966); and Dewey, *Human Nature and Conduct: An Introduction to Social Psychology* (1922; reprint, New York: Modern Library, 1930).

85. See, for example, John C. Almack, James F. Bursch, and James C. DeVoss, "Discussion of Democracy, Determinism and the IQ," *School and Society* 18 (1923): 292–95; Guy M. Whipple, "The Intelligence Testing Program and Its Objectors—Conscientious and Otherwise," *School and Society* 17 (1923): 561–604; and Robert S. Woodworth, "Aristocracy and Democracy Seen Through Intelligence Tests," *New York Times,* January 28, 1923, p. X13.

86. Lewis M. Terman, "The Great Conspiracy, or The Impulse Imperious of Intelligence Testers, Psychoanalyzed and Exposed by Mr. Lippmann," *New Republic* 33 (1922–23): 116–20.

87. Terman, "Great Conspiracy"; and Henry L. Minton, *Lewis M. Terman: Pioneer in Psychological Testing* (New York: New York University Press, 1988).

88. Terman, "Great Conspiracy," p. 117.

89. Psychologist Edgar A. Doll had made an important critique of the average age calculation as early as 1919. See Doll, "The Average Mental Age of Adults," *Journal of Applied Psychology* 3 (1919): 317–28.

90. "Tigert Raps Dr. Hopkins' Aristocracy of Brains," *Washington Post,* September 25, 1922, p. 2. See also "American College Democracy and the Aristocracy Acceptable," *Christian Science Monitor,* June 2, 1922, p. 18; Silas Bent, "University Head Derides Delusion of Democracy," *New York Times,* November 26, 1922, p. 111; Holmes, "Responsibility of the College"; Melvin Rigg, "Democracy in Education," *School and Society* 22 (1925): 339–40.

91. Bent, "University Head Derides Delusion of Democracy"; and George B. Cutten, "Nature's Inexorable Law—Inequality," *New York Times,* July 1, 1923, p. BR7.

92. Cutten, "Reconstruction of Democracy," p. 480.

93. See "Aristocracy of Character," *Los Angeles Times,* December 20, 1922, p. 114; "Colleges Must Keep Faith in Men and in Democracy, Says Meiklejohn," *Christian Science Monitor,* February 1, 1923, p. 14; "Democracy Held Success, Not a Popular Delusion," *New York Times,* December 17, 1922, p. 104; "Supra-Natural

Selection," *New York Times,* November 27, 1922; and "Tigert Raps Dr. Hopkins' Aristocracy of Brains."

94. "Tigert Raps Dr. Hopkins' Aristocracy of Brains," p. 2.

95. "Democracy Held Success," p. 104. See also Rigg, "Democracy in Education," although Rigg does concede the need for a highly differentiated curriculum.

96. See Hart's analysis of articles indexed in the *Reader's Guide to Periodical Literature* for the period 1905–1930, Table 3, in Hart, "Changing Social Attitudes and Interests."

97. F. H. Hankins, "Individual Differences and Democratic Theory," *Political Science Quarterly* 38 (1923): 388–412, quote on pp. 395–96.

98. See Kurt Danziger, *Naming the Mind: How Psychology Found its Language* (London: Sage, 1997), chap. 5.

99. "Intelligence, Limited," *Saturday Evening Post,* November 24, 1923, pp. 19, 120, 125–26; quote on p. 19.

100. See Chapman, *Schools as Sorters,* chap. 7.

101. See Petrina, "The 'Never-to-Be-Forgotten Investigation'"; and Robert S. Lynd and Helen M. Lynd, *Middletown: A Study of Modern American Culture* (New York: Harcourt, Brace, 1929). See also Susan P. Benson, *Counter Cultures: Sales Women, Managers, and Customers in American Department Stores, 1890–1940* (Urbana: University of Illinois Press, 1986); Richard Fox and T. J. Jackson Lears, eds., *The Culture of Consumption: Critical Essays in American History, 1880–1920* (New York: Pantheon, 1983); and Roland Marchand, *Advertising the American Dream: Making Way for Modernity, 1920–1940* (Berkeley: University of California Press, 1985).

102. See Chapman, *School as Sorters,* chap. 7; and Michael M. Sokal, "James McKeen Cattell and American Psychology in the 1920s," in *Explorations in the History of Psychology in the United States,* ed. Josef Brozek (Lewisburg: Bucknell University Press, 1984), pp. 273–323.

103. Chapman notes that the spread of testing was also facilitated by financial support from foundations, and the propagandizing/information dissemination efforts of the U. S. Bureau of Education, national educational organizations, educational periodicals, and educational research bureaus. See Chapman, *Schools as Sorters,* p. 152.

104. See "School Jobs for Soldiers," *New York Times,* February 23, 1919, p. 9; and Chapman, *Schools as Sorters,* pp. 72–77.

105. For two surveys of psychologists' opinions of intelligence testing, see Frank N. Freeman, "A Referendum of Psychologists: A Survey of Opinion on the Mental Tests," *Century Magazine* 107 (1923–24): 237–45; and Lewis M. Terman, collection of responses to a questionnaire on the mental test, Terman Papers.

106. For the numbers of tests produced, see Chapman, *Schools as Sorters,* pp. 147–49. See also Michael M. Sokal, "The Origins of the Psychological Corporation," *Journal of the History of the Behavioral Sciences* 17 (1981): 54–67; and for acute observations on the general phenomenon of the power of tests and forms to establish micro realms of power, see Michel Foucault, *Discipline and Punish: The Birth of the Prison,* trans. Alan Sheridan (New York: Vintage, 1979), pp. 135–69, 184–94. On the phenomenal growth of research bureaus in school systems during the early 1920s, see W. S. Deffenbaugh, "Research Bureaus in City

School Systems," in Department of the Interior, Bureau of Education, City School Leaflet no. 5 (Washington, D.C.: Government Printing Office, 1923); and Elise H. Martens, "Organization of Research Bureaus in City School Systems," in Department of the Interior, Bureau of Education, City School Leaflet no. 13 (Washington, D.C.: Government Printing Office, 1924). Martens found that in the fifty cities responding in her survey as having research bureaus in 1921–22, 100 percent reported mental and educational testing as one of their activities.

107. Elizabeth Frazer, "On the Job: The White-Collar World," *Saturday Evening Post* 195 (1923): 27, 132–37.

108. Ibid., p. 132.

109. See Chapman, *Schools as Sorters,* esp. chap. 7; Cremin, *American Education;* Fass, "The I.Q."; Porter, *Trust in Numbers,* pp. 209–10; and Tyack, *The One Best System.*

110. On the rise of white-collar work in America, see Alfred D. Chandler, *The Visible Hand: The Managerial Revolution in American Business* (Cambridge: Harvard University Press, 1977); Alan Trachtenberg, *The Incorporation of America: Culture & Society in the Gilded Age* (New York: Hill & Wang, 1982); and Olivier Zunz, *Making America Corporate, 1870–1920* (Chicago: University of Chicago Press, 1990).

111. Frazer, "On the Job," p. 133.

112. See, for example, Henry E. Garrett, "Jews and Others: Some Group Differences in Personality, Intelligence, and College Achievement," *Personnel Journal* 7 (1929): 341–48. See also Nicholas Lemann, *The Big Test: The Secret History of the American Meritocracy* (New York: Farrar, Straus, Giroux, 1999), esp. chap. 12. For complications in this story, see Michael Ackerman, "Mental Testing and the Expansion of Educational Opportunity," *History of Education Quarterly* 35 (1995): 279–300; Marcia G. Synnott, "The Admission and Assimilation of Minority Students at Harvard, Yale, and Princeton, 1900–1970," *History of Education Quarterly* 19 (1979): 285–304; Harold S. Wechsler, "Eastern Standard Time: High School-College Collaboration and Admissions to College, 1880–1930," in *A Faithful Mirror: Reflections on the College Board and Education in America,* ed. Michael C. Johanek (New York: College Board, 2001), pp. 45–79; and Wechsler, "The Rationale for Restriction: Ethnicity and College Admission in America, 1910–1980," *American Quarterly* 36 (1984): 643–67.

113. See Chapman, *Schools as Sorters,* pp. 155–64; and Williams, "From Polemics to Practice." Chapman's data was drawn from W. S. Deffenbaugh, "Uses of Intelligence and Achievement Tests in 215 Cities," in Department of the Interior, Bureau of Education, City School Leaflet no. 20 (Washington, D.C.: Government Printing Office, 1925); and "Cities Reporting the Use of Homogeneous Grouping and of the Winnetka Technique and the Dalton Plan," in Department of the Interior, Bureau of Education, City School Leaflet no. 22 (Washington, D.C.: Government Printing Office, 1926). "Cities Reporting" shows the overwhelming importance of IQ and teacher judgment in grouping. See also "The Mental Test's Use," *New York Times,* November 11, 1934, p. X9.

114. Frank N. Freeman, "Sorting the Students," *Educational Review* 68 (1924): 169–74.

115. See, for example, "Mentality Tests for Children Prove Value," *Los Angeles Times,* September 24, 1922, p. 112; "Will Class Pupils in Local Schools by Mental Tests," *Washington Post,* June 28, 1923, p. 2; M. B. Levick, "The New Superman in the Making," *New York Times,* April 27, 1924, pp. SM4, 10; "Colleges to Try Intelligence Tests," *New York Times,* November 11, 1925, p. 22; R. L. Duffus, "New Methods Remaking Old Colleges," *New York Times,* January 15, 1928, pp. 7, 17; "School for Superior Children," *Los Angeles Times,* July 5, 1929, p. A10; "'Bright Pupil' Test Wins Out," *Los Angeles Times,* May 4, 1930, pp. A1, 7; "Pupils' Classwork No Index of I.Q.," *New York Times,* May 18, 1938, p. 23; and "Making the School System Run Smoothly," *Washington Post,* March 10, 1940, p. 7. For more on Benson, see "Child Prodigy is Discovered," *Los Angeles Times,* September 19, 1922, p. 111; "Local Prodigy Will Enter College," *Los Angeles Times,* September 26, 1926, p. B2; Anne Austin, "Story of Elizabeth," *Los Angeles Times,* September 17, 1926, p. 9; September 18, 1926, p. 8; September 25, 1926, p. 7; September 28, 1926, p. 8; September 30, 1926, p. 11; and Elizabeth Benson, *The Younger Generation* (New York: Greenberg, 1927).

116. See, for example, "Keep Up Mental Tests in Army," *New York Times,* May 21, 1919, p. 10; "Better Recruits," *Washington Post,* October 20, 1927; "Intelligence Test for Soldiers," *New York Times,* April 22, 1928, sect. 10, p. 4; and "Army 'Laboratory' Set Up at Fort Dix," *New York Times,* September 15, 1940, p. 47

117. "Army's Strength is Put at 137,529," *New York Times,* November 27, 1929, p. 12.

118. See, for example, Ernest Greenwood, "Grading Human Beings: V. The Federal Executive Civil Service," *Independent* 115 (1925): 645–46, 663; L. L. Thurstone, "Intelligence Tests in the Civil Service," *Personnel Journal* 2 (1923–24): 431–41; "Problem of Right Man For Right Job Sifted," *New York Times,* May 31, 1925, p. XX13; and "Intelligence Tests that May Cost 900 Their Jobs," *United States News,* November 26, 1934, pp. 9, 13.

119. On testing of police, see "Police Get Mental Tests," *Los Angeles Times,* December 21, 1923, p. 118; "Inefficient Police Blamed for Crime—Lack of Brains and Inability to Catch Crooks Charged in Report to Commission," *New York Times,* February 28, 1924, p. 1; "'Intelligence Test' Questions Puzzle Philadelphia Police," *Christian Science Monitor,* January 6, 1930, p. 3; and "Brain Test for Police," *New York Times,* March 6, 1934, p. 42. On voting and intelligence testing, see "Tests of Voters is Advocated," *Christian Science Monitor,* December 20, 1927, p. 4; "Intelligence Test for Voters Urged," *New York Times,* January 6, 1928, p. 11 "Intelligence Tests for Voters," *New York Times,* November 26, 1928, p. 20; "Intelligence Test of Voters," *Washington Post,* November 27, 1928, p. 6; and William B. Munro, "Intelligence Tests for Voters: A Plan to Make Democracy Foolproof," *Forum* 80 (Dec. 1928): 823–30. For examples of other uses of intelligence for screening, see "Predicts Mental Test for Marriage," *New York Times,* October 19, 1923, p. 7; "Tests to Weed Out Dangerous Drivers," *Washington Post,* November 15, 1925, SM10; "Science Will Enter Prisons," *Los Angeles Times,* February 27, 1931, p. 9; "Intelligence Test Given Jurors in St. Louis,"

Washington Post, May 22, 1932, p. R8; and "Court Seeks Curb on Moron Drivers," *New York Times,* March 12, 1936, p. 23.

120. On psychology and industry, see especially Loren Baritz, *Servants of Power: A History of the Use of Social Science in American Industry* (Middletown: Wesleyan University Press, 1960); Leonard W. Ferguson, *The Heritage of Industrial Psychology* (Hartford: Finlay, 1963–68); Richard Gillespie, *Manufacturing Knowledge: A History of the Hawthorne Experiments* (New York: Cambridge University Press, 1991); Matthew Hale, "History of Employment Testing," in *Ability Testing: Uses, Consequences, and Controversies,* Part II, pp. 3–38; and Donald S. Napoli, *Architects of Adjustment: The History of the Psychological Profession in the United States* (Port Washington: Kennikat Press, 1981).

121. See Baritz, *Servants of Power,* pp. 71–72. Arthur W. Kornhauser and Forrest A Kingsbury, however, suggest that even in the early 1920s industrial adoption of psychological testing was limited, while Douglas Freyer, though agreeing intelligence testing was limited, estimated that almost three hundred firms might be doing testing in 1935. See Kornhauser and Kingsbury, *Psychological Tests in Business* (Chicago: University of Chicago Press, 1924), pp. 164–65; and Douglas Freyer, "Intelligence Testing in Industry," *Personnel Journal* 13 (1935): 321–23. For industry's continued interest in intelligence testing, see, for example, "Memorandum of first meeting of Committee on Psychological Testing of Employees [sic]," 5 November 1934, in Personnel Department—General Office Files; Records of the Pennsylvania Railroad; File 112.31 (1): Intelligence Tests for Employees, 1919–1935; Box 820; Hagley Museum and Library, Wilmington, DE (PRR Papers).

122. J. P. Lamb, "Intelligence Tests in Industry: Experience at the Cheney Brothers Plant Where Such Tests have been Used for Three Years," *Industrial Management* 58 (1919): 21–23. See also J. Crosby Chapman, "Mental Tests in Industry," *Personnel* 1 (1919): 1, 9.

123. C. S. Yoakum, "Can Executives be Picked by Mental Tests?" *Forbes Magazine* 9 (1922): 259–60; and Millicent Pond and Marion A. Bills, "Intelligence and Clerical Jobs: Two Studies of Relation of Test Score to Job Held," *Personnel Journal* 12 (1933): 41–43. See also W. V. Bingham, "Success in Business Proves Intelligence," *New York Times,* May 25, 1924, p. XX4; and Percy S. Straus, "Personnel Work in Retail Distribution," speech before the Philadelphia Chamber of Commerce, November 19, 1934, in Records of the Pennsylvania Railroad; File 112.31 (1): Intelligence Tests for Employees, 1919–1935, PRR Papers.

124. Donald A. Laird, *The Psychology of Selecting Employees,* 3rd ed. (New York: McGraw-Hill, 1937), chap. 16. See also Walter V. Bingham, *Aptitudes and Aptitude Testing* (New York: Harper & Bros., 1937).

125. See, for example, Morris Viteles, "Tests in Industry," *Journal of Applied Psychology* 5 (1921): 57–63.

126. Henry C. Link, "Psychological Tests in Industry," *Annals of the American Academy of Political and Social Science* 110 (1923): 32–44, quotes on pp. 41, 43. See also Link, "What is Intelligence?" *Atlantic Monthly* 132 (1923): 374–85.

127. Link, *Employment Psychology: The Application of Scientific Methods to*

the Selection, Training and Rating of Employees (New York: Macmillan, 1928), p. 139; and Morris Viteles, *Industrial Psychology* (New York: W. W. Norton, 1932), esp. pp. 121–33.

128. F. W. Hankins to J.F.D., 10 October 1934; and M.W.C. to J.F.D., 20 October 1934; both in File 112.31 (1), Box 820, PRR Papers.

129. "Intelligence and its Measurement: A Symposium," *Journal of Educational Psychology* 12 (1921): 123–47, 195–216, 271–75. See also "Mentality Tests: A Symposium," *Journal of Educational Psychology* 7 (1916): 229–40, 278–86, 348–60; and Guy M. Whipple, ed., *Intelligence Tests and Their Uses: Twenty-First Yearbook of the National Society for the Study of Education* (Bloomington: Public School Publishing Co, 1922).

130. File: Theories of Intelligence, I–III, Box: Vineland M934, Goddard Papers. For the lay analogue, see "What is Intelligence?" *Forum* 77 (1927): 602–604.

131. Frank N. Freeman, "The Meaning of Intelligence," in *Intelligence—Its Nature and Nurture: The Thirty-Ninth Yearbook of the National Society for the Study of Education,* ed. Guy M. Whipple (Bloomington: Public School Publishing Co., 1940), pp. 11–20.

132. See, for example, Edwin G. Boring, "Intelligence as the Tests Test It," *New Republic* 34 (1923): 35–36.

133. Freeman, "The Meaning of Intelligence," p. 18.

134. Charles Spearman, "'General Intelligence,' Objectively Determined and Measured," *American Journal of Psychology* 15 (1904): 201–93.

135. See Edward L. Thorndike, *An Introduction to the Theory of Mental and Social Measurements,* 2nd ed. (New York: Teachers College, 1916); and Thorndike, E. O. Bregman, M. V. Cobb, and Ella Woodyard, *The Measurement of Intelligence* (New York: Teachers College, 1926). For background, see Geraldine Jonçich Clifford, *Edward L. Thorndike: The Sane Positivist* (Middletown: Wesleyan University Press, 1984), esp. chaps. 13, 14, 23.

136. L. L. Thurstone, *The Nature of Intelligence* (New York: Harcourt, Brace, 1924); and Godfrey H. Thomson, *Instinct, Intelligence, and Character: An Educational Psychology* (New York: Longmans, Green, 1924). See also Truman L. Kelley, *Crossroads in the Mind of Man: A Study of Differentiable Mental Abilities* (Stanford: Stanford University Press, 1928).

137. See Dale Stout, "A Question of Statistical Inference: E. G. Boring, T. L. Kelley, and the Probable Error," *American Journal of Psychology* 102 (1989): 549–62; Stout, "Statistics in American Psychology: The Social Construction of Experimental and Correlational Psychology, 1900–1930" (Ph.D. dissertation: University of Edinburgh, 1988); and A. D. Lovie, "A Short History of Statistics in Twentieth-Century Psychology," in *New Developments in Statistics for Psychology and the Social Sciences,* vol. 2 (New York: Routledge, 1991), pp. 234–50.

138. Charles Spearman, *The Abilities of Man, Their Nature and Measurement* (New York: Macmillan, 1927).

139. Carl Campbell Brigham, "Intelligence Tests," *Princeton Alumni Weekly* 26 (May 5, 1926): 788–92; and Brigham, "The Scholastic Aptitude Test of the College Entrance Examination Board," in *The Work of the College Entrance Examination Board* (Boston: Ginn and Co., 1926).

140. See Danziger, *Constructing the Subject;* Stephen M. Stigler, *The History of Statistics: The Measurement of Uncertainty before 1900* (Cambridge: Harvard University Press, 1986); Theodore M. Porter, *The Rise of Statistical Thinking, 1820–1900* (Princeton: Princeton University Press, 1986); and Stout, "Statistics in American Psychology."

141. On these statistical issues see Truman L. Kelley, *Statistical Method* (New York: Macmillan, 1924); and for secondary overviews, Dale, "Statistics in American Psychology"; and Richard T. Von Mayrhauser, "The Mental Testing Community and Validity: A Prehistory," *American Psychologist* 47 (1992): 244–53.

142. On the intelligence of African Americans, see, for example, W. J. Mayo, "The Mental Capacity of the American Negro," *Archives of Psychology,* no. 28 (1913): 1–70; A. C. Strong, "Three Hundred Fifty White and Colored Children Measured by the Binet-Simon Scale of Intelligence: A Comparative Study," *Pedagogical Seminary* 20 (1913): 485–515; Robert S. Woodworth, "Comparative Psychology of the Races," *Psychological Bulletin* 13 (1916): 388–96; S. L. Pressey and G. F. Teter, "Comparison of Colored and White Children by Means of a Group Scale of Intelligence," *Journal of Applied Psychology* 3 (1919): 277–82; S. M. Derrick, "A Comparative Study of the Intelligence of Seventy-Five White and Fifty-Five Colored College Students by the Binet-Simon Scale," *Journal of Applied Psychology* 4 (1920): 316–29; and R. A. Schwegler and Edith Winn, "A Comparative Study of the Intelligence of White and Colored Children," *Journal of Educational Psychology* 2 (1920): 838–48. For reviews of the literature, see Charles S. Johnson and Horace M. Bond, "The Investigation of Racial Differences Prior to 1910," *Journal of Negro Education* 3 (1934): 328–39; J. St. Clair Price, "Negro-White Differences in General Intelligence," *Journal of Negro Education* 3 (1934): 424–52; and Audry M. Shuey, *The Testing of Negro Intelligence* (New York: Social Science Press, 1966). On immigrants see, in addition to Brigham, Clifford Kirkpatrick, *Intelligence and Immigration* (Baltimore: Williams & Wilkins, 1926). Garrett used a similar approach to argue for the *superiority* of Jewish Americans. See Garrett, "Jews and Others"; also Moshe Brill, "Studies of Jewish and Non-Jewish Intelligence," *Journal of Educational Psychology* 27 (1936): 331–52; and Irma H. Cohen, *The Intelligence of Jews as Compared with Non-Jews* (Columbus: Ohio State University Press, 1927). For the contrast with studies of intelligence differences between men and women, see Lewis Terman, "Women Equal in Intelligence," *World's Work,* September 30, 1922; "Radcliffe Girls Excel Harvard Men in Series of Tests for Intelligence," *New York Times,* December 27, 1923, p. 1; "Marks and Intelligence," *New York Times,* November 15, 1928, p. 24; G. M. Kuznets and Olga McNemar, "Sex Differences in Intelligence-Test Scores," in *Intelligence: Its Nature and Nurture,* chap. 6; Goodwin Watson, "Intelligence Test Has Limited Scope," *New York Times,* October 5, 1930, p. E7; and for an important secondary account, Lorraine Daston, "The Naturalized Female Intellect," *Science in Context* 5 (1992): 209–35.

143. See Franz Boas, *The Mind of Primitive Man* (New York: Macmillan, 1911); and Carl N. Degler, *In Search of Human Nature: The Decline and Revival of Darwinism in American Social Thought* (New York: Oxford University Press, 1991). Certain psychologists were also critical from the start; see, for example, Edwin G. Boring, "Facts and Fancies of Immigration," *New Republic,* April

25, 1923, pp. 245–46; Boring, Letter to Carl C. Brigham, 23 March 1923; Folder 83, Box 5, Robert M. Yerkes Papers, Yale University Archives, New Haven, CT; Raymond G. Fuller, "Intelligence of Our Immigrants," *New York Times,* March 18, 1923, p. BR18; Gustave A. Feingold, "Intelligence of the First Generation of Immigrant Groups," *Journal of Educational Psychology* 15 (1924): 65–72; Howard H. Long, "A Study of American Intelligence," *Opportunity* 1 (1923): 222–23; and "Merit Test Urged for Immigrants: Prof. Woodworth Says Regulation Should Be Based on Individuals, Not Groups," *New York Times,* July 10, 1924, p. 21.

144. On this general phenomenon, see Cravens, *Triumph of Evolution,* chap. 7; and Degler, *In Search of Human Nature,* chap. 7.

145. Carl C. Brigham, "Intelligence Tests of Immigrant Groups," *Psychological Review* 37 (1930): 158–65. See also Brigham, *A Study of Error* (New York: College Entrance Examination Board, 1932).

146. Otto Klineberg, "An Experimental Study of Speed and Other Factors in 'Racial' Differences," *Archives of Psychology* 15 (1928): 5–111.

147. Degler, *In Search of Human Nature;* Kevles, *In the Name of Eugenics;* Otto Klineberg, *Race Differences* (New York: Harper and Bros., 1935); Klineberg, *Negro Intelligence and Selective Migration* (New York: Columbia University Press, 1935); and Klineberg, "The Question of Negro Intelligence," *Opportunity* 9 (1931): 366–67.

148. W.E.B. Du Bois, "The Talented Tenth" (1903), in *Writings* (New York: Library of America, 1986).

149. See, for example, Charles S. Johnson, "Mental Measurements of Negro Groups," *Opportunity* 1 (1923): 21–25; Howard H. Long, "Race and Mental Tests," *Opportunity* 1 (1923): 22–27; Horace Mann Bond, "Intelligence Tests and Propaganda," *Crisis* 28 (1924): 61–64; Bond, "What the Army 'Intelligence' Tests Measured," *Opportunity* 2 (1924): 197–202; Long, "On Mental Tests and Racial Psychology—A Critique," *Opportunity* 3 (1925): 134–38; J. St. Clair Price, "Negro-White Differences in Intelligence," *Opportunity* 7 (1929): 341–43; Price, "The Intelligence of Negro College Freshmen," *School and Society* 30 (1929): 749–54; Melville J. Herskovits, *The Negro and the Intelligence Test* (Hanover: Sociological Press, 1929); and Albert Sidney Beckham, "Race and Intelligence," *Opportunity* 10 (1932): 240–42. And for historical overviews, see William B. Thomas, "Black Intellectuals' Critique of Early Mental Testing: A Little-Known Saga of the 1920s," *American Journal of Education* 90 (1982): 258–92; and Wayne J. Urban, "The Black Scholar and Intelligence Testing: The Case of H. M. Bond," *Journal of the History of the Behavioral Sciences* 25 (1989): 323–34.

150. Bond, "Intelligence Tests and Propaganda," p. 64.

151. See Cravens, *Triumph of Evolution,* chaps. 5, 7; and Degler, *In Search of Human Nature,* chaps. 7, 8.

152. Hamilton Cravens, *Before Head Start: The Iowa Station and America's Children* (Chapel Hill: University of North Carolina Press, 1993), esp. chaps. 4 and 6.

153. See Guy M. Whipple, ed., *Nature and Nurture: Their Influence Upon Intelligence—Twenty-Seventh Yearbook of the National Society for the Study of Education* (Bloomington: Public School Publishing Co., 1928); and Whipple, *In-*

telligence—Its Nature and Nurture. For a thoughtful analysis of this issue, see Cravens, *Before Head Start,* chap. 6.

154. See, for example, Frank S. Freeman, "Intelligence Tests and the Nature-Nurture Controversy," *School and Society* 30 (1929): 830–35; and Freeman, "The Effect of Environment on Intelligence," *School and Society* 31 (1930): 623–32.

155. "Mentality Tests for Children Prove Value," *Los Angeles Times,* September 24, 1922, p. 112; "Intelligence Tests in the Schools," *Washington Post,* September 19, 1924, p. 6; "Columbia Defends Its Mental Tests," *New York Times,* October 12, 1930, p. N18; and "Making the School System Run Smoothly," *Washington Post,* March 10, 1940, p. 7

156. Ernest Greenwood, "Grading Human Beings," *Independent* 115 (1925): 493–95, 530–31, 538, 557–58, 565, 584–85, 594, 613–14, 624, 645–46, 663, 681–82, 689, 711–12, 720, 737–39.

157. "Intelligence Tests in the Schools," *Washington Post,* September 12, 1924, p. 6.

158. "Californians Will Resist 'Mental Test,'" *Christian Science Monitor,* October 13, 1924, p. 14. See also "Grouping Pupil by Test Opposed," *Christian Science Monitor,* March 16, 1927, p. 4; and "Holds Mental Test is Unfair to Pupils," *New York Times,* November 18, 1930, p. 23.

159. See, for example, T.C.C., "Attempt to Measure Intellectual Capacity Seen as Mere Assumption," *Christian Science Monitor,* February 4, 1924, p. 10.

160. Dewey, "Mediocrity and Individuality."

161. "Mental Tests in the Schools," *New York Times,* September 13, 1921, p. 14.

162. Rollo G. Reynolds, "Intelligence Tests in Turn are Tested," *New York Times,* March 15, 1931, p. 55; and Reynolds, "Grouping of Pupils Tried in New Way," *New York Times,* March 22, 1931, p. E7.

163. Angelo Patri, "The I.Q.," *Washington Post,* February 2, 1922, p. 8.

164. Herbert S. Langfeld, "The Value of Intelligence Tests," *Forum* 76 (1926): 276–79; and Frank E. Hill, "Intelligence Tests Limited in Results," *New York Times,* December 9, 1934, p. XX4. See also W. A. Macdonald, "Brigham Adds Fire to 'War of I.Q.'s,'" *New York Times,* December 4, 1938, p. D10.

165. For an interesting example of the attempt to find a middle way, see Vernon L. Parrington, *Beginnings of Critical Realism in America, 1860–1920* (New York: Harcourt, Brace, 1930), pp. xxviii–xxix.

166. Austin G. Schmidt, "Can Intelligence Be Measured?" "What Does a Mental Scale Measure?" "Do We Measure Native Ability?" "How Well Do Tests Measure?" "Classroom Uses of Tests," and "Intelligence Tests," *America: A Catholic Review of the Week* 30 (1923): 93–94, 117–18, 141–42, 193–94, 217–18, 241–42; quote on p. 93. See also F. P. Donnelly, S.J., "A Cult of the Average," and a letter to the editor in response to Donnelly, Paul H. Furfey, "The Test Movement," *America: A Catholic Review of the Week* 30 (1924): 484–85, 571.

167. Schmidt, "Classroom Uses of Tests," p. 218.

168. Schmidt, "What Does a Mental Scale Measure?" p. 118; "Do We Measure Native Ability?" p. 141; and "How Well Do Tests Measure?" p. 193.

169. On scientism in Progressive culture, see Paul F. Boller, "The New Science and American Thought," in *The Gilded Age: A Reappraisal,* ed. H. Wayne Mor-

gan (Syracuse: Syracuse University Press, 1970); Raymond E. Callahan, *Education and the Cult of Efficiency: A Study of the Social Forces that Have Shaped the Administration of the Public Schools* (Chicago: University of Chicago Press, 1962); Jordan, *Machine-Age Ideology;* David F. Noble, *America by Design: Science, Technology, and the Rise of Corporate Capitalism* (New York: Knopf, 1977); Purcell, *Crisis of Democratic Theory;* Ross, *Origins of American Social Science,* chap. 10; and Cecilia Tichi, *Shifting Gears: Technology, Literature, Culture in Modernist America* (Chapel Hill: University of North Carolina Press, 1987).

170. See, for example, Patrick J. Ryan, "Unnatural Selection: Intelligence Testing, Eugenics, and American Political Cultures," *Journal of Social History* 30 (1997): 669–85.

171. Franz Samelson, "On the Science and Politics of the IQ," *Social Research* 42 (1975): 467–92; Sokal, "James McKeen Cattell and American Psychology in the 1920s" and Sokal, "The Origins of the Psychological Corporation."

172. "Browsing Among Brows," *New York Times,* October 6, 1927, p. 24, col. 6.

173. On the culture of America's managerial class, see John C. Burnham, "The Cultural Interpretation of the Progressive Movement," in *Paths Into American Culture: Psychology, Medicine, and Morals* (Philadelphia: Temple University Press, 1988), pp. 208–28; John Higham, "The Reorientation of American Culture in the 1890s," in *Writing American History: Essays on Modern Scholarship* (Bloomington: Indiana University Press, 1970), pp. 73–102; David A. Hollinger, "Inquiry and Uplift: Late Nineteenth-Century American Academics and the Moral Efficacy of Scientific Practice," in *The Authority of Experts,* ed. Thomas L. Haskell (Bloomington: Indiana University Press, 1984), pp. 142–56; Magali S. Larson, "The Production of Expertise and the Constitution of Expert Power," in *Authority of Experts,* pp. 28–80; and Wiebe, *Search for Order.*

174. On this issue generally, see Porter, *Trust in Numbers.*

175. For an excellent examination of politics and the human sciences in the post–World War II era, see Ellen Herman, *The Romance of American Psychology: Political Culture in the Age of Experts* (Berkeley: University of California Press, 1995). And for an important case study of the tangle of science and politics, see Sheila Jasanoff, "Science, Politics, and the Recognition of Expertise at EPA," *Osiris* 7 (1992): 195–217. For theoretical overviews, see Foucault, *Discipline and Punish;* and Bruno Latour, *The Politics of Nature: How to Bring the Sciences into Democracy,* trans. Catherine Porter (Cambridge: Harvard University Press, 2004).

Epilogue

1. Harold Brodkey, "A Story in an Almost Classical Mode," in *Stories in an Almost Classical Mode* (New York: Knopf, 1988), p. 221.

2. Richard J. Herrnstein and Charles A. Murray, *The Bell Curve: Intelligence and Class Structure in American Life* (New York: Free Press, 1994); and Steven Fraser, ed., *The Bell Curve Wars: Race, Intelligence, and the Future of America* (New York: Basic Books, 1995).

3. See Howard Gardner, *Frames of Mind: The Theory of Multiple Intelligences* (New York: Basic Books, 1983); and Daniel Goleman, *Emotional Intelligence: Why It Can Matter More Than IQ* (New York: Bantam, 1995).

4. See, for example, Pierre Bourdieu, *Homo Academicus,* trans. Peter Collier (Stanford: Stanford University Press, 1988).

5. See, for example, ibid., *The Logic of Practice* (Stanford: Stanford University Press, 1992).

6. I thank Sheila Jasanoff for helping me clarify this point.

7. Thomas Jefferson, *Notes on the State of Virginia* (1781–85), in *The Portable Thomas Jefferson,* ed. Merrill D. Peterson (New York: Penguin Books, 1977), pp. 193–98.

8. See Nicholas Lemann, *The Big Test: The Secret History of the American Meritocracy* (New York: Farrar, Straus, Giroux, 1999).

9. Diana Jean Schemo, "Head of U. of California Seeks to End S.A.T. Use in Admissions," *New York Times,* February 17, 2001, p. A1; and Michelle Maitre, "UC Adopts Standardized Test Changes; Atkinson's Criticisms Spurred Revamp for Incoming Class of 2006," *Oakland Tribune,* July 18, 2003.

10. See, for example, Erica Goode, "Brain Scans Reflect Problem-Solving Skills," *New York Times,* February 17, 2003, Section A; p. 14; Gina Kolata, "A Revolution at 50; Genetic Revolution: How Much, How Fast?" *New York Times,* February 25, 2003, Section F, p. 6; and Nicholas Wade, "Should We Improve Our Genome?" *New York Times,* November 11, 2003, Section F, p. 13.

11. All such survey data must be approached cautiously, as it is always difficult to know precisely what the respondents meant by their answers to any given question.

12. Orville G. Brim, David C. Glass, John Neulinger, and Ira J. Firestone, *American Beliefs and Attitudes about Intelligence* (New York: Russell Sage Foundation, 1969), esp. tables 2.2, 2.3, 2.7.1, 3.12, 5.2, 5.10, 12.1.

13. The issue of who believes what about intelligence and its relation to both test performance and innate ability remains enormously vexed. Claude Steele has done important work on African American attitudes toward intelligence and its tests; see for example Claude M. Steele and Joshua Aronson, "Stereotype Threat and the Intellectual Test Performance of African-Americans," *Journal of Personality and Social Psychology* 69 (1995): 797–811.

14. Other approaches include lowering the entry requirements for positions and raising the general standards of the citizenry—through such schemes as universal basic education—so that allocations according to the market, politics, social justice, kinship or friendship networks, or even a lottery would be possible.

15. Though it must be pointed out that both nations have found ways at various moments to exclude entire groups from the pool of potential choices, on the basis, for example, of their gender, ethno-racial status, or age (though often the justifications for such exclusions have themselves been couched in the merit-based language of suitable ability).

16. See Robert J. Sternberg, *Successful Intelligence: How Practical and Creative Intelligence Determine Success in Life* (New York: Simon & Schuster, 1996).

17. See Lorraine Daston, "The Naturalized Female Intellect," *Science in Context* 5 (1992): 209–35; and Lewis M. Terman and Maud A. Merrill, *Measuring*

Intelligence: A Guide to the Administration of the New Revised Stanford-Binet Tests of Intelligence (Boston: Houghton Mifflin, 1937), p. 22. In the 1916 version, girls performed very slightly better than boys across the ages five to thirteen; see Terman et al., *The Stanford Revision and Extension of the Binet-Simon Scale for Measuring Intelligence* (Baltimore: Warwick & York, 1917), pp. 62–83; Leta S. Hollingworth, "Sex Differences in Mental Traits," *Psychological Bulletin* 13 (1916): 377–84; and Terman, "Women Equal in Intelligence," *World's Work*, September 30, 1922.

18. Hamilton Cravens, *Before Head Start: The Iowa Station & America's Children* (Chapel Hill: University of North Carolina Press, 1993); and Carl N. Degler, *In Search of Human Nature: The Decline and Revival of Darwinism in American Social Thought* (New York: Oxford University Press, 1991), esp. chaps. 7–8.

19. Elazar Barkan, *The Retreat of Scientific Racism: Changing Concepts of Race in Britain and the United States Between the World Wars* (Cambridge: Cambridge University Press, 1992); and John P. Jackson, *Social Scientists for Social Justice: Making the Case against Segregation* (New York: New York University Press, 2001).

20. See Arthur R. Jensen, "How Much Can We Boost IQ and Scholastic Achievement?" *Harvard Educational Review* 39 (1969): 1–123; N. J. Block and Gerald Dworkin, eds., *The IQ Controversy: Critical Readings* (New York: Pantheon, 1976); Daniel J. Kevles, *In the Name of Eugenics: Genetics and the Uses of Human Heredity* (Berkeley: University of California Press, 1985), chap. 18; and William H. Tucker, *The Science and Politics of Racial Research* (Urbana: University of Illinois Press, 1994), chap. 5.

21. James Q. Wilson and Richard J. Herrnstein, *Crime and Human Nature* (New York: Simon and Schuster, 1985); Herrnstein and Murray, *The Bell Curve;* Russell Jacoby and Naomi Glauberman, eds., *The Bell Curve: History, Documents, Opinions* (New York: Times Books, 1995), esp. pp. 325–74.

22. See Block and Dworkin, *The IQ Controversy;* and Leon J. Kamin, *The Science and Politics of IQ* (Potomac: Erlbaum, 1974).

23. See, for example, the essays by Henry Lewis Gates, Jr., Randall Kennedy, and Orlando Patterson in Fraser, ed., *The Bell Curve Wars.*

24. See Sophia Rosenfeld, "Deaf Men on Trial: Language and Deviancy in Late Enlightenment France," *Eighteenth-Century Life* 21 (1997): 157–75.

Index